Disclaimer

The publisher of this book is by no way associated with the National Institute of Standards and Technology (NIST). The NIST did not publish this book. It was published by 50 page publications under the public domain license.

50 Page Publications.

Book Title: Chemicals in Arctic Seabirds: I. Annotated Bibliography

Book Author: Stacy S. Vander-Pol; Rebecca S. Pugh; Paul R. Becker

Book Abstract: The purpose of the work resulting in this report was to compare literature values of contaminant concentrations in the 39 species of Arctic seabirds that live or breed in Alaska. Contaminant data from literature tables are currently being entered into the, Arctic Seabird Contaminant Database and Annotated Bibliography (ASCODAB) for a comprehensive review. This report focuses on the 446 references in the annotated bibliography. A review of the bibliography including language of publication, publisher, contaminants reported, tissues examined, location of study, and other keywords precedes the annotated bibliography. Indices for authors, publishers and other keywords are included.

Citation: NIST Interagency/Internal Report (NISTIR) - 7196

Keyword: AK;annotated bibliography;Arctic;contaminants;review;seabirds

NISTIR 7196

Chemicals in Arctic Seabirds :
I. Annotated Bibliography

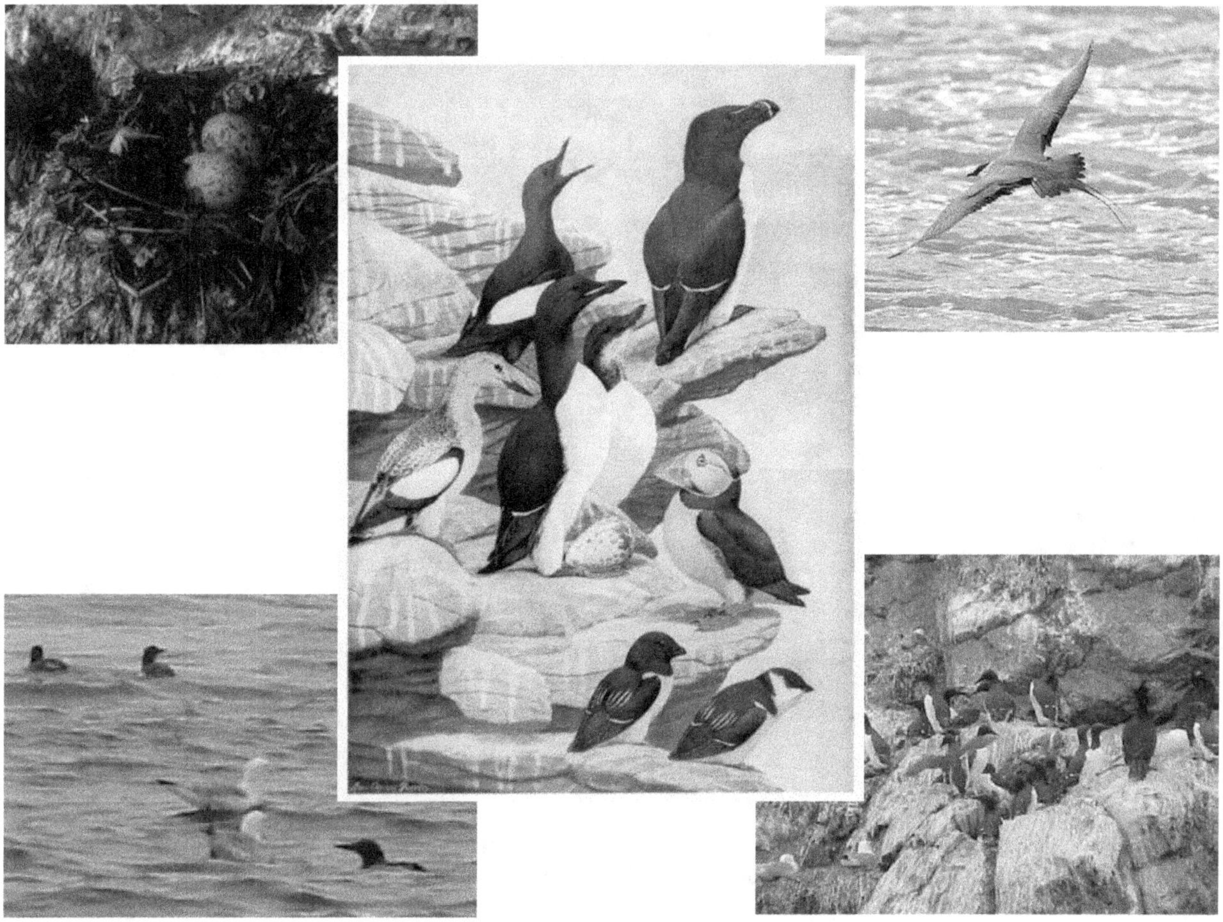

Stacy S. Vander Pol
Rebecca S. Pugh
Paul R. Becker

U.S. Department of Commerce
National Institute of Standards and Technology
Chemical Science and Technology Laboratory
Hollings Marine Laboratory
Charleston, SC 29412

NISTIR 7196

Chemicals in Arctic Seabirds:
I. Annotated Bibliography

Stacy S. Vander Pol
Rebecca S. Pugh
Paul R. Becker

U.S. Department of Commerce
National Institute of Standards and Technology
Chemical Science and Technology Laboratory
Hollings Marine Laboratory
Charleston, SC 29412

January 2005

TABLE OF CONTENTS

TABLE OF TABLES..ii
ACKNOWLEDGEMENTS..iii
DISCLAIMER..iii
BACKGROUND...1
BREAKDOWN OF REFERENCES...10
    Number of references..10
    Publications...11
    Authors..11
    Geography...11
    Tissues...14
    Chemicals..15
    Other Keywords..17
FUTURE WORK...18
    Suggestions for future research...18
    Expansion of ASCDAB...18
LITERATURE CITED..18
ANNOTATED BIBLIOGRAPHY...18

TABLE OF TABLES

TABLE 1. Arctic Seabird Species..2
TABLE 2. Additional Bird Species Found in References..3
TABLE 3. References Written in Languages Other Than English...........................11
TABLE 4. Locations of Studies..12
TABLE 5. Tissue Types Examined for Chemicals...14
TABLE 6. Chemicals Reported...15

ACKNOWLEDGEMENTS

The authors are very grateful for the tremendous assistance provided by Bev Baker (Marine Resources Library) in obtaining copies of most of the references.

DISCLAIMER

Certain commercial equipment or instruments are identified in this paper to specify adequately the experimental procedures. Such identification does not imply recommendations or endorsement by the National Institute of Standards and Technology nor does it imply that the equipment or instruments are the best available for the purpose.

BACKGROUND

Due to the amount of time seabirds spend in the air or on land, humans may associate with seabirds more than any other marine organism. While only 9 of the 156 existing families of birds are considered to be seabirds (Ainley, 1980), seabirds play a major role in the marine ecosystem. The purpose of the work resulting in this report was to compare literature values of chemical concentrations in the 39 species of Arctic seabirds that live or breed in Alaska (see Table 1). Although these species served as the basis for the literature search, the papers also contained 326 other bird species with chemical values (Table 2). All of these additional species are also indexed in this annotated bibliography.

Chemical data from literature tables for the 39 Alaskan seabird species are currently being entered into the Arctic Seabird Chemical Database and Annotated Bibliography (ASCDAB) for a comprehensive review. The review and chemical data of the ASCDAB will be available in the future. While the literature search and review was intended to be comprehensive, it is possible that references were overlooked or mistakes were made in entering the data. If you have any references you would like to contribute to the final database or would like to correct any information, please contact the authors. Copies of this report and electronic copies of the bibliography (Pro-Cite 5.0 for Windows or text document) may be obtained by contacting: Stacy S. Vander Pol, NIST – Charleston, Hollings Marine Laboratory, 331 Ft. Johnson Rd., Charleston, SC 29412. Telephone: 843-762-8994; Fax: 843-762-8742; E-mail: stacy.vanderpol@nist.gov

Table 1. Arctic seabird species for which chemical data are entered in the database. The number of references for each species that are included in the annotated bibliography are listed. When multiple names are used for the same species, the name in bold was chosen for inclusion in the bibliography. The list was compiled from U.S. Fish and Wildlife Service (2001) and AMAP (1998; identified with *).

Scientific Name	Common Name	# of References
Aethia cristatella	crested auklet	1
Aethia or *Cyclorrhynchus psittacula*	parakeet auklet	2
Aethia pusilla	least auklet	1
Aethia pygmaea	whiskered auklets	0
Brachyramphus brevirostris	Kittlitz's murrelet	1
Brachyramphus marmoratus	marbled murrelet	6
Cepphus columba	pigeon guillemot	12
Cepphus or *Uria grylle*	black guillemot	41
Cerorhinca monocerata	rhinoceros auklet	16
Fratercula cirrhata	tufted puffin	9
Fratercula corniculata	horned puffin	5
Fulmarus glacialis	northern fulmar	50
Larus argentatus	herring gull	183
Larus canus	common or mew gull	25
Larus glaucescens	glaucous-winged gull	18
Larus hyperboreus	glaucous gull	52
Larus philadelphia	Bonaparte's gull	7
Larus sabini	Sabine's gull	2
Oceanodroma furcata	fork-tailed storm petrel	9
Oceanodroma leucorhoa	Leach's storm-petrel	23
*Phalacrocorax aristotelis**	European shag*	25
Phalacrocorax auritus	double-crested cormorant	76
Phalacrocorax pelagicus	pelagic cormorant	12
Phalacrocorax penicillatus	Brandt's cormorant	8
Phalacrocorax urile	red-faced cormorant	5
*Polysticta stelleri**	Steller's eider*	3
Ptychoramphus aleuticus	Cassin's auklet	6
Rissa brevirostris	red-legged kittiwake	1
Rissa tridactyla	black-legged kittiwake	68
*Somateria mollissima**	common eider*	62
*Somateria spectabilis**	king eider*	16
Stercorarius longicaudus	long-tailed jaeger or skua	4
Stercorarius parasiticus	Arctic skua or parasitic jaeger	9
Stercorarius pomarinus	pomarine jaeger or skua	7
Sterna aleutica	Aleutian tern	1
Sterna paradisaea	Arctic tern	22
Synthliboramphus antiquus	ancient murrelet	10
Uria aalge	common murre or guillemot	99
Uria lomvia	thick-billed murre or Brünnich's guillemot	52

Table 2. Additional bird species found in the Arctic seabird references. The number of references included in the annotated bibliography are listed. When multiple names are used for the same species, text in bold was chosen for inclusion in the bibliography.

Scientific Name	Common Name	# of References
Accipiter cooperii	Cooper's hawk	1
Accipiter gentilis	northern goshawk	1
Accipiter nisus	sparrowhawk	1
Accipiter striatus	sharp-shinned hawk	3
Actitis macularia	spotted sandpiper	4
Aechmophorus occidentalis	western grebe	5
Agelaius phoeniceus	red-winged blackbird	4
Aix sponsa	wood duck	2
Alca torda	razorbill	26
Alcedo atthis	kingfisher	1
Alectoris graeca	chukar	2
Alle or *Plautus alle*	little auk or dovekie	19
Anas acuta	northern pintail	9
Anas americana	American wigeon	6
Anas clypeata	shoveler or northern shoveler	5
Anas crecca	common or green-winged teal	7
Anas discors	blue-winged teal	3
Anas penelope	Eurasian or European wigeon	1
Anas platyrhynchos	mallard or pekin duck	21
Anas poecilorhyncha	spot-billed duck	1
Anas rubripes	American black duck	4
Anas strepera	gadwall	5
Anous tenuirostris	black, lesser or white-capped noddy	1
Anser albifrons	white-fronted goose	2
Anser anser	goose	1
Anser fabalis	bean goose	1
Aphriza virgata	surfbird	1
Aquila chrysaetos	golden eagle	1
Ardea or *Casmerodius alba*	great egret	2
Ardea cinerea	grey heron	9
Ardea herodias	great blue heron	13
Arenaria melanocephala	black turnstone	1
Asio otus	long-eared owl	1
Asio flammeus	short-eared owl	1
Athene or *Speotyto cunicularia*	burrowing owl	2
Aythya affinis	lesser scaup	5
Aythya americana	redhead duck	2
Aythya collaris	ring-necked duck	4
Aythya ferina	pochard	2
Aythya fuligula	tufted duck	4
Aythya marila	greater scaup	6
Aythya valisineria	canvasback duck	3
Bartramia longicauda	Bartram's or upland sandpiper	1
Bonasa umbellus	ruffed grouse	1
Botaurus lentiginosus	American bittern	1

Table 2 (Cont.). Additional bird species found in the Arctic seabird references. The number of references included in the annotated bibliography are listed. When multiple names are used for the same species, text in bold was chosen for inclusion in the bibliography.

Scientific Name	Common Name	# of References
Branta bernicla	brant	1
Branta canadensis	Canada goose	13
Bubo bubo	eagle owl	3
Bubo virginianus	great horned owl	1
Bubulcus ibis	cattle egret	2
Bucephala albeola	bufflehead	6
Bucephala clangula	common goldeneye	10
Bucephala islandica	Barrow's goldeneye	4
Buteo buteo	buzzard	1
Buteo jamaicensis	red-tailed hawk	1
Buteo regalis	Ferruginous hawk	1
Buteo swainsoni	Swainson's hawk	1
Calidris or *Erolia alpina*	dunlin	2
Calidris maritima	purple sandpiper	4
Calidris mauri	western sandpiper	1
Calidris or *Erolia minutilla*	least sandpiper	3
Calidris ptilocnemis	rock sandpiper	2
Calidris tenuirostris	great knot	3
Callipepla or *Lophortyx californica*	California quail	1
Calonectris	shearwater	1
Calonectris diomedea	Cory's shearwater	4
Catharacta	skuas	1
Catharacta hamiltoni	Tristan skua	1
Catharacta lonnbergi	brown skua	2
Catharacta skua	great skua	11
Cathartes aura	turkey vulture	2
Catoptrophorus semipalmatus	willet	1
Ceryle alcyon	belted kingfisher	3
Charadrius hiaticula	ringed plover	2
Charadrius semipalmatus	semipalmated plover	1
Charadrius vociferus	killdeer	1
Chen caerulescens	snow goose	2
Chen rossii	Ross' goose	1
Chlidonias niger	black tern	3
Circus cyaneus	northern harrier or marsh hawk	1
Clangula hyemalis	long-tailed duck or oldsquaw	23
Colinus virginianus	northern bobwhite or bobwhite quail	2
Columba livia	rock dove or feral or common pigeon	4
Corvus caurinus	northwestern crow	3
Corvus corax	common raven	6
Corvus cornix	crow	1
Coturnix coturnix	common or Japanese quail	4
Crocethia alba	sanderling	3
Cyanocitta stelleri	Steller's jay	3

Table 2 (Cont.). Additional bird species found in the Arctic seabird references. The number of references included in the annotated bibliography are listed. When multiple names are used for the same species, text in bold was chosen for inclusion in the bibliography.

Scientific Name	Common Name	# of References
Cygnus cygnus	whooper swan	1
Cygnus olor	mute swan	3
Daption capense	cape pigeon	2
Dendragapus or *Falcipennis canadensis*	spruce grouse	1
Dendragapus obscurus	blue grouse	1
Diomedea bulleri	Buller's albatross	1
Diomedea cauta	shy or white-capped albatross	3
Diomedea cauta salvini	Salvin's albatross	1
Diomedea chlororhynchos	yellow-nosed albatross	3
Diomedea chrysostoma	grey-headed albatross	2
Diomedea epomophora	royal albatross	4
Diomedea exulans	wandering albatross	3
Diomedea immutabilis	Laysan albatross	5
Diomedea melanophris	black-browed albatross	3
Diomedea nigripes	black-footed albatross	9
Egretta alba	great white egret	1
Egretta garzetta	little egret	3
Egretta thula	snowy egret	2
Egretta tricolor	Louisiana heron	1
Eremophila alpestris	horned lark	1
Eudyptes chrysocome	rockhopper penguin	2
Eudyptes chrysolophus	macaroni Penguin	1
Eudyptes pachyrhynchus	Fiordland crested penguin	1
Eudyptes sclateri	erect-crested penguin	1
Eudyptula albosignata	white-flippered penguin	1
Eudyptula minor	blue penguin	2
Falco columbarius	pigeon hawk or merlin	2
Falco mexicanus	prairie falcon	1
Falco peregrinus	peregrine falcon	5
Falco rusticolus	gyrfalcon	1
Falco sparverius	American kestrel	3
Fratercula arctica	Atlantic or common puffin	34
Fregata magnificens	magnificent frigatebird	1
Fregata minor	great frigatebird	1
Fregetta tropica	black-bellied storm-petrel	1
Fulica americana	American coot	2
Fulica atra	coot	5
Fulmarus	fulmars	1
Capella or *Gallinago gallinago*	common snipe	2
Gallinula chloropus	common moorhen	1
Gallinula comeri	gough gallinule or moorhen	1
Gallus domesticus	chicken	2
Gallus gallus	domestic fowl	1
Gavia	loons	1

Table 2 (Cont.). Additional bird species found in the Arctic seabird references. The number of references included in the annotated bibliography are listed. When multiple names are used for the same species, text in bold was chosen for inclusion in the bibliography.

Scientific Name	Common Name	# of References
Gavia adamsii	yellow-billed loon or white-billed diver	3
Gavia arctica	Arctic loon or black-throated diver	6
Gavia immer	common loon	8
Gavia pacifica	Pacific loon	1
Gavia stellata	red-throated loon or diver	10
Grus canadensis	sandhill crane	1
Grus grus	crane	1
Haematopus bachmani	black oystercatcher	5
Haematopus ostralegus	oystercatcher	8
Haliaeetus albicilla	white-tailed sea eagle	9
Haliaeetus leucocephalus	bald eagle	12
Halobaena caerulea	blue petrel	1
Heterscelus incanus	wandering tattler	3
Himantopus himantopus	black-winged stilt	2
Hirundo rustica	barn swallow	1
Histrionicus histrionicus	harlequin duck	9
Hydrobates pelagicus	European storm-petrel	5
Lagopus lagopus	willow ptarmigan	3
Lagopus mutus	rock ptarmigan	3
Larus	gulls	1
Larus atricilla	laughing gull	2
Larus audouinii	Audouin's gull	3
Larus cachinnans	yellow-legged gull	1
Larus californicus	California gull	6
Larus crassirostris	black-tailed gull	3
Larus delawarensis	ring-billed gull	9
Larus dominicanus	southern black-backed gull	1
Larus fuscus	lesser black-backed gull	12
Larus genei	slender-billed gull	2
Larus glaucoides	Iceland gull or Thayer's gull	6
Larus marinus	great black-backed gull	16
Larus melanocephalus	Mediterranean Gull	1
Larus minutus	little gull	2
Larus novaehollandiae	silver or red-billed gull	3
Larus occidentalis	western gull	4
Larus pipixcan	Franklin's Gull	3
Larus ridibundus	black-headed gull	24
Leucocarbo atriceps	king shag	1
Leucocarbo ranfurlyi	Bounty Island shag	1
Limnodromus griseus	short-billed dowitcher	1
Limosa fedoa	marbled godwit	1
Limosa haemastica	hudsonian godwit	1
Limosa lapponica	bar-tailed godwit	3
Lonchura domestica	Bengalese finch	1

Table 2 (Cont.). Additional bird species found in the Arctic seabird references. The number of references included in the annotated bibliography are listed. When multiple names are used for the same species, text in bold was chosen for inclusion in the bibliography.

Scientific Name	Common Name	# of References
Lophodytes cucullatus	hooded merganser	3
Macronectes giganteus	southern giant petrel	2
Macronectes halli	northern giant-petrel	3
Megadyptes antipodes	yellow-eyed penguin	1
Melanitta	scoters	1
Melanitta fusca	velvet or white-winged scoter	10
Melanitta nigra	black or common scoter	10
Melanitta perspicillata	surf scoter	8
Meleagris gallopavo	turkey	1
Mergus merganser	common merganser or goosander	13
Mergus serrator	red-breasted merganser	14
Milvus lineatus	black-eared kite	1
Morus or *Sula serrator*	Australasian gannet	2
Mycteria americana	woodstork	1
Nesocichla eremita	Tristan thrush	1
Numenius americanus	long-billed curlew	1
Numenius phaeopus	whimbrel	1
Nycticorax nycticorax	black-crowned night-heron	14
Oceanites nereis	grey-backed storm-petrel	1
Oceanites oceanicus	Wilson's petrel	2
Oceanodroma homochroa	ashy petrel	2
Loomelania or *Oceanodroma melania*	black petrel	2
Halocyptena or *Oceanodroma microsoma*	least petrel	2
Oceanodroma monorhis	Swinhoe's storm petrel	1
Oxyura jamaicensis	ruddy duck	4
Pachyptila crassirostris	fulmar prion	1
Pachyptila desolata	Antarctic prion	1
Pachyptila turtur	fairy prion	2
Pachyptila vittata	broad-billed prion	2
Pagodroma	tubenoses	1
Pagodroma nivea	snow petrel	1
Pagophila eburnea	ivory gull	7
Pandion haliaetus	osprey	6
Passer montanus	tree sparrow	1
Pelecanoides georgicus	South Georgian diving-petrel	1
Pelecanoides urinatrix	diving petrel	2
Pelecanus erythrorhynchos	white pelican	7
Pelecanus occidentalis	brown pelican	6
Pelecanus onocrotalus	great white pelican	2
Perdix perdix	grey partridge	2
Phaethon lepturus	white-tailed tropicbird	2
Phaethon rubricauda	red-tailed tropicbird	1
Phalacrocorax	cormorants	1
Phalacrocorax atriceps	blue-eyed or imperial shag	1

Table 2 (Cont.). Additional bird species found in the Arctic seabird references. The number of references included in the annotated bibliography are listed. When multiple names are used for the same species, text in bold was chosen for inclusion in the bibliography.

Scientific Name	Common Name	# of References
Phalacrocorax bougainvillii	Guanay cormorant	2
Phalacrocorax capillatus	Japanese cormorant	1
Phalacrocorax carbo	great, common, or black cormorant	21
Phalacrocorax colensoi	Auckland Island shag	1
Phalacrocorax filamentosus	Temminck's cormorant	1
Phalacrocorax melanoleucos	little shag or little pied cormorant	1
Phalacrocorax nigrogularis	socotra cormorant	1
Phalacrocorax olivaceus	neotropic or olivaceous cormorant	1
Phalacrocorax pygmeus	pygmy cormorant	2
Phalaropus fulicarius	red phalarope	3
Lobipes or *Phalaropus lobatus*	northern or red-necked phalarope	6
Phasianus colchicus	pheasant	2
Philomachus pugnax	ruff	1
Phoebetria fusca	sooty albatross	2
Phoebetria palpebrata	light-mantled sooty albatross	3
Phoenicurus auroreus	Daurian redstart	1
Phoenicurus phoenicurus	redstart	1
Pica pica	black-billed magpie	3
Platalea leucorodia	spoonbill	1
Plegadis chihi	white-faced ibis	1
Pluvialis apricaria	golden plover	1
Pluvialis or *Squatarola squatarola*	black-bellied plover	1
Podiceps auritus	Slavonian or horned grebe	9
Podiceps caspicus	eared grebe	2
Podiceps cristatus	great crested grebe	5
Podiceps grisegena	red-necked grebe	7
Podiceps nigricollis	black-necked grebe	5
Procellaria aequinoctialis	white-chinned petrel	4
Procellaria cinerea	grey petrel	4
Procellaria westlandica	westland petrel	1
Pterodroma	petrels	1
Pterodroma axillaris	Chatham Island petrel	1
Lugensa or *Pterodroma brevirostris*	Kerguelen petrel	3
Pterodroma cahow	Bermuda petrel	2
Pterodroma cookii	Cook's petrel	1
Pterodroma incerta	Atlantic petrel	3
Pterodroma inexpectata	mottled petrel	2
Pterodroma lessonii	white-headed petrel	1
Pterodroma longirostris	Stejneger's petrel	1
Pterodroma macroptera	great-winged or grey-faced petrel	1
Pterodroma mollis	soft-plumaged petrel	2
Pterodroma nigripennis	black-winged petrel	1
Pterodroma pycrofti	Pycroft's petrel	1
Puffinus	shearwaters	1

Table 2 (Cont.). Additional bird species found in the Arctic seabird references. The number of references included in the annotated bibliography are listed. When multiple names are used for the same species, text in bold was chosen for inclusion in the bibliography.

Scientific Name	Common Name	# of References
Puffinus assimilis	little shearwater	1
Puffinus carneipes	flesh-footed shearwater	1
Puffinus creatopus	pink-footed shearwater	1
Puffinus gravis	great shearwater	4
Puffinus griseus	sooty shearwater	8
Puffinus huttoni	Hutton's shearwater	1
Puffinus lherminieri	Audubon's shearwater	1
Puffinus pacificus	wedge-tailed shearwater	3
Puffinus puffinus	Manx shearwater	8
Puffinus tenuirostris	slender-billed or short-tailed shearwater	6
Pygoscelis adeliae	Adelie penguin	1
Pygoscelis antarctica	chinstrap penguin	1
Pyrrhocorax pyrrhocorax	chough	1
Quelea quelea	red-billed quelea	1
Quiscalus mexicanus	great-tailed grackle	2
Quiscalus quiscalus	common grackle	1
Recurvirostra americana	American avocet	2
Recurvirostra avosetta	avocet	3
Rissa	kittiwakes	1
Rynchops niger	black skimmer	2
Scolopax minor	American woodcock	2
Somateria fischeri	spectacled eider	6
Spheniscus demersus	jackass penguin	1
Spheniscus magellanicus	magellanic penguin	1
Sterna	terns	1
Sterna albifrons	little tern	7
Hydroprogne or *Sterna caspia*	Caspian tern	13
Sterna dougallii	roseate tern	5
Sterna or *Thalasseus elegans*	elegant tern	1
Sterna forsteri	Forster's tern	11
Sterna fuscata	sooty tern	2
Sterna hirundo	common tern	39
Sterna nilotica	gull-billed tern	3
Sterna sandvicensis	sandwich tern	7
Sterna striata	white-fronted tern	1
Sterna vittata	Antarctic tern	1
Stictocarbo punctatus punctatus	blue shag	1
Stictocarbo punctatus steadi	spotted shag	2
Streptopelia risoria	ring dove	2
Sturnus vulgaris	starling	2
Sula bassana	gannet or northern gannet	14
Sula capensis	cape gannet	1
Sula dactylatra	masked booby	1
Sula leucogaster	brown booby	5

Table 2 (Cont.). Additional bird species found in the Arctic seabird references. The number of references included in the annotated bibliography are listed. When multiple names are used for the same species, text in bold was chosen for inclusion in the bibliography.

Scientific Name	Common Name	# of References
Endomychura or *Synthliboramphus craveri*	Craveri's murrelet	1
Synthliboramphus hypoleucus	Xantus's murrelet	1
Tachybaptus ruficollis	little grebe	1
Tachycineta bicolor	tree swallow	1
Tadorna tadorna	shelduck	5
Taeniopygia guttata	zebra finch	1
Thalassoica antarctica	Antarctic petrel	1
Totanus or *Tringa flavipes*	lesser yellowlegs	2
Tringa melanoleuca	greater yellowlegs	1
Tringa nebularia	greenshank	3
Tringa stagnatilis	marsh sandpiper	1
Tringa totanus	redshank	2
Pedioecetes or *Tympanuchus phasianellus*	sharp-tailed grouse	2
Vanellus vanellus	lapwing	1
Zenaida asiatica	white-winged dove	1
Zenaida macroura	mourning dove	3
Zonotrichia albicollis	white-throated sparrow	1

BREAKDOWN OF REFERENCES

Number of references

A comprehensive search of the literature resulted in the identification of 446 references containing chemical data on Arctic seabird species (see Table 1). Included in these references were 31 review papers. Whenever possible, the original papers were consulted when entering the data in the database. Supplemental papers accounted for 67 of the references, with 33 containing dosing studies and 44 dealing with oil pollution. An additional 56 references did not contain chemical tables for the Arctic seabird species. The search for references would also be greatly expanded if other languages were better represented; only 15 papers were written in languages other than English (Table 3).

Table 3. The number of references written in languages other than English.

Language	# of References
French	2
German	4
Italian	1
Japanese	2
Norwegian	1
Polish	2
Russian	3

Publications

The references were published in 146 different publications (see Publication Index), including: 89 journals (368 references), 14 books (19 references), 13 conference and symposium proceedings or abstracts (29 references), 25 agency reports, and 5 theses or dissertations. The journals with the most references were *Archives of Environmental Contamination and Toxicology*, *Environmental Pollution*, *Environmental Toxicology and Chemistry*, *Marine Pollution Bulletin*, *Science of the Total Environment*, and *Environmental Science and Technology*.

Authors

A total of 751 authors contributed to the references with almost 2/3 of these authoring only one paper (see Author Index). Several authors have contributed 20 or more papers on chemicals in Arctic seabirds, including Ross J. Norstrom, David B. Peakall, Janneche Utne Skaare, Geir Wing Gabrielsen, and Glenn A. Fox.

Geography

Any chemical data on the Arctic seabird species included in Table 1 were included in ASCDAB regardless of collection location. Most of the references examined birds from North America (230 references) and Europe (184 references). The majority of these studies were in Canada (89 references), USA (82 references), Norway (57 references), and Great Britain (26 references). Asia had fewer references (20) with most of these for Russia (12 references). Almost certainly more references exist, but finding and obtaining journals or reports from other countries (and in other languages) is more difficult and may represent an enormous bias. Surprisingly, New Zealand was the focus of two studies on Arctic seabirds and Chile was the location of one. The bodies of water most often studied were the Great Lakes (91 references), the Baltic Sea (39 references), and the North Sea (29 references).

Table 4. The locations of Arctic seabird species examined for chemicals broken down by continent with the number of references containing each location.

Location	# of References
Arctic Ocean	2
Asia	20
China	1
China Sea	1
Japan	5
Korea	3
Russia	12
Sea of Japan	1
Tokyo Bay	1
Atlantic Ocean	19
Bering Sea	9
Europe	184
Baltic Sea	39
Barents Sea	19
Belgium	12
Brittany	2
Danube Delta	1
Denmark	4
Finland	13
France	2
Germany	22
Great Britain	26
Iceland	4
Ireland	6
Italy	3
Netherlands	5
North Sea	29
Norway	57
Poland	6
Scotland	18
Shetland	4
Svalbard	21
Sweden	20
Switzerland	1
Indian Ocean	3
New Zealand	2
North America	230
Bay of Fundy	12
Canada	89
Alberta	3
British Columbia	16
Labrador	1
Manitoba	5

Table 4 (Cont.). The locations of Arctic seabird species examined for contaminants broken down by continent with the number of references containing each location.

Location	# of References
New Brunswick	14
Newfoundland	13
Northwest Territories	10
Nova Scotia	7
Nunavut	8
Ontario	8
Quebec	12
Saskatchewan	6
Yukon	4
Great Lakes	91
Greenland	19
Gulf of Alaska	7
Mexico	3
Baja	1
Northwater Polynya	4
USA	82
AK	26
CA	13
CO	1
CT	1
FL	1
GA	1
IA	1
ID	1
IL	1
LA	1
MA	3
ME	5
MI	2
MN	4
MO	1
MT	1
NC	2
NH	1
NJ	10
NM	2
NV	1
NY	12
OH	1
OK	1
OR	4
PA	1
SC	3

Table 4 (Cont.). The locations of Arctic seabird species examined for contaminants broken down by continent with the number of references containing each location.

Location	# of References
SD	5
TX	1
UT	1
VA	3
VT	1
WA	5
WI	11
Pacific Ocean	21
South America	1
Chile	1

Tissues

Chemicals were examined in 35 tissue types of Arctic seabirds (Table 5). The most common tissues examined were egg, liver, muscle, kidney, and feather.

Table 5. Tissue types examined for chemicals in Arctic seabirds with number of references containing each tissue.

Tissue	# of References
adrenal cortex	1
bile	1
blood	33
bone	12
brain	32
carcass	39
egg	190
eggshell	1
embryo	1
esophagus	1
eyeball	1
fat	42
feather	51
feces	5
gallbladder	2
gizzard	1
gland	11
gonad	6
gut	2
heart	7

Table 5 (Cont.). Tissue types examined for contaminants in Arctic seabirds with number of references containing each tissue.

Tissue	# of References
intestine	8
kidney	65
liver	174
lung	4
muscle	81
ovary	2
pancreas	3
skin	4
spleen	2
stomach	1
stomach contents	2
stomach oil	2
trachea	1
viscera	1
yolk	1

Chemicals

The chemicals reported were grouped according to chemical class and specific chemical. The chemical classes included organochlorines (274 references), metals (heavy metals and trace elements; 159 references), organometallic compounds (23 references), and hydrocarbons (11 references). The abbreviations for the chemicals and full names are listed in Table 6 with the count of references citing each chemical. Polychlorinated biphenyls (PCBs), DDTs, and mercury were the chemicals most reported in Arctic seabirds.

Table 6. Chemicals reported in Arctic seabird tissues as listed in the keyword index, with additional description and number of references found containing the chemical. The group headings are in bold.

Chemical	Additional Explanation	# of References
Ag	silver	7
Al	aluminum	7
aldrin		6
As	arsenic	20
B	boron	3
Ba	barium	3
benzenes	di-, tri-, tetra-, and/or pentachlorobenzene	24
Br	bromine	2
$C_{10}H_6N_2Br_4Cl_2$	a novel heterocyclic compound	2
Ca	calcium	7

15

Table 6 (Cont.). Contaminants reported in Arctic seabird tissues as listed in keyword index with additional description and number of references found containing the contaminant. The group headings are in bold.

Contaminant	Additional Explanation	# of References
Cd	cadmium	72
chlordanes	oxychlordane, *cis*-, *trans*-, -chlor, -nonachlor, heptachlor, and/or heptachlor epoxide	96
Co	cobolt	6
Cr	chromium	24
Cs	caesium	2
Cu	copper	43
DDTs	DDT, DDE, and/or DDT	185
dieldrin		85
EDCs	endocrine disrupting chemicals	2
endrin		7
ethylenes	tri- and tetrachloroethylene	1
Fe	iron	24
HBCD	hexabromocylodecane	1
HCB	hexachlorobenzene	95
HCHs	α-, β-, and/or γ-hexachlorocyclohexane	79
HDBPs	halogenated dimethyl bipyrroles	1
Hg	mercury	133
hydrocarbons		11
K	potassium	1
Li	lithium	2
malathion		1
metals		159
methylparathion		1
Mg	magnesium	8
mirex		56
Mn	manganese	24
Mo	molybdenum	4
Na	sodium	3
Ni	nickel	11
OCS	octachlorostyrene	14
oil		44
organic Hg	di- and methylmercury	20
organochlorines		274
organohalogens		1
organometallics		23
organophosphates		1
organolead	Me_3Pb^+, Me_2Pb^+, $MeEt_2Pb^{2+}$, Me_2Pb^{2+}, Et_3Pb^+, Et_2Pb^+, $MeEtPb^{2+}$	1
organotins	mono-, di-, tri-, -butyl-, and/or -phenol-tin	3
P	phosphorus	5
PAHs	polycyclic aromatic hydrocarbons	10
Pb	lead	52

Table 6 (Cont.). Contaminants reported in Arctic seabird tissues as listed in keyword index with additional description and number of references found containing the contaminant. The group headings are in bold.

Contaminant	Additional Explanation	# of References
PBBs	polybrominated biphenyls	1
PBDEs	polybrominated diphenyl ethers	10
PCBs	polychlorinated biphenyls	226
PCDDs	polychlorinated dibenzo-p-dioxins	64
PCDEs	polychlorinated diphenyl ethers	1
PCDFs	polychlorinated dibenzofurans	43
PCNs	polychlorinated naphthalenes	3
perfluorinated acids	FOSA, PFCA, PFHS, and/or PFOA	2
PFOS	perfluorooctane sulfonate	4
phenols		2
phosmethylan		1
phthalates		1
plastic		2
polyisobutylene		1
pulp mill		4
radionuclides		3
Rb	rubidium	2
S	sulfur	2
Sb	antimony	1
Se	selenium	51
Sn	tin	2
Sr	strontium	6
Th	thorium	1
Ti	titanium	4
Tl	thallium	1
toxaphene		18
V	vanadium	5
Zn	zinc	47
Zr	zirconium	1

Other Keywords

Most of the references did not simply measure the concentration of chemicals in Arctic seabird tissues, but rather examined the issue in more depth by examining the effects of the chemicals on the seabirds (173 references), biomagnification of the chemicals in the food web (121 references), bioaccumulation of the chemicals in the seabirds (69 references), toxicology of the chemical to the seabirds (111 references), reproductive success of the population (55 references), and/or variations in concentration of the chemicals spatially (117 references), temporally (90 references), among age classes of the seabirds (35 references), and between the genders (37 references; see Keyword Index).

FUTURE WORK

Suggestions for future research

A large amount of work has been conducted on chemicals in Arctic seabirds, but the bulk of this work was conducted on a few species in a few regions. Several species have only a few, if any, references on chemicals. The number of tissues examined could also be expanded to help answer the questions about body burden in seabirds. Further research should be conducted on current-use chemicals. Asian seabird populations may also benefit from more research.

Expansion of ASCDAB

The database should continue to be updated as new references are published. In addition, the database should be expanded to include other Arctic seabird species, particularly common tern (*Sterna hirundo*), Atlantic puffin *(Fratercula arctica)*, black-headed gull (*Larus ridibundus*), and razorbill (*Alca torda*). However, this is a great undertaking requiring numerous hours to complete and may not be completed in the near future.

LITERATURE CITED

AMAP. 1998. AMAP Assessment Report: Arctic Pollution Issues. Arctic Monitoring and Assessment Programme (AMAP), Oslo, Norway.

Ainley, David G. 1980. Birds as Marine Organisms: A Review. CalCOFI Rep. 21:48-53.

U.S. Fish and Wildlife Service. Region 7, Division of Migratory Bird Management. 2001. Migratory Birds – Seabird Species. Updated January 2001. http://www.r7.fws.gov/mbm/seabirds/species/

ANNOTATED BIBLIOGRAPHY

Allen, Janette R. and Thompson, Anu. 1996. PCBs and organochlorine pesticides in shag (*Phalacrocorax aristotelis*) eggs from the central Irish Sea: A preliminary study. Mar. Pollut. Bull. 32(12):890-892.
Rec #: 318
KEYWORDS: organochlorines; PCBs; HCB; HCHs; dieldrin; endrin; aldrin; DDTs; egg; Irish Sea; Europe; biomagnification; effects; abnormalities; shag; *Phalacrocorax aristotelis*
ABSTRACT: In 1994, five newly laid shag (*Phalacrocorax aristotelis*) eggs were collected from separate nests on the Calf of Man in the Central Irish Sea. The eggs were analyzed for ΣPCB, HCB, α-HCH, lindane, dieldrin, endrin, aldrin and ΣDDT. The pattern of individual

Annotated Bibliography (Cont.).

PCB congeners was also analyzed.

Anderson, Daniel W.; Hickey, Joseph J.; Risebrough, Robert W.; Hughes, Donald F.; and Christensen, Robert E. 1969. Significance of chlorinated hydrocarbon residues to breeding pelicans and cormorants. Can. Field. Nat. 83:91-112.
Rec #: 548
KEYWORDS: organochlorines; DDTs; dieldrin; aldrin; endrin; chlordanes; toxaphene; HCHs; PCBs; eggshell thickness; effects; reproductive success; double-crested cormorant; *Phalacrocorax auritus*; white pelican; *Pelecanus erythrorhynchos;* WI; MN; ND; USA; Manitoba; Saskatchewan; Canada; North America
ABSTRACT: This paper reports levels of chlorinated hydrocarbons present in eggs and spring food of double-crested cormorants (*Phalacrocorax auritus*) and white pelicans *(Pelecanus erythrorhynchos)* and the effects the residues might have upon the reproductive physiology of these species. On the basis of research concerning fat-kinetics in certain migratory birds (Hanson 1962, Weise 1963, Brenner 1967, and others), there is good reason to believe that residues in the eggs provide a reliable index to fat-stored contamination in the female, especially in birds with small clutches. The precise relationships have yet to be studied in detail (Lockie 1967, Stickel, 1968).

Antoine, Nathalie; Jansegers, Isabelle; Holsbeek, Ludo; Joiris, Claude; and Bouquegneau, Jean Marie. 1992 . Heavy metal contamination of seabirds and porpoises in the North Sea. Bulletin De La Societe Royale Des Sciences De Liege. 61(1-2):163-176.
Rec #: 458
KEYWORDS: metals; Cd; Zn; Cu; Cr; Ti; Fe; Pb; North Sea; Europe; common murre; *Uria aalge*; black-legged kittiwake; *Rissa tridactyla*; black-headed gull; *Larus ridibundus*; black scoter; *Melanitta nigra*; liver; muscle; biomagnification
ABSTRACT: In the North Sea, the contamination level of Cd in porpoises, Zn and Cu in seabirds, and Hg in both appear to be very high. The other studied metals (Cr, Ti, Fe, Pb) concentrations are comparable to literature. Seabirds and porpoises are located at the top of the food chains; however, their use as bioindicators of the contamination level of the ecosystem they are feeding on remains debatable. Actually many differences appear in both the contamination level and the storage and detoxification mechanisms from 1 species to another.

Appelquist, Helge; Asbirk, Sten; and Drabaek, Iver. 1984. Mercury monitoring: Mercury stability in bird feathers. Mar. Pollut. Bull. 15(1):22-24.
Rec #: 54
KEYWORDS: ultraviolet radiation; metals; Hg; common murre; *Uria aalge*; black guillemot; *Cepphus grylle*; feather; Greenland; North America; Denmark; Baltic Sea; Europe
ABSTRACT: The influence of ultraviolet light, heating, freezing and weathering on the mercury concentration in the primary feathers from Guillemot (*Uria aalge*) and Black Guillemot (*Cepphus grylle*) has been examined. Even within 8 months of exposure variation in mercury concentration due to either loss of mercury or weight loss of the feathers has been found to be less than 10 % relative.

Appelquist, Helge; Drabaek, Iver; and Asbirk, Sten. 1985. Variation in mercury content of guillemot feathers over 150 years. Mar. Pollut. Bull. 16(6):244-248.

Annotated Bibliography (Cont.).

Rec #: 55
KEYWORDS: metals; Hg; feather; black guillemot; *Cepphus grylle*; common murre; *Uria aalge*; thick-billed murre; *Uria lomvia*; Baltic Sea; Faroe Islands; North Sea; Kattegat; Europe; Greenland; North America; spatial variations; temporal trends
ABSTRACT: The feathers of 277 black (*Cepphus grylle*), common (*Uria aalge*), and Brünnich's guillemot (*Uria lomvia*) from the Baltic, Kattegat, Faroe Islands and Greenland were analysed for mercury. The levels were found to be higher in the Baltic and the Kattegat compared to the Faroe Islands and Greenland. In common guillemots from the Baltic a decrease was indicated after 1969. In general the mercury levels were higher for black guillemots living close to the coast compared to the Uria sp. living off-shore. It is concluded that chronological series based on feathers from sea birds and museum collections may contribute to the elucidation of the long term trend of mercury pollution at sea.

Arcos, J. M.; Ruiz, X.; Bearhop, S.; and Furness, R. W. 2002. Mercury levels in seabirds and their fish prey at the Ebro Delta (NW Mediterranean): The role of trawler discards as a source of contamination. Mar. Ecol. Prog. Ser. 232:281-290.
Rec #: 389
KEYWORDS: metals; Hg; Mediterranean Sea; Europe; Audouin's gull; *Larus audouinii*; yellow-legged gull; *Larus cachinnans*; common tern; *Sterna hirundo*; shag; *Phalacrocorax aristotelis*; liver; kidney; feather; biomagnification
ABSTRACT: Hg concentrations were determined in internal tissue and feathers from corpses of Audouin's (*Larus audouinii*) and yellow-legged gulls (*L. cachinnans Michaellis*), common terns (*Sterna hirundo*), European shags (*Phalacrocorax aristotelis Desmarestii*), and fish representative of trawler discards, collected at the Ebro Delta (northwestern Mediterranean) from Mar. to July (seabird breeding season), 1997-1999. Hg concentrations were significantly lower in epipelagic (Clupeiforms) vs. demersal fish. When representation of each species in discards was accounted for, the mean Hg concentration from this resource was more than double that of epipelagic fish (main natural prey for most area seabirds). Shag, the only species with direct access to benthic fish since it can dive to the seabed, had high Hg concentrations even though they do not feed on discards. The other seabirds exhibited Hg concentrations in accordance with their seasonal use of discards. Audouin's gull, which extensively exploits discards during the breeding season, had the highest concentrations in those tissues reflecting Hg intake during the breeding season (liver and first primary feathers). The common tern makes little use of discards and exhibited the lowest Hg concentrations. For samples reflecting Hg intake in winter (mantle feathers), when only the yellow-legged gull exploits discards extensively, this species exhibited highest concentrations. Audouin's gull and common tern had similarly low Hg concentrations in this period. It was concluded that consumption of discarded demersal fish strongly affected Hg contamination of surface-feeding seabirds.

Atwell, Lisa; Hobson, Keith A.; and Welch, Harold E. 1998. Biomagnification and bioaccumulation of mercury in an Arctic marine food web: Insights from stable nitrogen isotope analysis. Can. J. of Fish. Aquat. Sci. 55(5):1114-1121.
Rec #: 62
KEYWORDS: metals; Hg; stable isotopes; Northwest Territories; Canada; North America; biomagnification; black-legged kittiwake; *Rissa tridactyla*; glaucous gull; *Larus hyperboreus;* northern fulmar; *Fulmarus glacialis;* little auk; *Alle alle*; thick-billed murre; *Uria lomvia*; black

Annotated Bibliography (Cont.).

guillemot; *Cepphus grylle*; common eider; *Somateria mollissima*; Arctic tern; *Sterna paradisaea*; muscle

ABSTRACT: Several recent studies have shown that the use of $\delta^{15}N$ analysis to characterize trophic relationships can be useful for tracing biocontaminants in food webs. In this study, concentration of total mercury was measured in tissues from 112 individuals representing 27 species from the arctic marine food web of Lancaster Sound, Northwest Territories. Samples ranged from particulate organic matter through polar bears (*Ursus maritimus*). Using delta^{15}N values to identify trophic position, we found that total mercury in muscle tissue biomagnified in this food web. Polar bears were a notable exception, having a lower mean mercury concentration than their main prey, ringed seals (*Phoca hispida*). Most vertebrates showed greater variance in mercury concentration than invertebrates, and there was a trend in seabirds toward increased variability in mercury concentration with trophic position. Within species, we found no evidence of bioaccumulation of mercury with age in the muscle tissue of clams (*Mya truncata*) or ringed seals. Because stable nitrogen isotopes illustrated the relationship in this biome between trophic position and mercury level on a continuous, quantitative scale, we were able to determine that $\log^{10}(Hg)$ ($\mu g/g$ dry weight) $= 0.2(\delta^{15}N) - 3.3$. The measurement of $\delta^{15}N$ values and mercury concentration allowed us to quantitatively assess mercury biomagnification within this extensive arctic marine food web.

Bailey, Edgar P. and Davenport, Glenn H. 1972. Die-off of common murres on the Alaska Peninsula and Unimak Island. Condor. 74(2):215-219.
Rec #: 337
KEYWORDS: common murre; *Uria aalge*; AK; USA; North America; oil; effects
NOTES: SUPPLEMENTAL
ABSTRACT: Stormy weather prevailed in the Aleutian Islands and southern part of the Alaska Peninsula between 19 and 23 April 1970, climaxed by an intense disturbance with winds reaching 84 mph at Cold Bay on 22 April. Two days later reports from an aircraft north of Cold Bay indicated hundreds of dead and dying sea birds on the Bering Sea beach. On 25 April, airborne observers from Cold Bay surveyed beaches northward up the peninsula while another aircraft from Kodiak Island flew southward to Cold Bay. Thousands of dead and distressed murres were observed along the beaches between Ilnik Lake and Moffet Point. Aerial reconnaissance widened, and for 5 days, counts were conducted, sampling roughly 450 miles of coastline. Mortality was restricted to Common Murres (*Uria aalge*), and dead birds were sighted from Egegik Bay at the north end of the Alaska Peninsula to the western end of the Unimak Island. Maximum concentrations of dead murres existed in the Port Moller area. Total mortality probably exceeded 100,000 murres. Contrary to early reports, we found no evidence of oil spills, or evidence that any other species of bird or mammal suffered during the period of mortality in murres. No hydrocarbons were detected in murres or sand and water samples by six different laboratories. Test for other toxins proved inconclusive, and there was no evidence of disease. All murre specimens were emaciated and considerably underweight. Although paradoxical that only Common Murres were affected, the die-off most likely resulted from starvation precipitated by severe weather.

Barrett, R. T.; Skaare, J. U.; and Gabrielsen, G. W. 1996. Recent changes in levels of persistent organochlorines and mercury in eggs of seabirds from the Barents Sea. Environ. Pollut. 92(1):13-

Annotated Bibliography (Cont.).

18.
Rec #: 42
KEYWORDS: egg; metals; Hg; organochlorines; PCBs; DDTs; HCHs; chlordanes; Norway; Svalbard; Europe; Russia; Asia; Barents Sea; shag; *Phalacrocorax aristotelis*; common eider; *Somateria mollissima*; herring gull; *Larus argentatus*; glaucous gull; *Larus hyperboreus*; black-legged kittiwake; *Rissa tridactyla*; common murre; *Uria aalge*; thick-billed murre; *Uria lomvia*; razorbill; *Alca torda*; Atlantic puffin; *Fratercula arctica*; temporal trends; spatial variations
ABSTRACT: Eggs of ten seabird species were collected from six regions in North Norway, Svalbard and NW Russia in 1993, and were analyzed for organochlorines (OCs) and mercury. Significant declines in levels of PCBs, p,p'-DDE, HCB, beta -HCH, gamma -HCH and oxychlordane were documented in nearly half the data set since a similar study in 1983 in six of the seabird species breeding in North Norway. Only four of the 90 paired data sets increased significantly, and the remainder remained unchanged. There was very little change in mercury levels. The decline in OCs corresponds to similar declines found in Canada and the Baltic Sea, and to declines documented in marine fish in a Norwegian fjord. They can all be attributed to the reduction in use and spread of contaminants, both in Norway and internationally. No consistent regional differences in residue levels were found.

Barrett, R. T.; Skaare, J. U.; Norheim, G.; Vader, W.; and Frøslie, A. 1985. Persistent organochlorines and mercury in eggs of Norwegian seabirds 1983. Environ. Pollut. 39(1):79-93.
Rec #: 37
KEYWORDS: egg; metals; Hg; organochlorines; PCBs; DDTs; HCB; HCHs; bioaccumulation; Norway; Europe; shag; *Phalacrocorax aristotelis*; herring gull; *Larus argentatus*; black-legged kittiwake; *Rissa tridactyla*; common murre; *Uria aalge*; razorbill; *Alca torda;* Atlantic puffin; *Fratercula arctica*; spatial variations
ABSTRACT: Eggs of shags *Phalacrocorax aristotelis*, herring gulls *Larus argentatus*, kittiwakes *Rissa tridactyla*, common guillemots *Uria aalge*, razorbills *Alca torda* and puffins *Fratercula arctica* were collected from four regions in North Norway in 1983 and were analysed for organochlorines and mercury. The main organochlorines found were PCB, p,p'-DDE and HCB, but they were at levels well below those at which breeding could be expected to be affected. Mercury levels were also low. There appeared to be no statistically significant consistent changes in levels of any of the contaminants since a similar study in 1972, nor were there any consistent regional differences.

Becker, P. H.; Henning, D.; and Furness, R. W. 1994. Differences in mercury contamination and elimination during feather development in gull and tern broods. Arch. Environ. Contam. Toxicol. 27(2):162-167.
Rec #: 124
KEYWORDS: metals; Hg; herring gull; *Larus argentatus*; black-headed gull; *Larus ridibundus*; common tern; *Sterna hirundo*; bioaccumulation; Wadden Sea; Germany; Europe; feather
ABSTRACT: Eggs, feathers (down, body feathers from side/shoulder and back) and some dead chicks (liver) from broods of three species, herring gull (*Larus argentatus*), black-headed gull (*Larus ridibundus*), and common tern (*Sterna hirundo*) from the German North Sea coast were collected to study intersibling differences in mercury contamination and elimination into the growing feathers. The mercury contamination in eggs, feathers, and liver of the terns was about four times that of the gulls; black-headed gulls had lowest mercury concentrations. The body

Annotated Bibliography (Cont.).

feathers grown when the chicks became older had lower mercury levels than down in the more contaminated species (11 % lower in herring gulls, 49 % in common terns), indicating the advancing decontamination of the body by the plumage development. The elimination of mercury was greater in chicks with higher mercury levels. Down of the first hatched herring gull and common tern chick contained more mercury than down of the siblings hatched later, because of its higher burden derived from the first laid egg.

Becker, Peter H. 1992. Egg mercury levels decline with the laying sequence in Charadriiformes. Bull. Environ. Contam. Toxicol. 48(5):762-767.
Rec #: 118
KEYWORDS: metals; Hg; egg; toxicology; Wadden Sea; Germany; Europe; herring gull; *Larus argentatus*; common tern; *Sterna hirundo*; oystercatcher; *Haematopus ostralegus*; elimination
ABSTRACT: The author reports on investigations of intraclutch variation in mercury levels in three Charadriiform-species, Herring Gull, Common Tern and Oystercatcher (*Haematopus ostralegus*). The results confirm those previously reported in gulls and point to the importance of the egg in reducing the females' mercury burden.

Becker, Peter H.; Büthe, Annegret; and Heidmann, W. 1985. Pollutants in eggs of birds breeding at the German North Sea coast. I. Organochlorines. J. Ornithol. 126(1):29-51.
Rec #: 233
KEYWORDS: egg; effects; reproductive success; biomagnification; North Sea; Germany; Europe; organochlorines; DDTs; PCBs; HCHs; HCB; shelduck; *Tadorna tadorna*; oystercatcher; *Haematopus ostralegus*; ringed plover; *Charadrius hiaticula*; herring gull; *Larus argentatus*; black-headed gull; *Larus ridibundus*; common tern; *Sterna hirundo*; sandwich tern; *Sterna sandvicensis*
ABSTRACT: Chloroganic residues (gamma -HCH, HCB, DDT and metabolites, PCBs; mg/kg wet weight) were investigated in the eggs of seven species differing in migration and nutrition (*Tadorna tadorna, Haematopus ostralegus, Charadrius hiaticula, Larus argentatus, L. ridibundus, Sterna hirundo, S. sandvicensis*). Eggs laid at the Elbe (region V) or the Helgolaender Bucht (region VI) were more contaminated than in other regions. In the case of HCB the influx into the North Sea by the Elbe and a reduction due to the currents from West to Northeast could be traced by the contamination of bird eggs. The residue levels as a rule corresponded to the trophic level of the species' food. Unexpectedly, the highest levels of gamma -HCH were detected in eggs of the Shelduck and the Ringed Plover. PCBs were the compounds found in greatest quantity. A comparison with earlier results shows a decline of DDT and metabolites' levels. The authors estimate the endangering of the species by the given concentrations of organochlorines. The residues of gamma -HCH and DDT seemed to be below the critical quantities affecting breeding success. PCBs, however, partly reached concentrations which may have negative effects on the bird populations of the North Sea coast.

Becker, Peter H.; Furness, Robert W.; and Henning, Diana. 1993. The value of chick feathers to assess spatial and interspecific variation in the mercury contamination of seabirds. Environ. Monit. Assess. 28(3):255-262.
Rec #: 120
KEYWORDS: feather; egg; metals; Hg; bioaccumulation; spatial variations; Germany; North Sea; Europe; herring gull; *Larus argentatus*; common tern; *Sterna hirundo;* black-headed gull;

Annotated Bibliography (Cont.).

Larus ridibundus
NOTES: FIGURES ONLY
ABSTRACT: In 1991 we compared eggs, down and body feathers of chicks of Common Terns, Herring Gulls and Black-headed Gulls in their utility to assess contamination with mercury. Like eggs feathers distinctly show interspecific and intersite differences in contamination. Highest levels were found in Common Terns, lowest in Black-headed Gulls. Chicks hatched at the Elbe had much higher mercury levels than those in other areas of the German North Sea coast. Conversion ratios between tissues have to be used with caution.

Becker, Peter H. and Sperveslage, Hans. 1989. Organochlorines and heavy metals in herring gull (*Larus argentatus*) eggs and chicks from the same clutch. Bull. Environ. Contam. Toxicol. 42(5):721-727.
Rec #: 515
KEYWORDS: metals; Hg; Pb; Cd; organochlorines; HCHs; DDTs; dieldrin; HCB; PCBs; chlordanes; herring gull; *Larus argentatus*; egg; carcass; Germany; North Sea; Europe; bioaccumulation; age variations
ABSTRACT: Most of the organochlorine and heavy metal contaminants found in herring gull chicks and eggs were found in higher concentrations in the chicks than the eggs. An exception was PCBs which was lower in the chicks compared to the eggs. PCBs were the contaminants found in the highest concentration

Berge, John Arthur; Brevik, Einar M.; Bjørge, Arne; Følsvik, Norunn; Gabrielsen, Geir Wing; and Wolkers, Hans. 2004. Organotins in marine mammals and seabirds from Norwegian territory. J. Envrion. Monit. 6(2):108-112.
Rec #: 430
KEYWORDS: glaucous gull; *Larus hyperboreus*; organometallics; organotins; Barents Sea; Svalbard; Norway; Europe; liver
ABSTRACT: An increasing number of studies indicate that marine mammals and some seabirds are exposed to organotins. However, results from northern and Arctic areas are few. Here results from analysis of tributyltin (TBT), dibutyltin (DBT), monobutyltin (MBT), triphenyltin (TPhT), diphenyltin (DPhT) and monophenyltin (MPhT) in harbor porpoise (*Phocoena phocoena*), common seal (*Phoca vitulina*), ringed seal (*Phoca hispida*) and glaucous gull (*Larus hyperboreus*) from Norwegian territory are presented. Relatively high concentrations of DBT, TBT and MBT were observed in muscle, kidney and liver from harbor porpoises caught in northern Norway in 1988, just before restrictions on the use of tributyltin (TBT) (mainly on small boats) were introduced in several European countries. The concentrations in harbor porpoise muscle tissue were reduced significantly 11 years later, possibly as a result of the introduced restrictions. Considerably lower concentrations of butyltins were observed in the seals compared to porpoises. The lowest levels of organotins were found in ringed seals from Spitsbergen, where only traces of dibutyltin (DBT) and monobutyltin (MBT) were observed. Traces of DBT and MBT were also found in some individual glaucous gulls from Bear Island. The sum of the degradation products MBT and DBT in liver samples from all analyzed species were generally higher than TBT itself. Triphenyltin (TPhT) was observed in all porpoise samples and in livers of common seals. Also the sum of the degradation products MPhT and DPhT in liver samples from porpoise and common seals were higher than TPhT. No traces of phenyltins were found in ringed seals from Spitsbergen or in glaucous gulls from Bear Island.

Annotated Bibliography (Cont.).

The limited data available indicate low to moderate exposure to organotins in northern areas (Spitsbergen and Bear Island). Marine mammals are however more exposed further south along the Norwegian Coast.

Bergström, Rune and Norheim, Gunnar. 1986. Persistent chlorinated hydrocarbons in eggs of seabirds from the coast of Telemark in south-eastern Norway. Fauna (Blindern). 39(2):53-57.
Rec #: 176
KEYWORDS: Norway; Europe; egg; black-headed gull; *Larus ridibundus;* lesser black-backed gull; *Larus fuscus*; herring gull; *Larus argentatus*; organochlorines; HCB; PCBs; DDTs; OCS; temporal trends
ABSTRACT: The Frierfiord area in Telemark county (SE-Norway) is heavily burdened with pollution. Emissions from magnesium production include a release of persistent hydrocarbons. The most important of these are hexachlorobenzene (HCB), octachlorostyrene (OCS) and decachlorobiphenyl (DCB). Not only can they be detected in eggs of the Blackheaded Gull *Larus ridibundus* that breeds within the Frierfiord, but they are also found in eggs of the Lesser Blackbacked Gull *Larus fuscus* and the Herring Gull *Larus argentatus* breeding in outward coastal areas southwest of the fiord. From 1969 to the present eggs of the Herring Gull have shown a decrease in the levels of OCS and of HCB in particular. No reduction has been found in the level of DCB. All eggs have also been analysed for concentrations of DDE and PCB.

Bignert, A.; Litzén, K.; Odsjö, T.; Olsson, M.; Persson, W.; and Reutergårdh, L. 1995. Time-related factors influence the concentrations of sDDT, PCBs and shell parameters in eggs of Baltic guillemot (*Uria aalge*), 1861-1989. Environ. Pollut. 89:27-36.
Rec #: 298
KEYWORDS: organochlorines; DDTs; PCBs; Baltic Sea; Sweden; Europe; common murre; *Uria aalge*; egg; temporal trends; eggshell thickness; effects
ABSTRACT: Three hundred and nine eggs of guillemot (*Uria aalge*) have been used to study the time trends in DDT and PCB pollution of the Baltic during the period 1969-1989. As a comparison, herring (*Clupea harengus*) from the Baltic have been used. Eggshell thickness has been studied in 370 eggs collected between 1861-1989. The inter-relationships between various morphological and chemical parameters were studied with respect to the date of egglaying (first-laid eggs and replacement eggs) and year of collection. The following morphological and chemical parameters have been studied: weight, length and breadth of eggs; weight, thickness, thickness index and density index of eggshells; fat content and concentrations of sDDT and PCBs in the eggs.
sDDT and PCBs concentrations have decreased in the Baltic since the 1970s. Eggshell thickness is significantly lower in recent material compared to historical material collected before 1946; it has increased since the 1970s but it is still lower than before 1946.
Compared to the first-laid eggs, replacement eggs had significantly lower values for morphological data. The concentrations of sDDT and PCBs were significantly higher in replacement eggs. Fat content was similar in the two groups. The results show that date of egglaying has to be considered in monitoring studies and that such considerations will improve the interpretation of data and reduce the number of samples needed.

Bignert, Anders; Olsson, Mats; Persson, Wawa; Jensen, Sören; Zakrisson, Susanne; Litzén, Kerstin; Eriksson, Ulla; Häggberg, Lisbeth; and Alsberg, Tomas. 1998. Temporal trends of

Annotated Bibliography (Cont.).

organochlorines in Northern Europe, 1967-1995. Relation to global fractionation, leakage from sediments and international measures. Environ. Pollut. 99(2):177-198.
Rec #: 311
KEYWORDS: organochlorines; DDTs; PCBs; HCB; HCHs; temporal trends; Baltic Sea; Sweden; Europe; common murre; *Uria aalge;* egg; biomagnification
ABSTRACT: The time trend monitoring of organochlorine pollution was carried out in Sweden since the late 1960s. This report presents data on concentrations of DDT, PCB, HCHs and HCB in biota samples collected and analysed annually. All the matrices and compounds studied show a significant decrease over time. The data cover severely polluted Swedish marine and fresh water in southern Sweden as well as locally unpolluted waters in remote northern Arctic regions of Sweden. A total of 13 time series representing different locations and species are presented for the different pollutants. The period studied covers the time when pollution was serious as well as the time of recovery. All monitoring activities were carried out at the same laboratories over the entire study period, which means that comparability over time is good in the sets of data presented. The various time trends show a convincing agreement with trends and annual change over time, although the concentrations differ between the species and locations investigated, the highest concentrations being in the south. Since the annual changes are normally similar regardless of locations and species, spatial variations in concentrations remain over time, although concentrations are lower today. The onset of changes in concentrations over time can be related to international measures or other circumstances that lowered releases into the environment. Similarities in the annual changes, as well as the time when changes began, are discussed with respect to suggested hypotheses on the fate of the investigated organochlorines. It was not possible to verify that the oxygenation of anoxic sediments mobilised old pollution in Baltic sediments. Neither was it possible to conclude that eutrophication has caused a measurable effect on the rate and timing of the decreases. Finally, long-range transport to Arctic regions seems to be due more to a one step transport than to the `Grass-hopper' effect. The comprehensive database used, clearly shows how important it is to have datasets big enough to describe between-year variation before attempting to evaluate the time trend. In addition, if between-year variation is not known, it is then also difficult to evaluate spatial variation on the basis of single year observations.

Bishop, C. A.; Ng, P.; Norstrom, R. J.; Brooks, R. J.; and Pettit, K. E. 1996. Temporal and geographic variation of organochlorine residues in eggs of the common snapping turtle (*Chelydra serpentina serpentina*) (1981-1991) and comparisons to trends in the herring gull (*Larus argentatus*) in the Great Lakes basin in Ontario, Canada. Arch. Environ. Contam. Toxicol. 31(4):512-524.
Rec #: 92
KEYWORDS: organochlorines; egg; PCBs; DDTs; HCB; PCDDs; PCDFs; dieldrin; mirex; herring gull; *Larus argentatus*; Great Lakes; Quebec; Canada; North America; temporal trends; spatial variations
NOTES: NO TABLES FOR HERRING GULL; ONLY FIGURES
ABSTRACT: Common snapping turtle (*Chelydra serpentina serpentina*) eggs from five sites within the Great Lakes basin, and from a reference site in north-central Ontario were collected during 1981-1991 and analyzed for four organochlorine pesticides, polychlorinated biphenyls (PCBs) including six non-ortho PCBs, polychlorinated dibenzodioxins (PCDDs), and polychlorinated dibenzofurans (PCDFs). The pattern of geographic variation was consistent over

Annotated Bibliography (Cont.).

time in eggs with Cootes Paradise/Hamilton Harbour and Lynde Creek eggs on Lake Ontario containing the highest concentrations and most PCDD and PCDF congeners among all sites. Eggs from Cranberry Marsh on Lake Ontario contained organochlorine concentrations similar to those from Big Creek Marsh and Rondeau Provincial Park on Lake Erie except PCDDs and PCDFs which occurred at higher concentrations and more congeners were detectable in Cranberry Marsh eggs. Concentrations of most contaminants in turtle eggs from Algonquin Park, the reference site, have significantly decreased in the past decade. Dieldrin concentrations, however, increased in Algonquin Park eggs from 1981 to 1989. Significant decreases in concentrations of hexachlorobenzene, mirex and PCBs occurred between turtle eggs collected in 1981/84 and 1989 at Big Creek Marsh and Rondeau Provincial Park, whereas there was no significant change in concentrations of p,p'-DDE and dieldrin. In Lake Ontario eggs, concentrations of PCBs, p,p'-DDE and dieldrin increased significantly between 1984 and 1991. Differences were also found in patterns of temporal variation in contamination between herring gulls (*Larus argentatus*) and snapping turtles which were attributed to differences in diet. Elevated and continued contamination in turtle eggs from Lake Ontario is probably due to a combination of local sources of chemicals and consumption of large migratory fish that spawn in wetlands inhabited by these turtles.

Bishop, C. A; Weseloh, D. V; Burgess, N. M.; Struger, J.; Norstrom R. J.; Turle, R.; and Logan, K. A. 1992. An atlas of contaminants in eggs of fish-eating colonial birds of the Great Lakes (1970-1988). Technical Report Series No. 153. Burlington, Ontario, Canada: Canadian Wildlife Service Ontario Region. 135 pp.
Rec #: 366
KEYWORDS: herring gull; *Larus argentatus*; double-crested cormorant; *Phalacrocorax auritus*; Caspian tern; *Sterna caspia*; common tern; *Sterna hirundo*; black-crowned night-heron; *Nycticorax nycticorax*; Forster's tern; *Sterna forsteri*; ring-billed gull; *Larus delawarensis*; organochlorines; chlordanes; benzenes; PCBs; PCDDs; PCDFs; egg; Great Lakes; North America; temporal trends; spatial variations
ABSTRACT: During 1970-1988, Canadian Wildlife Service (Ontario) collected eggs from 7 species fish-eating colonial birds from 67 colonies throughout the Great Lakes to measure the levels of 39 contaminants and the lipid concentrations present. There are 4491 data points presented in the atlas. Data were generated as part of a monitoring program started in 1970 to understand the temporal and spatial trends in environmental contaminant levels in biota of the Great Lakes. The species sampled include: Herring Gull (*Larus argentatus*), Double-crested Cormorant (*Phalacrocorax auritus*), Caspian Tern (*Sterna caspia*), Common Tern (*Sterna hirundo*), Black-crowned Night-Heron (*Nycticorax nycticorax*), Forster's Tern (*Sterna forsteri*), and Ring-billed Gull (*Larus delawarensis*). The purpose was to measure the levels of the following compounds: organochlorine pesticides, chlorinated benzenes, polychlorinated biphenyls, dioxins and furans, and moisture. Volume I contains contaminant data summarized by location. Volume II contains contaminant data summarized by compound.

Bjerk, John Erik and Holt, Gunnar. 1971. Residues of DDE and PCB in eggs from herring gull (*Larus argentatus*) and common gull (*Larus canus*) in Norway. Acta Vet. Scand. 12(3):429-441.
Rec #: 377
KEYWORDS: herring gull; *Larus argentatus*; common gull; *Larus canus*; organochlorines; DDTs; PCBs; Norway; Europe; egg; eggshell thickness; spatial variations; effects

Annotated Bibliography (Cont.).

ABSTRACT: A total of 294 eggs from herring gull (*Larus argentatus*) were collected at eight localities in Norway in 1969. Nine eggs from common gull (*Larus canus*) were also collected from one of the sites. The eggs were analyzed for DDE and PCBs and were found to range from 0.2 to 5.4 ppm in DDE and 0.2 and 3.8 ppm in PCB 10 with significant differences between the localities. Eggshell thickness had a higher correlation with PCB than DDE, but neither was significant.

Blight, Louise K. and Burger, Alan E. 1997. Occurrence of plastic particles in seabirds from the eastern North Pacific. Mar. Pollut. Bull. 34(5):323-325.
Rec #: 174
KEYWORDS: plastic; Pacific Ocean; Canada; British Columbia; USA; WA; OR; North America; black-footed albatross; *Diomedea nigripes*; fork-tailed storm petrel; *Oceanodroma furcata*; Leach's storm petrel; *Oceanodroma leucorhoa*; northern fulmar; *Fulmarus glacialis*; Stejneger's petrel; *Pterodroma longirostris*; sooty shearwater; *Puffinus griseus*; tufted puffin; *Fratercula cirrhata*; horned puffin; *Fratercula corniculata*; common murre; *Uria aalge*; Xantus' murrelet; *Synthliboramphus hypoleucus*; rhinoceros auklet; *Cerorhinca monocerata*
NOTES: SUPPLEMENTAL
ABSTRACT: We found plastic particles in the stomachs of 8 of the 11 species of seabirds caught as bycatch in the pelagic waters of the eastern North Pacific (41-50 degree N, 131-134 degree W). Plastic was found in all surface-feeding birds (two storm-petrel, one albatross, one petrel and one fulmar species) and in 75 % of shearwaters. Densities in some storm-petrels, shearwaters and the petrel were possibly sufficient to impede digestion, but were negligible in other birds. Plastic was also found in two diving species (puffins) but absent in three others (murre, auklet and murrelet). Of 353 anthropogenic items examined, 29 % were industrial pellets and 71 % were fragments of discarded products ("user" plastic), with user plastic making up 60 % of total mass. Our study is evidence of widespread plastic pollution affecting birds in a previously unsampled sector of the North Pacific

Bogan, J. A. and Bourne, W. R. P. 1972. Organochlorine levels in Atlantic seabirds. Nature. 240(5380):358.
Rec #: 11
KEYWORDS: Atlantic Ocean; Scotland; Norway; Barents Sea; Europe; organochlorines; PCBs; DDTs; muscle; liver; fat; stomach oil; black-legged kittiwake; *Rissa tridactyla*; northern fulmar; *Fulmarus glacialis*; storm-petrel; *Hydrobates pelagicus*; Atlantic puffin; *Fratercula arctica*; glaucous gull; *Larus hyperboreus*; common murre; *Uria aalge*
NOTES: NO TABLES
ABSTRACT: Normal seabirds from a representative range of species occurring between Scotland and the Artic contained polychlorinated biphenyls (PCB) and smaller amounts of p,p'-DDE (I) with the PCB/DDE ratio usually being 2-10 in all species except the kittiwake (Rissa tridactyla), where the ratio was always higher, but at least 60. The total organochlorine content in muscle and liver was generally low (0.1-1.0 ppm) in the auks and shearwaters examined, and was greater in more pelagic species, with levels >10 ppm being found in gulls and skuas feeding largely around trawlers and seabird breeding stations. The organochlorine content apparently was not inherited, and the PCB content of the parent was not passed on to the chicks through the feeding of stomach oil. Normal birds which appeared to have been feeding largely on other seabirds' eggs had an average content of 24 ppm PCB and 17 ppm DDE in the liver and one

Annotated Bibliography (Cont.).

contained 535 ppm PCB and 67 ppm DDE in the fat, but a bird containing 311 ppm PCB and 67 ppm DDE in the liver showed weakness and failure of coordination of the limbs, as did experimentally poisoned finches which had a mean liver PCB content of 345 ppm.

Bondarev, A. Ya.; Denisova, A. V.; Kishchinskii, A. A.; and Ryazhenov, N. I. 1976. Chlorinated organic pesticides in tissues of some wild animals. Nauchnye Osnovy Okhrany Prirody. 7:110-117.
Rec #: 420
KEYWORDS: organochlorines; DDTs; HCHs; fat; Steller's eider; *Polysticta stelleri*; spectacled eider; *Somateria fischeri*; long-tailed duck; *Clangula hyemalis*; bean goose; *Anser fabalis*; ruff; *Philomachus pugnax*; Russia; Asia
ABSTRACT: The fat of the birds *Polysticta stelleri*, *Somateria fischeri*, *Clangula hyemalis*, *Anser fabalis*, and *Philomachus pugnax* from the lower reaches of the Indigirka River contained 4 (1-16) µg HCH (I) [58-89-9] and 2 (1-120) µg total DDT (II) [50-29-3]/kg. The content of I in wolves from Altai ranged from 0 to 90 µg/kg brain and 140 µg/kg fat in pups. The maximum levels of II were 14 µg/kg in the brain of a pup and 40 µg/kg in the fat of an adult male. The presence of I and II in the polar sea birds *C. hyemalis* and *P. stelleri*, which never come to areas treated with pesticides, indicates global spread of pesticides.

Borgå, K.; Gabrielsen, G. W.; and Skaare, J. U. 2001. Biomagnification of organochlorines along a Barents Sea food chain. Environ. Pollut. 113:187-198.
Rec #: 344
KEYWORDS: biomagnification; organochlorines; HCHs; HCB; chlordanes; DDTs; PCBs; mirex; liver; *Uria lomvia;* thick-billed murre; *Cepphus grylle;* black guillemot; *Rissa tridactyla;* black-legged kittiwake; *Larus hyperboreus*; glaucous gull; Barents Sea; Europe
ABSTRACT: To trace the biomagnification of organochlorines in marine food chains near Svalbard, which may lead to the high organochlorine concentrations in top predators from the area, we compared concentrations and patterns of organochlorines in selected taxa. The pelagic crustaceans, *Calanus* spp. (copepods), *Thysanoessa* spp. (euphausiids), *Parathemisto libellula* (amphipod), and the fish species, *Boreogadus saida* (polar cod) and *Gadus morhua* (cod) were selected to represent the lower trophic levels in the food web. Four seabird species were chosen at the higher trophic levels, *Uria lomvia* (Brünnich's guillemot), *Cepphus grylle* (black guillemot), *Rissa tridactyla* (black-legged kittiwake) and *Larus hyperboreus* (glaucous gull). We found low concentrations of the organochlorines Σhexachlorocyclohexanes (HCHs), hexachlorobenzene (HCB), ΣChlordanes, ΣDDTs, and Σpolychlorinated biphenyls (PCBs) in crustaceans (11-50 ng g^{-1} lipid weight) and fish (15-222 ng g^{-1} lipid weight). In seabirds, the organochlorine concentrations biomagnified one to three orders of magnitude dependent on species and compound class. Glaucous gulls had the highest concentrations of all organochlorines. The organochlorine levels in all taxa except glaucous gull were comparable to those recorded in similar species in the Canadian Arctic. The organochlorine pattern changed from crustaceans and fish to seabirds. Moving up the food chain, the relative contribution of ΣHCHs, HCB and ΣChlordanes decreased, and the relative contribution of Σ DDTs, ΣPCBs, persistent compounds and metabolites increased. The results reflected trophic transfer of organochlorines along the food chain as well as different elimination potentials due to direct diffusion in crustaceans and fish, and higher contaminant metabolic activity in seabirds.

Annotated Bibliography (Cont.).

Borgå, Katrine; Gabrielsen, Geir W.; and Skaare, Janneche U. 2003. Comparison of organochlorine concentrations and patterns between free-ranging zooplankton and zooplankton sampled from seabirds' stomachs. Chemosphere. 53(6):685-689.
Rec #: 474
KEYWORDS: organochlorines; PCBs; HCB; HCHs; chlordanes; DDTs; stomach contents; thick-billed murre; *Uria lomvia*; Barents Sea; Europe; biomagnification
ABSTRACT: To study the use of predators' stomach contents to measure the organochlorine load in free-ranging prey, the 2 zooplankton species *Thysanoessa inermis* (euphausiids) and *Themisto libellula* (amphipods) were collected simultaneously from the water column and from the seabird Brünnich's guillemot (*Uria lomvia*)'s stomach (from crop to proventriculus). The organochlorine compounds' concentrations and relative proportion of PCB-153 generally did not differ between the 2 sampling methods (least square mean of .sum. organochlorines=5.9 and 6.8 ng/g wet weight for water column and stomach-sampled prey, respectively), indicating that the organochlorines in stomach-sampled prey were not yet affected by mechanical, chemical and bacterial degradation Although the sample size is restricted, similar organochlorine concentrations and pattern in free-ranging and stomach-sampled prey may suggest that Brünnich's guillemots feed randomly on the zooplankton population rather than at more (or less) contaminated individuals. Alternatively or in addition, the similar concentrations and pattern suggest that the contaminant levels in Barents Sea zooplankton do not influence their behavior to change the vulnerability to predation. The organochlorine concentrations and pattern in zooplankton collected from seabirds' stomach correspond with levels in free-ranging prey.

Borgå, Katrine; Gabrielsen, Geir WIng; Hop, Haakon; and Skaare, Janneche Utne. 1998. Organochlorines and trophic positions in a marine pelagic food chain leading to seabirds in the Norwegian Arctic. Organohalogen Compounds. 39:431-434.
Rec #: 316
KEYWORDS: organochlorines; HCHs; HCB; chlordanes; DDTs; PCBs; stable isotopes; biomagnification; Svalbard; Norway; Europe; liver; black-legged kittiwake; *Rissa tridactyla*; glaucous-winged gull; *Larus glaucescens*
NOTES: FIGURES ONLY
ABSTRACT: The bioaccumulation of organochlorines was studied in the food chain leading to glaucous gull. N isotope ratios were used to detect the trophic position of the organisms. The $\delta15$-N-values increased from 9.38 % in copepods over euphausiids, amphipods, Spider crab, polar cod, Atlantic cod, black-legged kittiwake to 15.8 % in seabirds (glaucous gull). In the lower end of the food chain (copepods to fish) the concentrations of the organochlorines (HCHs, HCB, chlordanes, DDTs, and PCBs) were low (10.1-85.8 ng/g lipid weight) and did not increase with the organisms' trophic position. In the upper end of the food chain (seabirds) the concentrations of organochlorines were 1-3 orders of magnitude higher than in crustaceans and fish, depending on the organochlorine group. The overall increase in organochlorine concentration with the organisms trophic position in the marine food chain was HCHs$\leq$HCB$\leq$chlordanes$\leq$DDTs$\leq$PCBs. This is consistent with the physicochemical properties of the organisms.

Borlakoglu, J. T. and Haegele, K. D. 1991. Comparative aspects on the bioaccumulation, metabolism and toxicity with PCBs. Comp. Biochem. Physiol. C. 100C(3):327-328.
Rec #: 205

Annotated Bibliography (Cont.).

KEYWORDS: bioaccumulation; organochlorines; PCBs; metabolism; Great Lakes; North America; Baltic Sea; Europe; Tokyo Bay; Japan; Asia
NOTES: SUPPLEMENTAL
ABSTRACT: The commercial advantage of PCBs was based on their properties of high chemical stability, low aqueous solubility, non-flammability, and excellent electrical insulation. This versatility led to their widespread use as heat transfer fluids in transformers and capacitors, flame retardant, lubricating and hydraulic oils, additives in the manufacture of plastics, vehicles for ink in copying paper, and even as immersion oils for microscopes. Serious environmental contamination with PCBs has been documented for industrialized areas such as the Great Lakes, the Baltic Sea and Tokyo Bay. However, their accumulation throughout the ecosystem is shown by the presence of PCBs in samples of air, snow, ice, fish, birds and mammals obtained from the polar regions, in the surface and sub-surface water and atmosphere of oceans, and in a wide range of plankton, fish, and marine and terrestrial mammals including humans. The concentration of PCBs in lower trophic levels of aquatic organisms such as plankton, crustacean and fish depends primarily on the PCB concentration in the sediments and particulate matter in the ambient water and is to a lesser degree influenced by the ingestion of food contaminated wit

Borlakoglu, J. T.; Wilkins, J. P. G.; and Walker, C. H. 1988. Polychlorinated biphenyls in fish-eating sea birds--Molecular features and metabolic interpretations. Mar. Environ. Res. 24:15-19.
Rec #: 303
KEYWORDS: organochlorines; PCBs; bioaccumulation; cormorant; *Phalacrocorax carbo;* common murre; *Uria aalge;* Atlantic puffin; *Fratercula arctica;* razorbill; *Alca torda;* shag; *Phalacrocorax aristotelis*; fat
NOTES: SUPPLEMENTAL
ABSTRACT: Polychlorinated biphenyls (PCBs) are an important group of environmental pollutants, there being a total of 209 theoretical congeners. Residue analysis of the adipose tissue of five species of fish-eating sea birds from British and Irish coastal waters revealed the presence of up to 60 different congeners. Comparison of the molecular structure for the persistent PCB congeners revealed the lack of meta-para unsubstituted adjacent carbon atoms. It has been shown that meta-para unsubstituted adjacent carbon atoms facilitate the metabolism of PCBs and it is hypothesised that the formation of hydroxy derivatives may depend upon such a requirement.

Borlakoglu, J. T.; Wilkins, J. P. G.; Walker, C. H.; and Dils, R. R. 1990. Polychlorinated biphenyls (PCBs) in fish-eating sea birds - 2. Molecular features of PCB isomers and congeners in adipose tissue of male and female puffins (*Fratercula arctica*), guillemots (*Uria aalge*), shags (*Phalacrocorax aristotelis*) and cormorants (*Phalacrocorax carbo*) of British and Irish coastal waters. Comp. Biochem. Physiol. C. 97C(1):161-171.
Rec #: 261
KEYWORDS: organochlorines; PCBs; DDTs; Great Britain; Ireland; Europe; fat; razorbill; *Alca torda*; Atlantic puffin; *Fratercula arctica*; common murre; *Uria aalge*; shag; *Phalacrocorax aristotelis;* cormorant; *Phalacrocorax carbo*; gender differences
ABSTRACT: The concentration of individual PCBs was measured in adipose tissue of male and female puffins (*Fratercula arctica*), shags (*Phalacrocorax aristotelis*), guillemots (*Uria aalge*) and cormorants (*Phalacrocorax carbo*) obtained from the Isle of May and the Saltees islands. The concentrations of total PCBs showed positive correlations with that of p,p'-DDE in the tissues. Enrichment factors were calculated by comparing the concentration of an individual PCB

Annotated Bibliography (Cont.).

in the tissue with its abundance in commercial mixtures of PCBs. Of the 47 individual PCBs identified five prominent PCBs had enrichment factors considerably > 1 and accounted for approximately 35 % of the total concentration of PCBs present.

Borlakoglu, J. T.; Wilkins, J. P. G.; Walker, C. H.; and Dils, R. R. 1990. Polychlorinated biphenyls (PCBs) in fish-eating sea birds -- 3. Molecular features and metabolic interpretations of PCB isomers and congeners in adipose tissues. Comp. Biochem. Physiol. C. 97C(1):173-177.
Rec #: 260
KEYWORDS: biomagnification; organochlorines; PCBs; fat; Great Britain; Europe; razorbill; *Alca torda*; Atlantic puffin; *Fratercula arctica*; common murre; *Uria aalge*; shag; *Phalacrocorax aristotelis;* cormorant; *Phalacrocorax carbo*; metabolism
NOTES: SUPPLEMENTAL
ABSTRACT: Enrichment factors have been calculated for several persistent PCB congeners in the adipose tissue for five species of fish-eating sea birds (female razorbills, puffins, guillemots, shags and cormorants) (*Alca torda, Fratercula, Uria aalge, Phalacrocorax* spp.). There were no significant differences between the five species in the enrichment factor of individual persistent PCBs compared with congener 153, indicating similar levels of diminished metabolism of this group of congeners. Of the 47 individual PCBs identified, ten congeners had enrichment factors of > 1 in all of the species and these accounted for up to 70 % of the concentration of total PCBs present.

Bouquegneau, J. M.; Das, K.; Debacker, V.; Gobert, S.; and Nellissen, J. P. 1996. Toxicological investigations on the heavy metals contamination of seabirds and marine mammals beached along the Belgian coast. Bulletin De La Societe Royale Des Sciences De Liege. 65(1):31-34.
Rec #: 88
KEYWORDS: review; metals; Cu; Zn; Cd; liver; Hg; metallothionein; Belgium; Norway; Scotland; Europe; toxicology; common murre; *Uria aalge*; effects
NOTES: NO TABLES
ABSTRACT: A short review of contaminants in stranded seabirds and marine mammals and the toxicity of these contaminants.

Bouquegneau, J. M.; Debacker, V.; Gobert, S.; and Havelange, S. 1996. Role of metallothioneins in metal regulation by the guillemot *Uria aalge*. Comp. Biochem. Physiol., C. 113C(2):135-139.
Rec #: 225
KEYWORDS: metals; Cu; Cd; Zn; liver; kidney; metallothionein; North Sea; Belgium; Europe; effects; common murre; *Uria aalge*
NOTES: FIGURES ONLY
ABSTRACT: Guillemots (*Uria aalge*), like other seabird species living in the North Sea, appear to be heavily contaminated by copper. Metallothioneins are present in both liver and kidney but, at least in the specimens stranded along the Belgian coast, fail to maintain constant the copper, zinc and cadmium load of the high molecular weight soluble proteins of both organs, stressing the potential toxic role of these metals, mainly copper.

Bourne, W. R. P. 1976. Seabirds and pollution. In: Johnston, R. (Ed.) Marine Pollution. London: Academic Press. 403-502.
Rec #: 540

Annotated Bibliography (Cont.).

KEYWORDS: review; oil; organochlorines; PCBs; PCDDs; PCDFs; DDTs; dieldrin; metals; Fe; Co; As; Hg; Ag; Pb; Ni; Cr; Zn; Ca; Cu; Sn; plastic; gland; bone; egg; carcass; feather; liver; fat; brain; muscle; effects; reproductive success; abnormalities; common murre; *Uria aalge*; razorbill; *Alca torda*; white-tailed sea eagle; *Haliaeetus albicilla*; northern fulmar; *Fulmarus glacialis*; gannet; *Sula bassana*; shag; *Phalacrocorax aristotelis*; great black-backed gull; *Larus marinus*; lesser black-backed gull; *Larus fuscus*; herring gull; *Larus argentatus*; black-legged kittiwake; *Rissa tridactyla*; Atlantic puffin; *Fratercula arctica*; common eider; *Somateria mollissima*; horned grebe; *Podiceps auritus*; western grebe; *Aechmophorus occidentalis*; cape gannet; *Sula capensis*; spotted shag; *Stictocarbo punctatus steadi*; white-tailed tropicbird; *Phaethon lepturus*; long-tailed duck; *Clangula hyemalis*; shelduck; *Tadorna tadorna*; greater scaup; *Aythya marila*; glaucous gull; *Larus hyperboreus*; brown pelican; *Pelecanus occidentalis*; double-crested cormorant; *Phalacrocorax auritus*; ashy petrel; *Oceanodroma homochroa*; osprey; *Pandion haliaetus*; Bermuda petrel; *Pterodroma cahow*; peregrine falcon; *Falco peregrinus*; sandwich tern; *Sterna sandvicensis*; spoonbill; *Platalea leucorodia*; cormorant; *Phalacrocorax carbo*; snow petrel; *Pagodroma nivea*; Wilson's petrel; *Oceanites oceanicus*; Manx shearwater; *Puffinus puffinus*; black guillemot; *Cepphus grylle*; thick-billed murre; *Uria lomvia*; turkey vulture; *Cathartes aura*; common raven; *Corvus corax*; common merganser; *Mergus merganser*; common loon; *Gavia immer*; common tern; *Sterna hirundo*; sooty tern; *Sterna fuscata*; Laysan albatross; *Diomedea immutabilis*; black-footed albatross; *Diomedea nigripes*; jackass penguin; *Spheniscus demersus*; magellanic penguin; *Spheniscus magellanicus*; white-flippered penguin; *Eudyptula albosignata*; blue penguin; *Eudyptula minor*; rockhopper penguin; *Eudyptes chrysocome*; Adelie penguin; *Pygoscelis adeliae*; chinstrap penguin; *Pygoscelis antarctica*; black-browed albatross; *Diomedea melanophris*; wandering albatross; *Diomedea exulans*; southern giant petrel; *Macronectes giganteus*; cape pigeon; *Daption capense*; broad-billed prion; *Pachyptila vittata;* fairy prion; *Pachyptila turtur*; mottled petrel; *Pterodroma inexpectata*; Atlantic petrel; *Pterodroma incerta*; Kerguelen petrel; *Pterodroma brevirostris*; grey petrel; *Procellaria cinerea*; great shearwater; *Puffinus gravis*; Cory's shearwater; *Calonectris diomedea*; sooty shearwater; *Puffinus griseus*; short-tailed shearwater; *Puffinus tenuirostris*; Audubon's shearwater; *Puffinus lherminieri*; Leach's storm-petrel; *Oceanodroma leucorhoa*; storm-petrel; *Hydrobates pelagicus*; fork-tailed storm petrel; *Oceanodroma furcata*; least petrel; *Oceanodroma microsoma*; black petrel; *Oceanodroma melania*; little auk; *Alle alle*; brown booby; *Sula leucogaster*; Australasian gannet; *Morus serrator*; masked booby; *Sula dactylatra*; white pelican; *Pelecanus erythrorhynchos*; Brandt's cormorant; *Phalacrocorax penicillatus*; pelagic cormorant; *Phalacrocorax pelagicus*; socotra cormorant; *Phalacrocorax nigrogularis*; Aukland Island shag; *Phalacrocorax colensoi*; blue-eyed shag; *Phalacrocorax atriceps*; magnificent frigatebird; *Fregata magnificens*; great skua; *Catharacta skua*; parasitic jaeger; *Stercorarius parasiticus*; great crested grebe; *Podiceps cristatus*; red-breasted merganser; *Mergus serrator*; red phalarope; *Phalaropus fulicarius*; purple sandpiper; *Calidris maritima*; yellow-billed loon; *Gavia adamsii*; Arctic loon; *Gavia arctica*; common goldeneye; *Bucephala clangula*; white-winged scoter; *Melanitta fusca*; black scoter; *Melanitta nigra;* red-throated loon; *Gavia stellata*; sooty albatross; *Phoebetria fusca*; diving petrel; *Pelecanoides urinatrix*; little gull; *Larus minutus*; little tern; *Sterna albifrons*; Antarctic tern; *Sterna vittata*; oystercatcher; *Haematopus ostralegus*; Bartram's sandpiper; *Bartramia longicauda*; coot; *Fulica atra*; gough gallinule; *Gallinula comeri*; Tristan thrush; *Nesocichla eremita*; Irish Sea; North Sea; Baltic Sea; Great Britain; Scotland; Shetland; Netherlands; Denmark; Ireland; Poland; Belgium; France;

Annotated Bibliography (Cont.).

Kattegat; Sweden; Europe; Ontario; New Brunswick; British Columbia; Canada; CA; NC; NJ; MA; USA; North America; Japan; New Zealand; Asia; Australia; Atlantic Ocean; Pacific Ocean; seasonal variations; spatial variations
ABSTRACT: A review examining the effects of marine pollution on seabirds. Four types of pollution (oil, toxic chemical, artefacts, and micro-organisms and parasites) are examined in different oceans and groups of birds. An appendix estimates the concentrations of chlorinated hydrocarbons in some 250 birds from different parts of the world.

Bourne, W. R. P. and Bogan, J. A. 1972. Polychlorinated biphenyls in North Atlantic seabirds. Mar. Pollut. Bull. 3(11):171-175.
Rec #: 423
KEYWORDS: organochlorines; PCBs; liver; muscle; fat; stomach oil; northern fulmar; *Fulmarus glacialis;* great shearwater; *Puffinus gravis;* Manx shearwater; *Puffinus puffinus*; storm-petrel; *Hydrobates pelagicus*; gannet; *Sula bassana*; shag; *Phalacrocorax aristotelis;* great skua; *Catharacta skua*; parasitic jaeger; *Stercorarius parasiticus*; glaucous gull; *Larus hyperboreus*; great black-backed gull; *Larus marinus*; lesser black-backed gull; *Larus fuscus;* herring gull; *Larus argentatus;* black-legged kittiwake; *Rissa tridactyla*; Arctic tern; *Sterna paradisaea*; black guillemot; *Cepphus grylle*; common murre; *Uria aalge;* thick-billed murre; *Uria lomvia;* razorbill; *Alca torda*; Atlantic puffin; *Fratercula arctica;* little auk; *Alle alle*; Atlantic Ocean; Norway; Scotland; Great Britain; Europe; spatial variations
ABSTRACT: Polychlorinated biphenyls are usually thought to be discharged in industrial effluents and therefore to be commonest in inshore waters close to centres of industry. It now appears that high concentrations are found in pelagic-feeding seabirds from the North Atlantic, far from possible industrial sources of these compounds.

Braestrup, Liselotte; Clausen, Jørgen; and Berg, Ole. 1974. DDE, PCB, and aldrin levels in arctic birds of Greenland. Bull. Environ. Contam. Toxicol. 11:326-332.
Rec #: 523
KEYWORDS: organochlorines; PCBs; DDTs; aldrin; HCHs; Greenland; North America; biomagnification; fat; *Somateria spectabilis;* king eider; *Somateria mollissima;* common eider; *Histrionicus histrionicus;* harlequin duck; *Clangula hyemalis;* long-tailed duck; *Calidris maritima;* purple sandpiper; *Uria lomvia;* thick-billed murre; *Phalacrocorax carbo*; cormorant; *Lagopus mutus;* rock ptarmigan; *Corvus corax;* common raven
ABSTRACT: The use of chlorinated pesticides for more than 30 years and of polychlorinated biphenyls (PCB) for more than 40 years has caused a world-wide accumulation of these components in different ecosystems (1,7,9,10,12,13,18,19,20). Since no studies have been made on theme components in Arctic ecosystems of Greenland, the present study was designed to elucidate this, especially concerning the concentration of polychlorinated hydrocarbons (PCHC) in birds. This is of interest since these often migrate or eat migrating fish, thus being affected by areas in which PCHC are used.

Braune, B. M.; Donaldson, G. M.; and Hobson, K. A. 2002. Contaminant residues in seabird eggs from the Canadian Arctic. II. Spatial trends and evidence from stable isotopes for intercolony differences. Environ. Pollut. 117:133-145.
Rec #: 340
KEYWORDS: organochlorines; PCBs; DDTs; chlordanes; dieldrin; mirex; benzenes; HCHs;

Annotated Bibliography (Cont.).

metals; Hg; Se; stable isotopes; egg; glaucous gull; *Larus hyperboreus*; black-legged kittiwake; *Rissa tridactyla*; thick-billed murre; *Uria lomvia*; black guillemot; *Cepphus grylle*; Nunavut; Canada; North America; spatial variations; biomagnification

ABSTRACT: Eggs of glaucous gulls (*Larus hyperboreus*), black-legged kittiwakes (*Rissa tridactyla*), thick-billed murres (*Uria lomvia*) and black guillemots (*Cepphus grylle*) were collected from several sites throughout the Canadian Arctic. Samples were analyzed for organochlorines as well as mercury and selenium. Glaucous gulls breeding at sites in the High Arctic showed higher levels of organochlorine contamination than those in the western Low Arctic. This was likely due to dietary differences among colonies as suggested by stable isotope data, although different overwintering areas may also play a role. Levels of ΣPCB, ΣDDT, ΣCHLOR, ΣCBz and dieldrin were significantly lower in thick-billed murres from Prince Leopold Island in the High Arctic compared with colonies in the eastern Low Arctic. This difference was likely due to the combined effects of different atmospheric deposition patterns in the High and Low Arctic and different overwintering areas since murres from Prince Leopold Island may winter farther north than murres from the other colonies sampled. Eggs from colonies at higher latitudes generally contained higher concentrations of mercury. The trophic and dietary differences/similarities suggested by stable-nitrogen and carbon isotope data in this study were useful in explaining the spatial patterns of contaminant concentrations observed among colonies of seabirds such as the glaucous gull and the black-legged kittiwake where variation in latitudinal atmospheric deposition patterns and different overwintering grounds did not appear to be confounding factors.

Braune, B. M.; Donaldson, G. M.; and Hobson, K. A. 2001. Contaminant residues in seabird eggs from the Canadian Arctic. Part I. Temporal trends 1975-1998. Environ. Pollut. 114:39-54. Rec #: 305
KEYWORDS: organochlorines; PCBs; DDTs; benzenes; chlordanes; dieldrin; mirex; HCHs; metals; Hg; Se; stable isotopes; biomagnification; temporal trends; egg; thick-billed murre; *Uria lomvia*; northern fulmar; *Fulmarus glacialis*; black-legged kittiwake; *Rissa tridactyla;* Nunavut; Canada; North America
ABSTRACT: Concentrations of total mercury, selenium and a suite of organochlorine compounds were measured in eggs of thick-billed murres (*Uria lomvia*), northern fulmars (*Fulmarus glacialis*) and black-legged kittiwakes (*Rissa tridactyla*) collected on Prince Leopold Island in Lancaster Sound, Nunavut, Canada, between 1975 and 1998. Mercury levels in thick-billed murre and northern fulmar eggs increased significantly during this period while selenium concentrations decreased significantly in northern fulmar eggs. Mercury and selenium concentrations in black-legged kittiwake eggs exhibited no significant temporal trends. Concentrations of ΣPCB, ΣDDT and total chlorobenzenes decreased over time for all three species and there was a shift in the PCB congener pattern as the hexachlorobiphenyl fraction of ΣPCB increased and the lower chlorinated biphenyl fraction decreased. Total chlordane, dieldrin and mirex concentrations decreased in kittiwake eggs while no significant trends were observed for the other two species. Increases in ΣHCH levels were detected in thick-billed murre eggs but not in northern fulmar and black-legged kittiwake eggs. Levels of the β-HCH isomer, however, increased significantly in murres and fulmars. Stable-nitrogen isotope analyses ($\delta^{15}N$) indicate that the temporal trends observed for contaminant concentrations in eggs were not the result of shifts in trophic level. Changing deposition patterns of xenobiotic compounds over the summer

Annotated Bibliography (Cont.).

and winter ranges of these birds provide a likely explanation for differing exposures through time.

Braune, B. M.; Wakeford, Brian; and Gaston, Anthony. 1999. Contaminants in Arctic seabird eggs. Synopsis of Research Conducted Under the Northern Contaminants Program 1998-1999. Ottawa: Department of Indian Affairs and Northern Development. 77-80.
Rec #: 310
KEYWORDS: organochlorines; PCBs; DDTs; benzenes; chlordanes; dieldrin; mirex; HCHs; metals; Hg; egg; black guillemot; *Cepphus grylle*; thick-billed murre; *Uria lomvia*; northern fulmar; *Fulmarus glacialis*; black-legged kittiwake; *Rissa tridactyla;* glaucous-winged gull; *Larus glaucescens*; Nunavut; Canada; North America; temporal trends
ABSTRACT: Five species of arctic seabird eggs were collected from Prince Leopold Island, Canada, in 1998 and analyzed for PCBs, DDE, chlordanes, dieldrin, mirex, chlorobenzenes, hexachlorobenzenes, and total mercury.

Braune, Birgit M. 1987. Comparison of total mercury levels in relation to diet and molt for nine species of marine birds. Arch. Environ. Contam. Toxicol. 16(2):217-224.
Rec #: 382
KEYWORDS: metals; Hg; double-crested cormorant; *Phalacrocorax auritus*; common eider; *Somateria mollissima*; black guillemot; *Cepphus grylle*; red-necked phalarope; *Phalaropus lobatus*; herring gull; *Larus argentatus*; Bonaparte's gull; *Larus philadelphia*; common tern; *Sterna hirundo*; Arctic tern; *Sterna paradisaea*; black-legged kittiwake; *Rissa tridactyla;* New Brunswick; Canada; North America; muscle; liver; kidney; brain; feather; biomagnification
ABSTRACT: Total Hg concentrations were analyzed for tissues of 9 species of marine birds from the Quoddy region, New Brunswick, Canada, including cormorants (*Phalacrocorax auritus*), eiders (*Somateria mollissima*), guillemots (*Cepphus grylle*), phalaropes (*Phalaropus lobatus*), gulls (*Larus argentatus, L. philadelphia*), terns (*Sterna hirundo, S. paradisaea*), and kittiwakes (*Rissa tridactyla*). There was a progressive decrease in Hg concentration from the innermost to the outermost primary feather in Bonaparte's gulls, herring gulls, black-legged kittiwakes, and Arctic terns. Primaries of common terns and black guillemots showed no significant trend. Cormorants, guillemots, and eiders, which feed on benthic organisms, and common terns, which feed predominantly on fish, had the highest tissue Hg levels, whereas birds such as kittiwakes and phalaropes, which consume mainly pelagic invertebrates, had the lowest Hg levels.

Braune, Birgit M. and Gaskin, David E. 1987. A Mercury Budget for the Bonaparte's Gull During Autumn Moult. Ornis Scand. 18(4):244-250.
Rec #: 212
KEYWORDS: energetics; metals; Hg; seasonal variations; bioaccumulation; elimination; feather; carcass; feces; Bonaparte's gull; *Larus philadelphia*; Bay of Fundy; New Brunswick; Canada; North America
ABSTRACT: A bioenergetics-based budget that predicts net total Hg loss during the period of autumn moult was calculated for adult Bonaparte's Gulls (*Larus philadelphia*) in the southwestern Bay of Fundy, Canada. Daily food consumption and, hence, ingestion of Hg measured in prey samples, was calculated from estimated energy requirements during the period of moult. The amount of Hg eliminated was estimated from analyses of Hg content in excreta

Annotated Bibliography (Cont.).

and feathers. Elimination of Hg via the feathers accounted for 68 % of the total loss from the body in females and 59 % in males during the period of autumn moult.

Braune, Birgit M. and Gaskin, David E. 1987. Mercury levels in Bonaparte's gulls (*Larus philadelphia*) during autumn molt in the Quoddy region, New Brunswick, Canada. Arch. Environ. Contam. Toxicol. 16(5):539-549.
Rec #: 213
KEYWORDS: metals; Hg; body burden; North America; Canada; New Brunswick; Bay of Fundy; Bonaparte's gull; *Larus philadelphia*; feather; muscle; brain; liver; kidney; carcass; gender differences; age variations; bioaccumulation; elimination
ABSTRACT: No significant between sex differences were detected in Hg concentrations in primary feathers, pectoral muscle, brain, liver, and kidney tissues of fall migrating juvenile and second-year Bonaparte's gulls (*Larus philadelphia*) collected in the Quoddy region. Adults showed sexual differences only in the first 5 primary feathers, and in muscle, kidney and brain. Differences in Hg concentrations among age groups were reflected in the primary feathers and body tissues, but as the molt progressed, Hg concentrations decreased as they converged toward a minimum asymptotic Hg level for each tissue. This suggests that the body burden of Hg was reduced through its redistribution from the body tissues into the growing feathers. Mercury concentrations in premolt head feathers (pre-egg-laying) did not vary significantly between adult females and males, whereas Hg concentrations in postmolt feathers (post-egg-laying) were significantly lower in females, suggesting that egg-laying was also a route for Hg elimination.

Braune, Birgit M. and Norstrom, Ross J. 1989. Dynamics of organochlorine compounds in herring gulls: III. Tissue distribution and bioaccumulation in Lake Ontario gulls. Environ. Toxicol. Chem. 8(10):957-968.
Rec #: 214
KEYWORDS: Great Lakes; North America; biomagnification; herring gull; *Larus argentatus*; carcass; egg; liver; organochlorines; PCBs; HCB; DDTs; mirex; HCHs; chlordanes; dieldrin; PCDDs; PCDFs
ABSTRACT: Apparent biomagnification factors (BMFs, wet weight basis) for organochlorine compounds in herring gulls (*Larus argentatus*) in Lake Ontario were shown to be related to chlorine substitution patterns in the case of polychlorinated biphenyls (PCBs), polychlorinated dibenzo-p-dioxins (PCDDs) and polychlorinated dibenzofurans (PCDFs). PCBs accumulated according to the availability of adjacent positions not substituted with chlorine: (a) para-meta unsubstituted (P group), mean BMF 20 plus or minus 6.2; (b) meta-ortho unsubstituted (M group), mean BMF 87 plus or minus 36; and (c) no adjacent unsubstituted positions (blocked, B group), mean BMF 154 plus or minus 39. Among the other organochlorines, DDT, cis- and trans-nonachlor, dieldrin and octachlorostyrene magnified to the least extent (BMF 3-19), and photomirex, mirex and DDE magnified to the greatest extent (BMF 85-100). Of the PCDDs and PCDFs, 2,3,7,8-TCDD and 2,3,4,7,8-PnCDF biomagnified to a greater degree than did other congeners (BMF 32 and 7, respectively).

Braune, Birgit M. and Simon, Mary. 2003. Dioxins, furans, and non-ortho PCBs in Canadian Arctic seabirds. Environ. Sci. Technol. 37(14):3071-3077.
Rec #: 454
KEYWORDS: organochlorines; PCDDs; PCDFs; PCBs; toxicology; TEQs; thick-billed murre;

Annotated Bibliography (Cont.).

Uria lomvia; northern fulmar; *Fulmarus glacialis*; black-legged kittiwake; *Rissa tridactyla*; Nunavut; Canada; North America; egg; liver; temporal trends

ABSTRACT: This is the first account of polychlorinated dibenzodioxins (PCDDs), polychlorinated dibenzofurans (PCDFs), and non-ortho polychlorinated biphenyls (PCBs) in Canadian Arctic seabirds. Livers and eggs of thick-billed murres, northern fulmars, and black-legged kittiwakes were collected in 1975 and 1993 from Prince Leopold Island in Lancaster Sound, Canada. Detectable concentrations of PCDDs, PCDFs, and non-ortho PCBs were found in all the Arctic seabird samples analyzed. Of the PCDD congeners assayed, only 2,3,7,8-substituted PCDDs were detected in the samples, whereas non-2,3,7,8-substituted PCDFs were found in addition to 2,3,7,8-substituted PCDFs in some of the samples. The predominant PCDD/F congener found in the livers of all three species was 2,3,4,7,8-pentachlorodibenzofuran, both in 1975 and 1993. Concentrations of most dioxins decreased in the fulmars and kittiwakes between 1975 and 1993 but increased in the murres. Of the non-ortho PCBs measured, PCB-126 occurred in the highest concentrations and contributed the majority of the non-ortho PCB-TEQ in all three species in both years. The highest concentrations of dioxins, as well as the highest TEQ values, were found in the northern fulmar livers in both 1975 and 1993. Concentrations of some of the PCDDs and PCDFs are among the highest reported for Canadian Arctic biota.

Braune, Birgit M.; Wong, Michael P.; Belles-Isles, Jean-Claude; and Marshall, W. Keith. 1991. Chemical residues in Canadian game birds. Technical Report Series No. 124. Ottawa, Ontario, Canada: Canadian Wildlife Service Ottawa Region. 375 pp.
Rec #: 419
KEYWORDS: review; organochlorines; aldrin; chlordanes; benzenes; HCHs; DDTs; dieldrin; endrin; PCDDs; PCDFs; PCBs; HCB; mirex; OCS; toxaphene; phenols; metals; Ag; As; Cd; Cs; Cr; Cu; Fe; Pb; Hg; Ni; Se; Sr; V; Zn; organometallics; organolead; organic Hg; Canada goose; *Branta canadensis*; snow goose; *Chen caerulescens*; mallard; *Anas platyrhynchos*; gadwall; *Anas strepera*; pintail; *Anas acuta*; green-winged teal; *Anas crecca*; blue-winged teal; *Anas discors*; American wigeon; *Anas americana*; shoveler; *Anas clypeata*; wood duck; *Aix sponsa*; redhead duck; *Aythya americana*; ring-necked duck; *Aythya collaris*; canvasback duck; *Aythya valisineria*; greater scaup; *Aythya marila*; lesser scaup; *Aythya affinis*; common goldeneye; *Bucephala clangula*; bufflehead; *Bucephala albeola*; long-tailed duck; *Clangula hyemalis*; common eider; *Somateria mollissima*; white-winged scoter; *Melanitta fusca*; surf scoter; *Melanitta perspicillata*; black scoter; *Melanitta nigra*; ruddy duck; *Oxyura jamaicensis*; common merganser; *Mergus merganser*; hooded merganser; *Lophodytes cucullatus*; red-breasted merganser; *Mergus serrator*; blue grouse; *Dendragapus obscurus*; spruce grouse; *Dendragapus canadensis*; ruffed grouse; *Bonasa umbellus*; willow ptarmigan; *Lagopus lagopus*; sharp-tailed grouse; *Tympanuchus phasianellus*; pheasant; *Phasianus colchicus*; chukar; *Alectoris graeca*; grey partridge; *Perdix perdix*; common moorhen; *Gallinula chloropus*; American coot; *Fulica americana*; American woodcock; *Scolopax minor*; common snipe; *Gallinago gallinago*; razorbill; *Alca torda*; common murre; *Uria aalge*; thick-billed murre; *Uria lomvia*; black guillemot; *Cepphus grylle*; pigeon guillemot; *Cepphus columba*; marbled murrelet; *Brachyramphus marmoratus*; ancient murrelet; *Synthliboramphus antiquus*; Cassin's auklet; *Ptychoramphus aleuticus*; rhinoceros auklet; *Cerorhinca monocerata*; Atlantic puffin; *Fratercula arctica*; tufted puffin; *Fratercula cirrhata*; common pigeon; *Columba livia*; mourning dove; *Zenaida macroura*; bone; brain; muscle; carcass; egg; fat; feather; gonad; intestine; kidney; liver; gland; Newfoundland; Nova Scotia; New Brunswick; Quebec; Ontario; Manitoba;

Annotated Bibliography (Cont.).

Saskatchewan; Alberta; British Columbia; Northwest Territories; Canada; North America
ABSTRACT: This report reviews data on chemical residues in Canadian game birds. Past efforts on surveying residues in game birds and the health hazards posed by toxic chemicals to game birds and their consumers are discussed. The majority of residue surveys conducted in Canada focused on mercury, DDT, and PCB contamination. Although information is available for a large spectrum of contaminants, data are often limited for compounds other than those mentioned above. Most of the information was collected prior to 1975 in areas of known or suspected contamination. Important data gaps, thus, exist.

Earlier studies reported high residue concentrations in tissues of game birds from a few areas (e.g., Alberta, New Brunswick). Chemical residues were generally found at low concentrations in birds from recent surveys. Nonetheless, there is evidence of important toxic chemical pollution in some areas (e.g., Hamilton Harbour, Ontario; Baie des Chaleurs, New Brunswick) Canadian game birds can also accumulate chemical residues along their migration routes or on their wintering grounds. Although levels of most organochlorines have declined since the mid1960s in the United States, unexpectedly high levels of contaminants are still found in game birds from a few wintering locations (e.g., Alabama).

Game birds can concentrate chemical residues to levels of concern for their health and that of their consumers. Although estimates of the controlled harvests are available, there is little information on the native use of wildlife as food resources. More information is needed to assess the health hazard from consumption of contaminated game by native populations.

Braune, Birgit Margret. 1985. Total mercury accumulation during Autumn moult in Bonaparte's gulls of the Quoddy region, New Brunswick, Canada. University of Guelph. Dept. of Zoology. Ph.D. Thesis. 232 pp.
Rec #: 383
KEYWORDS: metals; Hg; feather; liver; kidney; muscle; brain; heart; carcass; Bonaparte's gull; *Larus philadelphia*; double-crested cormorant; *Phalacrocorax auritus;* common eider*; Somateria mollissima;* black guillemot*; Cepphus grylle;* common tern; *Sterna hirundo;* Arctic tern*; Sterna paradisaea;* black-legged kittiwake*; Rissa tridactyla*; red-necked phalarope; *Phalaropus lobatus;* New Brunswick; Canada; North America; bioaccumulation; gender differences; age variations; elimination
NOTES: NOT ABLE TO OBTAIN
ABSTRACT: Bonaparte's Gulls (*Larus philadelphia*) used the Quoddy region as a major autumn staging ground during late July through December where they underwent postnuptial moult and feather renewal which lasted about 14 weeks. While in the region, the largest energy and total mercury contribution to the birds' diet came from fish, with smaller contributions from euphausiids, insects and other marine invertebrates. After feathers, liver and kidney accumulated the highest mercury concentrations, and muscle, brain, heart and whole body mercury levels were not significantly different from each other. Juvenile and second-year birds showed no significant difference between sexes in mercury concentrations in primary feathers nor in the soft tissues, whereas adult males had higher mercury levels than females in the first five primary feathers, and in muscle, kidney and brain. Inter-age differences in mercury concentrations were reflected in the primary feathers and the soft tissues, but as the body burden of mercury was redistributed into the newly-developing feathers during moult, mercury concentrations in the soft

Annotated Bibliography (Cont.).

tissues converged toward a minimum asymptotic mercury level for each tissue. After the completion moult, the new feathers contained most the body burden of mercury (93.0 % in adults). The calculated total mercury budget came within 33 % of the expected body burden of mercury for juveniles, within 20 % for second-year birds and within 10 % far adults. Other seabird species (*Phalacrocorax auritus, Somateria mollissima,Cepphus grylle, Sterna hirundo*) which fed predominantly on benthic organisms or fish had the highest mercury levels whereas species (*Rissa tridactyla, Phalaropus lobatus*) which consumed mainly pelagic invertebrates had the lowest levels. Four species (*Larus philadelphia, L. argentatus, R. tridactyla. Sterna paradisaea*) showed a progressive decrease in mercury concentrations through the primary feather sequence, while two other species (*S. hirundo, C. grylle*) showed no significant trend.

Briggs, Kenneth T.; Yoshida, Steven H.; and Gershwin, M. Eric. 1996. The influence of petrochemicals and stress on the immune system of seabirds. Regul. Toxicol. Pharmacol. 23(2):145-155.
Rec #: 265
KEYWORDS: oil; immunology; effects; review
NOTES: SUPPLEMENTAL
ABSTRACT: There is increasing attention directed to the role of environmental pollutants in altering immune function. Only with the identification of the responsible environmental toxicants, and an understanding of their mechanisms of action, can we hope to treat immunotoxic injuries. This situation is exemplified by the exposure of wild birds to oil spills, the subsequent potential for direct toxicity from the oil, and the secondary toxicity of stress-induced immune modulation. Immunosuppressive mechanisms related to oil ingestion and handling stress are implicated in the morbidity and mortality of seabirds during care and following reentry into the wild. This does suggest that improvements in the treatment of these affected animals will enhance their survival and well-being. However, a survey of the literature shows that the implementation of better techniques are hampered by inadequate information on the immunological consequences of oil contact with seabirds. Marine oil pollution is a constant occurrence and will continue as long as oil and oil products are important commodities transported by sea routes. Among the numerous negative consequences of oil pollution are its effects on marine wildlife. There is much evidence that oil spills are responsible for massive seabird deaths. However, the constant, low level releases of petrochemicals probably contribute to the harmful effects of oil pollution on seabird populations. In an attempt to rectify the damage inflicted on seabirds by accidental oil discharge, rehabilitation centers are established for the cleaning and care of affected wildlife. Unfortunately, there is evidence that the ingestion of oil by preening and the handling stress undergone by birds in these centers lowers their ability to survive and reproduce following release to their native habitats. Although the reasons for this are unclear, there is the suggestion that both oil and handling will induce immunosuppressive mechanisms that ultimately predispose birds to infections and immune-mediated diseases, as well as reproductive, behavioral, and other problems. Thus, there are questions concerning the effectiveness of intervention measures currently being used in the rehabilitation of seabirds.

Broman, Dag; Näf, Carina; Lundbergh, Ivar; and Zebühr, Yngve. 1990. An in situ study on the distribution, biotransformation and flux of polycyclic aromatic hydrocarbons (PAHs) in an aquatic food chain (seston-*Mytilus edulis* L.-*Somateria mollissima* L.) from the Baltic: An ecotoxicological perspective. Environ. Toxicol. Chem. 9(4):429-442.

Annotated Bibliography (Cont.).

Rec #: 485
KEYWORDS: hydrocarbons; PAHs; common eider; *Somateria mollissima*; gallbladder; liver; fat; egg; Baltic Sea; Europe; biomagnification; metabolism
ABSTRACT: This in situ study is focusing on the distribution, biotransformation, and flux of 19 polycyclic aromatic hydrocarbons (PAHs) in the food chain seston-blue mussel (*M. edulis*)-common eider duck (*S. mollissima*) as well as the distribution in the gallbladder, liver, adipose tissue, and egg of the duck. All samples were collected within the open northern Baltic proper coastal areas. Analyses were carried out by GC/mass spectrometry with electron-impact and negative-ion chemical ionization. With a multivariate statistical method, a significant change in the PAH composition through the food chain was found. This change probably depends on an increasing metabolic activity with increasing trophic level, due to a selective biotransformation capacity for different PAHs. Decreasing PAH concentrations with increasing trophic level were found. The PAH concentrations in the different eider duck organs were: gallbladder > adipose tissue ≥ liver. The theoretical inhalation of air-dispersed PAHs was of no significance compared to the exposure from food. The relatively high theoretical PAH flux through the food chain did not result in increasing concentrations with increasing trophic level, which indicates that PAHs are biotransformed quite fast. However, many intermediate metabolites of PAHs have a mutagenic and carcinogenic potential, which makes it important to observe these compounds when assessing ecotoxicological risks.

Broman, Dag; Näf, Carina; Rolff, Carl; Zebühr, Yngve; Fry, Brian; and Hobbie, John. 1992. Using ratios of stable nitrogen isotopes to estimate bioaccumulation and flux of polychlorinated dibenzo-p-dioxins (PCDDs) and dibenzofurans (PCDFs) in two food chains from the northern Baltic. Environ. Toxicol. Chem. 11(3):331-345.
Rec #: 483
KEYWORDS: organochlorines; PCDDs; PCDFs; stable isotopes; Baltic Sea; Sweden; Europe; common eider; *Somateria mollissima*; carcass; muscle; fat; viscera; feces; bile; biomagnification; elimination; metabolism
ABSTRACT: A method for the estimation of in situ biomagnification of organic contaminants uses ratios of naturally occurring stable isotopes of N to classify trophic levels of organisms from one littoral and one pelagic food chain in the northern Baltic proper. Results indicated a biomagnification of polychlorodibenzo-p-dioxins and -furans (PCDD/Fs), whereas the total concentration of 2,3,7,8-substituted PCDD/Fs, and particularly octachlorodibenzo-p-dioxin and -dibenzofuran (OCDD/F), decreased with increasing trophic level in the food chains. A calculated flux estimate of PCDD/Fs in juvenile eider ducks was supported by results from the biomagnification study; i.e., the most toxic 2,3,7,8-substituted isomers tended to accumulate in the tissue of the eider duck. Out of the total PCDD/Fs consumed by the eider ducks, only 10 % were recovered in the body and approximately 10 % were recovered in feces, whereas 80 % were not recovered, i.e., were excreted or remained as unidentified metabolites. Of the most toxic 2,3,7,8-substituted isomers, 57 % were recovered in the tissue, 30 % were recovered in the feces, and only 13 % were not recovered (i.e., metabolized and/or excreted).

Brown, N. J. and Brown, A. W. A. 1970. Biological fate of DDT in a sub-arctic environment. J. Wildl. Manage. 34(4):929-940.
Rec #: 384
KEYWORDS: organochlorines; DDTs; willow ptarmigan; *Lagopus lagopus;* Arctic tern; *Sterna*

Annotated Bibliography (Cont.).

paradisaea; Bonaparte's gull; *Larus philadelphia*; Manitoba; Canada; North America; fat; brain; muscle; kidney; liver; intestine; biomagnification
ABSTRACT: Residues of DDT plus metabolites were determined in 660 samples of soil, plants, and animals in an area at Fort Churchill treated with 22 airsprays at 0.22 lb./acre applied between 1947 and 1964. The average residues of DDT plus metabolites in the fat of birds ranged between 3 ppm in the willow ptarmigan (*Lagopus lagopus*) and 64 ppm in the arctic tern (*Sterna paradisaea*) as compared with 0.2-12 ppm in fat from birds taken in the surrounding unsprayed area. The proportion of DDE found in the various bird species was partly dependent on the DDE production of their plant or invertebrate food. Of the tissues assessed, the pectoral muscle had the lowest residue, except in the case of the Arctic tern and Bonaparte's gull (*Larus philadelphia*).

Brunström, Björn and Halldin, Krister. 2000. Ecotoxicological risk assessment of environmental pollutants in the Arctic. Toxicol. Lett. 112-113:111-118.
Rec #: 436
KEYWORDS: organochlorines; PCBs; toxicology; TEQs; egg; liver; glaucous gull; *Larus hyperboreus*; herring gull; *Larus argentatus*; egg; liver; Barents Sea; Svalbard; Norway; Europe; Canada; North America; temporal trends; reproductive success; effects
NOTES: FIGURES ONLY
ABSTRACT: A review and discussion with many references concentrations of such persistent organic pollutants (POPs) as polychlorinated biphenyls (PCBs) are high in certain Arctic animal species. The polar bear, Arctic fox, and glaucous gull may be exposed to PCB levels above lowest-observed-adverse-effect-level (LOAEL) values for adverse effects on reproduction in mammals and birds. However, the dioxin-like congeners seem to be major contributors to the reproductive effects of PCBs and the relative concentrations of these congeners are low in polar bears. Temporal trends for POPs in Arctic wildlife and the sensitivities of Arctic species to these compounds detect the risk for future adverse health effects.

Buckman, Andrea H.; Norstrom, Ross J.; Hobson, Keith A.; Karnovsky, Nina J.; Duffe, Jason; and Fisk, Aaron T. 2004. Organochlorine contaminants in seven species of Arctic seabirds from northern Baffin Bay. Environ. Pollut. 128(3):327-338.
Rec #: 439
KEYWORDS: Northwater Polynya; Canada; Greenland; North America; Atlantic Ocean; fat; liver; muscle; little auk; *Alle alle*; thick-billed murre; *Uria lomvia*; black guillemot; *Cepphus grylle*; black-legged kittiwake; *Rissa tridactyla*; ivory gull; *Pagophila eburnea*; glaucous gull; *Larus hyperboreus*; northern fulmar; *Fulmarus glacialis;* organochlorines; PCBs; HCB; benzenes; HCHs; chlordanes; DDTs; OCS; dieldrin; mirex; stable isotopes; biomagnification; gender differences
ABSTRACT: Organochlorine contaminants (OCs) were detected in liver and fat of seven species of seabirds (*Alle alle, Uria lomvia, Cepphus grylle, Rissa tridactyla, Pagophila eburnea, Larus hyperboreus,* and *Fulmarus glacialis*) collected in May/June 1998 from the Northwater Polynya in northern Baffin Bay. OC concentrations ranged over an order of magnitude between seabird species and OC groups, with PCBs having the highest concentrations followed by DDT, chlordane, HCH and ClBz. Positive relationships between $\delta^{15}N$ (estimator of trophic level) and OC concentrations (lipid basis) were found for all OC groups, showing that trophic position and biomagnification significantly influence OC concentrations in Arctic seabirds. Concentrations of

Annotated Bibliography (Cont.).

a number of OCs in particular species (e.g., HCH in *P. eburnean*) were lower than expected based on $\delta^{15}N$ and was attributed to biotransformation. *P. eburnea* and *F. glacialis*, which scavenge, and *R. tridactyla*, which migrate from the south, were consistently above the $\delta^{15}N$-OC regression providing evidence that these variables can elevate OC concentrations Stable isotope measurements in muscle may not be suitable for identifying past scavenging events by seabirds. OC relative proportions were related to trophic position and phylogeny, showing that OC biotransformation varies between seabird groups. Trophic level, migration, scavenging and biotransformation all play important roles in the OCs found in Arctic seabirds. Concentrations of organochlorines in high Arctic seabirds are influenced by trophic level, migration, scavenging and biotransformation.

Bull, K. R.; Murton, R. K.; Osborn, D.; Ward, P.; and Cheng, Lana. 1977. High levels of cadmium in Atlantic seabirds and sea-skaters. Nature. 269(5628):507-509.
Rec #: 388
KEYWORDS: metals; Cd; kidney; liver; northern fulmar; *Fulmarus glacialis*; Manx shearwater; *Puffinus puffinus*; Atlantic puffin; *Fratercula arctica*; Leach's storm-petrel; *Oceanodroma leucorhoa*; storm-petrel; *Hydrobates pelagicus*; razorbill; *Alca torda*; Scotland; Europe; spatial variations
ABSTRACT: The Cd concentrations of kidney and liver were determined in 21 Atlantic seabirds representing 6 species (fulmar, Manx shearwater, puffin, Leach's petrel, storm petrel, and razorbill); the Cd residues were higher in the kidney (range 14.6-240 mg/kg dry weight) than in the liver (range 1.4-57.0 mg/kg dry weight). Cd residues were higher in the pelagic species (shearwaters, puffins, and petrels) than in the razorbill. Samples of marine insects (*Halobates micans*), obtained from tropical areas of the Atlantic Ocean, contained a mean Cd concentration of 22.7 mg/kg dry weight High Cd concentrations found in seabirds appear to originate from natural, rather than industrial, sources; the birds have probably developed mechanisms which enable them to tolerate Cd.

Burger, J. 1990. Behavioral effects of early postnatal lead exposure in herring gull (*Larus argentatus*) chicks. Pharmacol. Biochem. Behav. 35(1):7-13.
Rec #: 196
KEYWORDS: effects; herring gull; *Larus argentatus*; metals; Pb; dosing study; NJ; USA; North America
NOTES: SUPPLEMENTAL
ABSTRACT: In this paper, I use the herring gull, *Larus argentatus*, as an animal model to examine effects of lead exposure on early development. Like humans, birds rely mainly on visual and vocal, rather than olfactory, modes of communication. Although on most days, begging behavior, balance and righting response did not differ significantly, over the 45 days of the experiment control birds performed better on more days than the lead-injected birds. Balance was disturbed by lead-injection for the first six days following injection. Individual recognition developed by day 5 in control birds, by day 10 for 0.1 Pb mg/g birds, and by day 14 for 0.2 Pb mg/g birds. Depth perception and thermoregulation behavior were also adversely affected by lead.

Burger, J. 1997. Heavy metals and selenium in herring gulls (*Larus argentatus*) nesting in colonies from eastern Long Island to Virginia. Environ. Monit. Assess. 48(3):285-296.

Annotated Bibliography (Cont.).

Rec #: 158
KEYWORDS: NY; NJ; VA; USA; North America; metals; Pb; Hg; Cd; Se; Cr; Mn; feather; herring gull; *Larus argentatus*; feather; spatial variations
ABSTRACT: With increasing interest in assessing the health or well-being of communities and ecosystems, birds are being used as bioindicators. Colonially nesting species breed mainly in coastal areas that are also preferred for human development, exposing the birds to various pollutants. In this paper concentrations of heavy metal and selenium in the feathers of Herring Gulls (*Larus argentatus*) nesting in several colonies from Massachusetts to Delaware are reported. There were significant differences among colonies for all metals, with metal concentrations being two to nearly five times higher at some colonies than others. Selenium showed the least difference, and cadmium showed the greatest difference among sites. Concentrations of lead were highest at Prall's Island; mercury was highest at Shinnecock, Huckleberry and Harvey, and manganese was highest at Captree.

Burger, J. and Gochfeld, M. 1995. Heavy metal and selenium concentrations in eggs of herring gulls (*Larus argentatus*): Temporal differences from 1989 to 1994. Arch. Environ. Contam. Toxicol. 29(2):192-197.
Rec #: 275
KEYWORDS: metals; Pb; Cd; Hg; Cr; Mn; Se; temporal trends; herring gull; *Larus argentatus*; egg; NY; USA; North America
ABSTRACT: Concentrations of five metals and selenium in the eggs of herring gulls (*Larus argentatus*) were examined at a breeding colony on western Long Island, New York from 1989 to 1994. There were significant yearly differences in lead, cadmium, mercury, selenium, chromium, and manganese. Chromium and cadmium were significantly higher in 1993 compared to the other years. Lead levels were highest in 1989, and were uniformly lower in the succeeding four years. Manganese showed no clear pattern. Selenium concentrations decreased from 1991 through 1994, whereas mercury increased from 1992 through 1994. Generally, concentrations of cadmium were similar to those reported for avian eggs from elsewhere; mercury and lead were within the range, but were at the high end; and chromium concentrations were higher than elsewhere. For all years combined, there was a positive correlation between lead and cadmium concentrations and between chromium and manganese, and a negative correlation between lead and mercury concentrations. In conclusion, egg contents can be used to monitor heavy metal concentrations, but consecutive years must be examined because concentrations can vary significantly among years. Ideally, data are needed for more than three years before trends, or lack thereof, can be determined.

Burger, Joanna. 2002. Food chain differences affect heavy metals in bird eggs in Barnegat Bay, New Jersey. Environ. Res. 90(1):33-39.
Rec #: 359
KEYWORDS: biomagnification; egg; herring gull; *Larus argentatus;* great black-backed gull; *Larus marinus;* black skimmer; *Rynchops niger;* Forster's tern; *Sterna forsteri;* common tern; *Sterna hirundo*; metals; As; Cd; Cr; Pb; Mn; Hg; Se; NJ; USA; North America
ABSTRACT: There is an abundance of field data on levels of mercury in a wide variety of birds and on a suite of heavy metals in single species of birds, but few studies examine a suite of metals in a suite of birds that represent different trophic levels. Thus it is often difficult to detect whether food chain differences exist and have ecological relevance for the birds. In this paper I

Annotated Bibliography (Cont.).

examine the levels of seven metals in the eggs of five species of marine birds that nest in Barnegat Bay, New Jersey to detect whether there are differences among species and whether such differences reflect food chain differences. There were significant differences among species for all metals, except cadmium, with black skimmers (*Rynchops niger*) having the highest levels of all metals except manganese and selenium. Metal concentrations in eggs mainly represented food chain differences. Mercury exhibited the greatest interspecific difference, with skimmer eggs having five times higher mercury levels than the eggs of great black-backed gulls (*Larus marinus*). Although there were significant interspecific differences in the other metals, they were generally less than an order of magnitude. There were few high, significant correlations among metals, although mercury was positively correlated with arsenic overall. Mean mercury levels exceeded the level known to adversely affect development in bird eggs for common (*Sterna hirundo*) and Forster's (*Sterna forsteri*) terns and for skimmers and exceeded the mean for eggs of fish-eating birds reported from 68 studies.

Burger, Joanna. 1995. Heavy metal and selenium levels in feathers of herring gulls (*Larus argentatus*): Differences due to year, gender, and age at Captree, Long Island. Environ. Monit. Assess. 38(1):37-50.
Rec #: 53
KEYWORDS: metals; Se; Hg; Pb; Cd; Cr; Mn; Se; bioaccumulation; effects; herring gull; *Larus argentatus*; feather; NY; USA; North America; gender differences; age variations
ABSTRACT: The concentrations of heavy metals (mercury, lead, cadmium, chromium, manganese) and selenium in the feathers of herring gulls (*Larus argentatus*) from a nesting colony at Captree, Long Island, New York were examined from 1989 to 1993 to determine if there were differences from year to year, and between males and females, adult and young, and dead versus live gulls. Variation in metal levels in regression models was explained by age (all metals), year (all except manganese), and whether the feathers were from live or dead birds (all except lead and chromium). The feathers of adults had significantly higher levels of mercury, lead and manganese than those of young, but lower levels of selenium and cadmium than those of young. Levels in down and fledgling feathers were similar for lead, cadmium and selenium, but fledgling feathers had higher levels for mercury, chromium, and manganese. There were no gender differences in metal levels for adult feathers except for lead (females had higher levels). Levels of mercury and manganese were higher in feathers of live adults whereas levels of cadmium and selenium were higher in the feathers of dead adults.

Burger, Joanna. 1994. Heavy metals in avian eggshells: Another excretion method. J. Toxicol. Environ. Health. 41(2):207-220.
Rec #: 90
KEYWORDS: effects; eggshell; egg; elimination; roseate tern; *Sterna dougallii;* herring gull; *Larus argentatus*; metals; Pb; Cd; Hg; Se; Mn; Cr; NY; USA; North America
ABSTRACT: Birds can rid their bodies of heavy metals through both excretion and deposition in feathers, and females can also eliminate heavy metals in the contents of their eggs. In this paper the levels of heavy metals (lead, cadmium, mercury, selenium, manganese, chromium) in the contents and shells of eggs of roseate terns (*Sterna dougallii*) and herring gulls (*Larus argentatus*) nesting at Cedar Beach, Long Island, are reported. For both species, metal concentrations were significantly higher in the contents compared to the shells for lead, mercury, selenium, and chromium. For herring gulls, metal levels were higher in the shells for cadmium

Annotated Bibliography (Cont.).

and manganese. Levels of cadmium, mercury, and selenium were significantly higher in roseate ten egg contents than for herring gulls. In eggshells, lead, cadmium, mercury, and selenium were significantly higher in roseate terns compared to herring gulls. For both species, eggshells account for about 7-8 % of the egg by weight, but less than 1 % of the egg burden for mercury, 1-5 % for lead, selenium, and chromium, and 7-11 % for manganese. For cadmium, shells account for only 5 % of the egg burden for roseate terns, but 29 % for herring gulls. These data suggest that, except for mercury, eggshells provide another method of excretion of metals in these two species of birds.

Burger, Joanna and Gochfeld, Michael. 1995. Behavior effects of lead exposure on different days for gull (*Larus argentatus*) chicks. Pharmacol. Biochem. Behav. 50(1):97-105.
Rec #: 277
KEYWORDS: effects; metals; Pb; temporal trends; herring gull; *Larus argentatus;* NJ; USA; North America; dosing study
NOTES: SUPPLEMENTAL
ABSTRACT: Lead exposure early in life affects behavioral, physiologic, and intellectual development in humans and other animals. In this article, we examine the effects of temporal differences in lead exposure on early development in herring gulls (*Larus argentatus*). Each of 72 1-day-old herring gull chicks was randomly assigned to one of six treatment groups to receive a lead nitrate concentration of 100 µg/g at age 2 or at age 6, a similar cumulative dose evenly divided on days 2, 4, and 6, or matched-volume saline injections on the same days. Behavioral tests were performed (some at 2- and others at 5-day intervals) to examine locomotion, balance, righting response, thermoregulation, and visual cliff. Most variation in weight was explained by testing age, although treatment affected weight gain for the lead-6 gulls, particularly after 20 days. Although treatment influenced balance and locomotion, the effect was small. The lead-6 birds were unable to remain on an incline as long as the lead-2, lead-246, and control birds. The overall score for balance improved with age for controls, showed little change for the lead-2 and lead-2-4-6 gulls, but showed a decrease in performance for the lead-6 birds. On the thermoregulation test, the lead-6 birds performed less well under both low- and high-temperature test conditions. Although the lead-2-4-6 birds had a lower score on the visual cliff tests than the other groups, the lead-6 gulls showed a significant delay in response and gave significantly fewer calls then the other groups. Overall, the data showed that the lead-6 group was more affected by the dose than the other groups, suggesting that 6 days of age may be a more critical period than earlier ages for some behaviors.

Burger, Joanna and Gochfeld, Michael. 1994. Behavioral impairments of lead-injected young herring gulls in nature. Fundam. Appl. Toxicol. 23(4):553-561.
Rec #: 285
KEYWORDS: metals; Pb; herring gull; *Larus argentatus*; effects; NY; USA; North America; dosing study
NOTES: SUPPLEMENTAL
ABSTRACT: Lead is ubiquitous in the environment, and trace amounts enter the food chain and bioaccumulate in organisms high on the food chain. Although lead levels have been examined in a variety of wild species, effects data are usually from laboratory studies. Thus the relevance of effects to survival and fitness are not directly determined. In the field we compared the behavior of lead-injected young herring gulls (*Larus argentatus*) to the behavior of their control siblings

Annotated Bibliography (Cont.).

who received an injection with no lead and to chicks from control nests that received no injections. Lead-injected chicks had significantly lower survival rates than all controls. Lead-injected chicks were less healthy than control chicks as measured by begging and walking scores and by the number of times they stumbled when walking. Control chicks had a higher degree of accuracy when pecking at their parents' bills to stimulate feeding compared to the lead-injected chicks. For all chicks, begging and walking scores improved with age. Behavioral deficits measured in the laboratory are homologous with those observed in the field.

Burger, Joanna and Gochfeld, Michael. 1988. Effects of lead on growth in young herring gulls (*Larus argentatus*). J. Toxicol. Environ. Health. 25(2):227-236.
Rec #: 82
KEYWORDS: metals; Pb; effects; herring gull; *Larus argentatus;* NJ; USA; North America; dosing study
NOTES: SUPPLEMENTAL
ABSTRACT: One-day-old herring gulls (*Larus argentatus*) were injected intraperitoneally with lead nitrate solution (0.1 or 0.2 mg Pb/g) or sterile saline to examine differences in growth rates. Despite the low levels of lead exposure, by d 8 there were significant differences in growth rates as a function of treatment. There were also, by d 8, significant differences in bill length, tarsus length, and wing bone length. Except for bill length, these differences persisted for the duration of the study. Developmental curves varied, with lead-treated birds reaching the same asymptote as control birds for bill length, but having a lower asymptote for tarsus length. For curves where asymptotes were reached, lead-treated birds required more days to reach it than control birds. The initial amount of food eaten each day was positively correlated with weight gain for control birds, but negatively correlated for lead-treated birds.

Burger, Joanna and Gochfeld, Michael. 1995. Growth and behavioral effects of early postnatal chromium and manganese exposure in herring gull (*Larus argentatus*) chicks. Pharmacol. Biochem. Behav. 50(4):607-612.
Rec #: 287
KEYWORDS: metals; Cr; Mn; herring gull; *Larus argentatus*; effects; NY; USA; North America; dosing study
NOTES: SUPPLEMENTAL
ABSTRACT: Organisms in marine environments are exposed to chromium and manganese, yet little is known of the effects of these metals on physiology and behavior. In this article we examine the effects of chromium and manganese on early neurobehavioral development in herring gulls, *Larus argentatus*. Each of 36 2-day-old herring gull chicks was randomly assigned to one of three treatment groups to receive either chromium nitrate, manganese acetate (50 mg/kg), or a control dose of sterile saline solution. The trios were not siblings, but were matched by age and weight. Behavioral tests examined food-begging, balance, locomotion, righting response, recognition, thermoregulation, and perception. There were significant differences in begging behavior by 5 days postinjection, and there were significant differences in weight gain throughout development until 50 days of age, when the experiment was terminated. Behavioral tests, administered from 18-48 days postinjection, indicated significant differences between control and the exposed groups for time to right themselves; thermoregulation behavior; and performance on a balance beam, inclined plane, actual cliff, and visual cliff (although not all components varied significantly). Of the 14 behavioral measures with significant differences,

Annotated Bibliography (Cont.).

control birds performed best on 12.

Burger, Joanna and Gochfeld, Michael. 1996. Heavy metal and selenium levels in birds at Agassiz National Wildlife Refuge, Minnesota: Food chain differences. Environ. Monit. Assess. 43(3):267-282.
Rec #: 279
KEYWORDS: metals; Pb; Cd; Hg; Se; Cr; Mn; double-crested cormorant; *Phalacrocorax auritus*; black-crowned night heron; *Nycticorax nycticorax*; Franklin's gull; *Larus pipixcan*; American bittern; *Botaurus lentiginosus*; American coot; *Fulica americana*; eared grebe; *Podiceps caspicus*; Canada goose; *Branta canadensis*; biomagnification; feather; egg; MN; USA; North America
ABSTRACT: The levels of heavy metals and selenium in the eggs and in breast feathers of adult double-crested cormorant (*Phalacrocorax auritus*), black-crowned night heron (*Nycticorax nycticorax*), and Franklin's gull (*Larus pipixcan*) nesting at Agassiz National Wildlife Refuge in Marshall County, northwestern Minnesota were examined. Also examined were metal levels in the feathers of fledgling night herons and gulls, in the feathers of adult and fledgling American bittern (*Botaurus lentiginosus*), in eggs of American coot (*Fulica americana*) and eared grebe (*Podiceps caspicus*), and in feathers of adult Canada geese (*Branta canadensis*). These species represent different levels on the food chain from primarily vegetation-eating species (geese, coot) to species that eat primarily fish (cormorant). A clear, positive relationship between level on the food chain and levels of heavy metals occurred only for mercury in feathers and eggs. Otherwise, eared grebes had the highest levels of all other metals in their eggs compared to the other species. No clear food chain pattern existed for feathers for the other metals. For eggs at Agassiz: 1) lead, selenium, and manganese levels were similar to those reported in the literature, 2) mercury levels were slightly higher for cormorants and night herons, 3) all species had higher chromium and cadmium levels than generally reported, and 4) eared grebes had significantly higher levels of cadmium than reported for any species from elsewhere. For adult feathers: 1) gulls had higher levels of lead than the other species, 2) cadmium levels were elevated in gulls and adult herons and cormorants, 3) mercury levels showed an increase with position on the food chain, 4) selenium and chromium levels of all birds at Agassiz were generally low and 5) manganese levels in adults were generally higher than in the literature for other species. Adults had significantly higher mercury levels than fledgling gulls, night herons, and bitterns.

Burger, Joanna and Gochfeld, Michael. 1993. Lead and behavioral development in young herring gulls: Effects of timing of exposure on individual recognition. Fundam. Appl. Toxicol. 21(2):187-195.
Rec #: 276
KEYWORDS: effects; metals; Pb; herring gull; *Larus argentatus*; NY; NJ; USA; North America; dosing study
NOTES: SUPPLEMENTAL
ABSTRACT: Lead exposure early in life affects behavioral, physiological, and intellectual development in humans and other animals. Recognition of parents or other caregivers and eventual bonding are essential aspects of behavioral development. In this paper young herring gulls, *Larus argentatus*, were used to examine the effect of timing of lead exposure on individual recognition behavior and development. Each of 60 1-day-old herring gull chicks was randomly assigned to a control group or to one of three treatment groups that received a single dose of lead

Annotated Bibliography (Cont.).

nitrate solution (100 mg/kg) at Day 2 or 6 of age or the same total dose divided in thirds on Day 2, 4, and 6. Matched controls were injected with isotonic saline on the same schedules. Variations in individual recognition of human attendants were largely explained by age and status (lead versus control), and some variation was explained by day of injection (exposure regime). Using a feeding paradigm, the percentage responding to their caretaker compared to another person was higher (70 % versus 38 %), and occurred earlier in controls compared to lead-injected birds. Lead-injected birds required longer to respond initially, took longer to choose, moved less distance per time, and took longer to eventually reach the food. These results were all significant by GLM models and Kruskal-Wallis tests. Among lead-injected birds there was a disjunction of effect related to dosing schedule: birds injected on Day 6 chose more correctly but were slower to respond, indicating that these behavioral traits were differently affected by timing of exposure. These data suggest that there is a critical period during development when individual recognition can be disrupted by lead.

Burger, Joanna and Gochfeld, Michael. 1993. Lead and cadmium accumulation in eggs and fledgling seabirds in the New York Bight. Environ. Toxicol. Chem. 12(2):261-267.
Rec #: 281
KEYWORDS: bioaccumulation; metals; Pb; Cd; egg; feather; NY; NJ; USA; North America; common tern; *Sterna hirundo*; roseate tern; *Sterna dougallii*; Forster's tern; *Sterna forsteri*; black skimmer; *Rynchops niger*; herring gull; *Larus argentatus*; spatial variations
ABSTRACT: We measured lead and cadmium concentrations in eggs and in the breast feathers of fledglings of common tern (*Sterna hirundo*), roseate tern (*S. dougallii*), Forster's tern (*S. forsteri*), black skimmer (*Rynchops niger*), and herring gull (*Larus argentatus*) nesting in mixed-species colonies in the New York Bight, USA, in 1989. Metal concentrations in fledgling feathers represent in part metals sequestered in the egg by females and accumulation from food brought back to chicks by parents, and thus may be a measure of local metal acquisition. There were significant interspecific differences in lead in eggs, and lead and cadmium in fledgling feathers. Herring gulls had the most lead in eggs, whereas the terns had the least. Cadmium concentrations were generally low in all examined eggs. Lead concentrations were high in fledgling feathers in some populations of all species. Cadmium was highest in fledgling feathers of herring gull and skimmers. Among fledgling terns, the roseate tern (a federally endangered species) had the highest concentrations. For all species except herring gull, the feathers of fledglings had higher levels of metals than did eggs.

Burger, Joanna; Reilly, Stephanie M.; and Gochfeld, Michael. 1992. Comparison of lead levels in bone, feathers, and liver of herring gull chicks (*Larus argentatus*). Pharmacol. Biochem. Behav. 41(2):289-293.
Rec #: 79
KEYWORDS: metals; Pb; bone; liver; feather; liver; herring gull; *Larus argentatus*; NY; USA; North America; dosing study
ABSTRACT: Bone, feathers, and liver were analyzed for lead in herring gull chicks (*Larus argentatus*) of two different ages. The highest levels were found in the bone, evidence of chronic exposure. No differences were found within the bones. Differences occurred between different bones, with the ribs having twice the amount of lead than any other bone. These studies indicate that type of bone affects lead levels; thus researchers should clearly state which parts of which bones are examined. It is also suggested that for humans consistent location should be used for

Annotated Bibliography (Cont.).

analysis by in vivo X-ray fluorescence.

Bustnes, J. O.; Bakken, V.; Erikstad, K. E.; Mehlum, F.; and Skaare, J. U. 2001. Patterns of incubation and nest-site attentiveness in relation to organochlorine (PCB) contamination in glaucous gulls. J. Appl. Ecol. 38(4):791-801.
Rec #: 381
KEYWORDS: effects; reproductive success; glaucous gull; *Larus hyperboreus*; Norway; Svalbard; Europe; organochlorines; PCBs; DDTs; HCB; chlordanes; blood; gender differences;
ABSTRACT: 1. Although experimental studies show that organochlorines (OC) can affect bird behavior, field assessments are invariably confounded by ecological differences between contaminated and uncontaminated sites. The behavior of individual birds in the field has rarely been related to the contaminant burden. 2. We examined individual patterns of incubation and nest-site attentiveness in relation to OC burden, measured as polychlorinated biphenyl (PCB) concentration in the blood, of 27 glaucous gulls, *Larus hyperboreus*, in two breeding areas at Bear Island, in the north-eastern Atlantic. 3. Blood PCB concentrations ranged from 52 ng g^{-1} to 1079 ng g^{-1} (wet weight). There were significant differences between the two breeding areas, and females had significantly lower concentrations than males. 4. Gull behavior differed significantly between breeding areas and sexes independently of PCB. Females incubated more than males (54 % vs. 46 %) but spent more time away from the nest site than males, both overall (23 % vs. 12 %) and when not incubating (50 % vs. 21 %). They were also absent for longer periods (4.5 vs. 2.8 h). Moreover, length of incubation bouts (6.4 vs. 4.4 h), the amount of time absent from the nest site when not incubating (51 % vs. 25 %) and length of absences (5.6 vs. 1.8 h) differed between breeding areas, probably due to different feeding specializations. 5. After controlling for these area and sex effects, the proportion of time absent from the nest site when not incubating, and the number of absences, were both significantly related to blood concentration of PCB. 6. Increased absence from the nest site in individual glaucous gulls with high blood concentrations of OC suggests effects on reproductive behavior. We speculate that endocrine disruption or neurological effects might be involved, leading to increased energetic costs during incubation and reduced reproductive output.

Bustnes, Jan O.; Bakken, Vidar; Erikstad, Kjell E.; Mehlum, Fritjof; and Skaare, Janneche U. 2001. Effects of long-transported organochlorines on the behavior of Arctic breeding glaucous gulls (*Larus hyperboreus*): Evidence using blood samples. Organohalogen Compounds. 53:204-206.
Rec #: 379
KEYWORDS: glaucous gull; *Larus hyperboreus*; organochlorines; PCBs; Barents Sea; Svalbard; Norway; Europe; blood; reproductive success; effects
NOTES: NO TABLES
ABSTRACT: The variation in the patterns of incubation and nest-site attentiveness of individual glaucous gulls, *Larus hyperboreus*, was examined in relation to their blood polychlorinated biphenyl concentration. Two different breeding areas were used, one at a seabird cliff, where birds fed predominantly on eggs and of other seabirds, and one close to sea level where the birds fed largely on fish. The nesting period is the most critical phase in avian reproduction, rendering both parents and offspring vulnerable to predation and starvation. Gull behavior differed considerably between breeding areas and sexes independently of PCB. Organochlorines may affect reproductive behavior through endocrine disruption, since several compounds are known

Annotated Bibliography (Cont.).

to alter hormone levels and may act as estrogen or thyroxine agonists or antagonists. Even if the levels of PCB found in the brains of glaucous gulls at Bear Island are considerably lower than those usually considered lethal in birds, the individuals with high levels are well within the range expected to produce behavioral aberrations.

Bustnes, Jan O.; Skaare, Janneche U.; Erikstad, Kjell E.; Bakken, Vidar; and Mehlum, Fritjof. 2001. Whole blood concentrations of organochlorines as a dose metric for studies of the glaucous gull *Larus hyperboreus*. Environ. Toxicol. Chem. 20(5):1046-1052.
Rec #: 326
KEYWORDS: organochlorines; HCB; HCHs; chlordanes; DDTs; PCBs; blood; glaucous gull; *Larus hyperboreus*; Norway; Europe; seasonal variations; temporal trends
ABSTRACT: In order to examine if whole blood concentrations of organochlorines (OCs) is an appropriate dosimetric parameter for use in ecotoxicological studies of free-living birds, a number of incubating glaucous gulls (*Larus hyperboreus*) were repeatedly sampled within and between subsequent breeding seasons. The wet weight concentrations of selected OCs, differing in persistence and fat solubility, were compared and it was assessed to what extent present concentrations could be predicted from concentrations previously measured in the individuals. There were only a few significant differences in the blood concentrations of the selected OCs within and between seasons. The most persistent compound, polychlorinated biphenyl (PCB)-153, showed a low interindividual variability, and between seasons, 70 % of the variance could be explained by the level in the previous year, while changes in body condition and blood lipid percentage were of less importance. For PCB-101, the predictability of the present blood concentration from the previous concentration was lower than for PCB-153, and changes in body condition and blood lipid percentage explained a higher proportion of the variance. The present level of -hexachlorocyclohexane (HCH) could not be predicted from the previous level. Sex did not explain any significant proportion of the variance in OC concentrations when previous level and changes in body mass and blood lipid were included in the statistical models. Thus, for the most persistent OCs, concentration in the blood of incubating glaucous gulls is representative for the interindividual differences over time and whole blood concentrations of OCs appear adequate as a dose metric in ecotoxicological studies.

Bustnes, Jan Ove; Bakken, Vidar; Skaare, Janneche Utne; and Erikstad, Kjell Einar. 2003. Age and accumulation of persistent organochlorines: A study of arctic-breeding glaucous gulls (*Larus hyperboreus*). Environ. Toxicol. Chem. 22(9):2173-2179.
Rec #: 440
KEYWORDS: glaucous gull; *Larus hyperboreus*; Norway; Europe; blood; organochlorines; HCB; chlordanes; DDTs; PCBs; age variations; bioaccumulation
ABSTRACT: The authors studied the relationship between increasing age and blood concentrations of 4 persistent organochlorines (OCs), hexachlorobenzene (HCB), oxychlordane, p,p'-dichlorodiphenyldichloroethylene (DDE), and 2,2',4,4',5,5'-hexachlorbiphenyl (PCB 153), in arctic-breeding glaucous gulls (*Larus hyperboreus*). The authors measured OC concentrations in 31 individuals of known age and took repeated blood samples of 64 individuals in different years, either 1 year apart or 3 or 4 years apart. The age of individuals was not related to the blood concentrations for any of the 4 compounds, and in birds whose values were measured repeatedly, there was no effect of the length of time (number of years) between sampling events on the relative change in OC concentrations. This indicates that steady-state levels were reached

Annotated Bibliography (Cont.).

before the age of first breeding. However, breeding area significantly influenced the changes in OC concentration between sampling events. In areas in which birds fed on prey from higher trophic levels, the OC concentrations showed large increases between sampling events; in areas in which birds fed at lower trophic levels, OC concentrations increased relatively little or not at all. This indicates that individual birds had different equilibrium concentrations, which are reached at different ages depending on the intake of OCs through the food. It also indicates that some individuals had not reached steady-state concentrations at the onset of reproduction. Changes in body condition and amount of blood lipids were of lesser importance than trophic level and influenced the concentrations of HCB and oxychlordane more strongly than DDE and PCB 153. Thus, steady-state concentrations of persistent OCs are reached early in life in most glaucous gulls, considering the long life span of the species.

Butler, R. G.; Peakall, D. B.; Leighton, F. A.; Borthwick, J.; and Harmon, R. S. 1986. Effects of crude oil exposure on standard metabolic rate of Leach's storm-petrel. Condor. 88(2):248-249.
Rec #: 139
KEYWORDS: oil; effects; metabolism; Leach's storm-petrel; *Oceanodroma leucorhoa*; dosing study
NOTES: SUPPLEMENTAL
ABSTRACT: In a previous report (Trivelpiece et al. 1984) the authors noted decreased survival and reduced growth rate in chicks of oil-dosed adult Leach's Storm-Petrels (*Oceanodroma leucorhoa*). It was suggested that these effects were related to impaired ability of oil-dosed adults to provide food for their young, possibly due to elevated metabolic demands following contamination. The authors report here the effects of oil exposure on the standard metabolic rate of adults of this species using two different experimental methods.

Butler, Ronald G.; Trivelpiece, Wayne; Miller, David; Bishop, Paul; D'Amico, Christopher; D'Amico, Melissa; Lambert, Gabrielle; and Peakall, David. 1979. Further studies of the effects of petroleum hydrocarbons on marine birds. Bull. Mt. Desert Isl. Biol. Lab. 19:33-35.
Rec #: 129
KEYWORDS: oil; effects; toxicology; dosing study; herring gull; *Larus argentatus*; Leach's storm petrel; *Oceanodroma leucorhoa*
NOTES: SUPPLEMENTAL
ABSTRACT: The effects of crude oil on *Larus argentatus* and *Oceanodroma leucorhoa* were studied to find out which components of the oil were directly responsible for the effects observed and whether the toxic effects could be modified by emulsification of the oil. High molecular weight aromatic compounds in the oil were found to be responsible for the physiological effects observed in the birds. The effects of some dispersants are discussed and the effect of crude oil on subsequent generations is also briefly mentioned.

Cahill, T. M.; Anderson, D. W.; Elbert, R. A.; Perley, B. P.; and Johnson, D. R. 1998. Elemental profiles in feather samples from a mercury-contaminated lake in central California. Arch. Environ. Contam. Toxicol. 35(1):75-81.
Rec #: 402
KEYWORDS: feather; metals; Hg; S; Ca; Ti; Sc; Cr; Fe; Ni; Zn; As; Se; Br; Rb; Sr; Pb; osprey; *Pandion haliaetus*; western grebe; *Aechmophorus occidentalis*; great blue heron; *Ardea herodias*; mallard; *Anas platyrhynchos*; turkey vulture; *Cathartes aura*; double-crested

Annotated Bibliography (Cont.).

cormorant; *Phalacrocorax auritus;* CA; ID; USA; North America; age variations; bioaccumulation; biomagnification; reproductive success; spatial variations
NOTES: NO TABLES FOR *PHALACROCORAX AURITUS*
ABSTRACT: Flight feathers from 6 bird species at Clear Lake were analyzed to determine the extent and distribution of Hg contamination from an abandoned Hg mine and associated levels of 14 other elements. Further samples were collected from adult and juvenile osprey (*Pandion haliaetus*), including juvenile osprey from 3 additional comparison sites; adult western grebes (*Aechmophorus occidentalis*); adult great blue herons (*Ardea herodias*); adult mallards (*Anas platyrhynchos*); adult turkey vultures (*Cathartes aura*); and juvenile double-crested cormorants (*Phalacrocorax auritus*). Samples were analyzed by a multielemental x-ray fluorescence method. The osprey from Clear Lake showed significantly elevated Hg concentrations relative to the comparison sites. Different species at Clear Lake had different Hg concentrations based on trophic status; osprey exhibited the highest Hg concentrations and the mallards showed the lowest. We quantified differences in elemental concentrations, including Hg, between adult and juvenile osprey from Clear Lake. Elements known to be nutrients, such as S and Zn, did not vary significantly among species or sites. Reproductive success of osprey at Clear Lake was monitored from 1992 to 1996 to determine if osprey reproduction was depressed. During this 5 years, the breeding population grew from 10 to 20 nesting pairs and the average reproductive rate was 1.4 fledglings/nesting attempt. Although the osprey showed the highest Hg levels of any species sampled, their reproduction does not appear to be depressed.

Calambokidis, John; Speich, Steven M.; Peard, John; Steiger, Gretchen H.; and Cubbage, James C. 1985. Biology of Puget Sound marine mammals and marine birds: Population health and evidence of pollution effects. NOAA Tech. Memo. NOS OMA 18. 170 pp.
Rec #: 101
KEYWORDS: effects; WA; USA; North America; glaucous-winged gull; *Larus glaucescens*; great blue heron; *Ardea herodias*; pigeon guillemot; *Cepphus columba*; egg; blood; liver; toxicology; reproductive success; eggshell thickness; abnormalities; parasites
NOTES: NO CONTAMINANT TABLES
ABSTRACT: The objective of the research was to determine whether detrimental effects possibly caused by toxic chemicals could be observed in Puget Sound marine mammals and marine birds. The study design was based on examination of a wide variety of indices of population and individual health and comparison of these indices from areas of suspected high contaminant levels (target areas) to those from areas of suspected low contaminant levels (reference areas) and to those reported by other researchers. Primary species considered here are harbor seal. Glaucous-winged Gull, Great Blue Heron, and Pigeon Guillemot; these species were chosen because they reside, feed, and breed in some of the most contaminated portions of Puget Sound. Three other mammal species (killer whale, harbor porpoise, and river otter) were chosen as secondary study species either because they seasonally occur in contaminated areas of Puget Sound or they were found through previous research to be experiencing problems that might be pollutant-related.

Caldwell, C. A.; Arnold, M. A.; and Gould, W. R. 1999. Mercury distribution in blood, tissues, and feathers of double-crested cormorant nestlings from arid-lands reservoirs in south central New Mexico. Arch. Environ. Contam. Toxicol. 36(4):456-461.
Rec #: 442

Annotated Bibliography (Cont.).

KEYWORDS: egg; blood; liver; muscle; feather; metals; Hg; double-crested cormorant; *Phalacrocorax auritus*; NM; USA; North America; spatial variations; bioaccumulation; biomagnification

ABSTRACT: Eggs, blood, liver, muscle, and feathers were analyzed for concentrations of total mercury in double-crested cormorant (*Phalacrocorax auritus*) nestlings from two reservoirs in south central New Mexico. Total mercury concentrations among eggs, tissues, and feathers were not significantly correlated. Concentrations of total mercury averaged 0.40 µg/g in liver and 0.18 µg/g in muscle tissues in both populations of nestlings. There were no significant changes in concentrations of total mercury in whole blood of nestlings collected 7-10 days and 17-22 days posthatch in Caballo Reservoir (0.36 µg/g and 0.39 µg/g and 0.34 µg/g, respectively). Total mercury concentrations were similar for blood, muscle, and liver in nestlings for both reservoirs. Total mercury concentrations were higher in eggs and tail, primary, and secondary features from nestlings at Caballo Reservoir compared to Elephant Butte Reservoir. Although there were no differences in concentrations of total mercury in fishes between the two reservoirs, bioaccumulation and biomagnification was evident in planktivorous and piscivorous fishes. The data demonstrate that feather analysis may not be a good predictor of tissue burden in nestlings from regions of low contamination.

Cameron, Marjorie and Weis, I. Michael. 1993. Organochlorine contaminants in the country food diet of the Belcher Island Inuit, Northwest Territories, Canada. Arctic. 46(1):42-48.
Rec #: 198

KEYWORDS: North America; Canada; Northwest Territories; organochlorines; PCBs; DDTs; HCHs; chlordanes; dieldrin; common eider; *Somateria mollissima*; carcass; biomagnification; spatial variations

ABSTRACT: An initial assessment of the country food diet at the Belcher Islands' community of Sanikiluaq, Northwest Territories, was made by interviewing 16 families during May - July 1989. Estimates of consumption per day were established over a two-week period for 10 of these families. This information was utilized along with previously published harvest data for the community to estimate country food consumption in grams/day and kg/year. Beluga (*Delphinapterus leucas*), ringed seal (*Phoca hispida*), arctic charr (*Salvelinus alpinus*), common eider (*Somateria mollissima*) and Canada goose (*Branta canadensis*) were found to be important components in the diet during this period. Results of analysis for organochlorine contaminants reveal that ringed seal fat and beluga muktuk (skin and fat layer) samples have the highest concentration of DDE and total PCBs among the country food species. Average DDE and total PCB values were 1504.6 µg/kg and 1283.4 µg/kg respectively in ringed seal fat and 184.3 µg/kg and 144.7 µg/kg respectively in beluga muktuk. Comparison of contaminants in seal fat indicates concentrations approximately two times higher in samples from the Belcher Islands than from sites in the Canadian Western Arctic, but lower than concentrations reported from various European sites. The daily consumption estimates in grams/day were used along with organic contaminant analysis data to calculate the estimated intake levels of 0.22 µg/kg body weight/day of total DDT and 0.15 µg/kg body weight/day of total PCBs during the study period. Although limited in sample size, studies such as this provide a framework from which to establish future consumption guidelines more applicable to arctic systems and native diets.

Camphuysen, Kees C. J.; Barreveld, Hein; Dahlmann, Gerhard; and Van Franeker, Jan Andries.

Annotated Bibliography (Cont.).

1999. Seabirds in the North Sea demobilized and killed by polyisobutylene $(C_4H_8)_n$ (PIB). Mar. Pollut. Bull. 38(12):1171-1176.
Rec #: 43
KEYWORDS: effects; toxicology; common murre; *Uria aalge*; northern fulmar; *Fulmarus glacialis*; black scoter; *Melanitta nigra*; polyisobutylene; North Sea; Netherlands; Europe
NOTES: SUPPLEMENTAL; NO TABLES
ABSTRACT: This paper reports on a mass stranding of seabirds in the North Sea in December 1998. Hundreds of birds were washed ashore alive in Zeeland (SW Netherlands), covered in a whitish, sticky substance, and were transported to a rehabilitation centre. About 10 days later, more (dead) casualties washed ashore further to the north on Texel and along the mainland coast, again covered in a glue-like substance. Common guillemots *Uria aalge*, northern fulmars *Fulmarus glacialis* and common scoters *Melanitta nigra* were the most numerous birds affected in this incident. Both strandings were temporarily (10 days) and geographically separated (ca. 120 km apart), but were apparently caused by a single source of pollution. The meteorology at the time was consistent with the course of a single incident. At least 1100 seabirds were affected by this substance, soon identified as polyisobutylene $(C_4H_8)_n$. PIB is known as a non-toxic, non-aggressive substance. Volunteers cleaning the birds in the rehabilitation centre reported serious discomfort and dizziness and the soft parts of the PIB-affected birds found dead (bill, eye, throat, feet, webs) appeared to dissolve in a few days time. Both effects cannot be attributed to PIB, and are therefore unexplained. Although the dumping of PIB in the marine environment is not explicitly prohibited under MARPOL, the effects on wildlife observed are enough to plead for counter-measures.

Carlberg, G. E. and Böler, J. B. 1985. Determination of persistent chlorinated hydrocarbons and inorganic elements in samples from Svalbard. Senter for Industriforskning, Report No. 83/11/01-1. 1-20.
Rec #: 541
NOTES: DO NOT HAVE
KEYWORDS: organochlorines; metals; Svalbard; Norway; Europe

Champoux, Louise. 1993. Contamination et ecotxicologie de la faune dans la region d'une usine de pate blanchie au chlore a Lat Turque (Quebec). Technical Report Series No. 187. Canadian Wildl. Serv. St. Foy, PQ. 63 pp.
Rec #: 552
KEYWORDS: pulp mill; organochlorines; PCBs; DDTs; HCHs; chlordanes; mirex; dieldrin; OCS; PCDDs; PCDFs; effects; toxicology; vitamin A; retinol; enzymes; thyroid hormones; porphyrin; EROD; reproductive success; Quebec; Canada; North America; egg; carcass; herring gull; *Larus argentatus*; hooded merganser; *Lophodytes cucullatus*; common merganser; *Mergus merganser*
ABSTRACT: Following the finding that dioxins and furans were present in the environment of bleached kraft pulp mills, a study was undertaken on the St. Maurice River in La Tuque. The objectives of this study were to document dioxin and furan bioaccumulation in piscivorous birds and mammals and to evaluate different biochemical bioindicators of wildlife health. The selected species were the Herring gull, the Hooded Merganser, the Common Merganser and the American Mink. Herring gull eggs and youngs were collected at two colonies, one contaminated and one control. Hooded and Common Merganser eggs were collected in nest boxes along the St.

Annotated Bibliography (Cont.).

Maurice River, while Common Merganser young were collected at two sites on the river, one upstream and one downstream of La Tuque. Mink were trapped around La Tuque and along the river. The sampled tissues were analysed for contaminants and vitamin A, hepatic enzymes, thyroid hormones and other biochemical biomarkers.

Different contamination patterns in dioxins, furans and PCBs were observed between species. Herring gull eggs and young were weakly contaminated (Eggs: 24 ppt of dioxins, 7 ppt of furans, 3,7 ppm of PCBs; youngs: 16 ppt of dioxins, 8 ppt of furans, 0,1 ppm of PCBs). Hooded and Common Merganser eggs were contaminated (50 and 129 ppt of dioxins, 108 and 374 ppt of furans, 0,4 and 13 ppm of PCBs, in the Hooded and the Common, respectively) , while Common Merganser youngs were contaminated with 2,3,7,8-TCDF (21 ppt). Some mink contained elevated levels of dioxins and furans (mean values, 91 ppt and 97 ppt). Herring gull eggs showed a significant difference in retinol/retinyl palmitate ratio between the two colonies. Herring gull youngs from the contaminated colony showed EROD induction, and significant differences in uroporphyrin, T4 and creatinin in comparison to the control colony. These results indicate that piscivorous aquatic birds from La Tuque area seem to be exposed to polyhalogenated aromatic hydrocarbons like dioxins and furans. However, these effects are relatively minor and do not permit identification of a direct causal link between contaminants and the observed effects, nor to predict health effects. At a medium or long term, some of the observed effects could lead to problems in metabolism, growth or reproduction. The reproductive success of the Herring gull seems acceptable. Levels of furans in Mergansers and mink are of concern.

Chan, H. M. 1997. A review of environmental contaminant levels in traditional food in northern Canada. Proceedings of the 23rd Annual Aquatic Toxicity Workshop: October 7-9, 1996, Calgary, Alberta. Can. Tech. Rep. Fish. Aquat. Sci. 35-39.
Rec #: 192
KEYWORDS: review; organochlorines; PCBs; chlordanes; toxaphene; metals; Hg; Canada; Northwest Territories; Yukon; Quebec; North America
ABSTRACT: The author conducted an extensive literature review on levels of environmental contaminants in northern Canada. The range of levels of four contaminants of major concern (chlordane, Hg, PCB and toxaphene) in 59 species of marine mammals, terrestrial mammals, birds, fish and plants are summarized. This data represents 58 % of the 101 species of fish, wildlife and plants mentioned in our dietary interviews conducted in the northern communities. Mathematic modeling of the distributions of the data showed that contaminant levels in most food groups are log-normally distributed and have a typical coefficient of variation of about 100 %. An example of using the mathematical model for dietary exposure assessment is presented. With the current knowledge of environmental contaminant levels in the northern traditional food system, it may be feasible to conduct preliminary risk assessment of dietary exposure of environmental contaminants when some diet information for a community is available.

Choi, J. W.; Matsuda, M.; Kawano, M.; Min, B. Y.; and Wakimoto, T. 2001. Accumulation profiles of persistent organochlorines in waterbirds from an estuary in Korea. Arch. Environ. Contam. Toxicol. 41:353-363.
Rec #: 369
KEYWORDS: organochlorines; PCDDs; PCDFs; PCBs; DDTs; HCHs; HCB; chlordanes; toxicology; TEQs; Korea; Asia; fat; black-necked grebe; *Podiceps nigricollis*; great knot; *Calidris tenuirostris;* sanderling; *Crocethia alba*; greenshank; *Tringa nebularia*; bar-tailed

Annotated Bibliography (Cont.).

godwit; *Limosa lapponica*; black-headed gull; *Larus ridibundus*; herring gull; *Larus argentatus*; common gull; *Larus canus*; black-tailed gull; *Larus crassirostris*; common tern; *Sterna hirundo*; little tern; *Sterna albifrons*; biomagnification

ABSTRACT: Persistent organochlorine pollutants (POPs), such as polychlorinated dibenzo-p-dioxins (PCDDs), polychlorinated dibenzofurans (PCDFs), polychlorinated biphenyls (PCBs), and organochlorine (OC) pesticides such as DDTs (dichlorodiphenyltrichloroethane), HCHs (hexachlorocyclohexane isomers), CHLs (chlordane compounds) and HCB (hexachlorobenzene), were measured in subcutaneous fat of resident and migratory birds collected from the Nakdong River estuary (NRE) in Korea. Black-tailed gull, a resident bird from the NRE, contained greater concentrations of PCDD/Fs and PCBs than the migratory birds collected in the estuary. For example, mean concentrations of PCDD/Fs in black-tailed gull (395.5 pg/g fat weight) were higher than those in migratory birds, such as greenshank (198.3 pg/g fat weight), common gull (90.9 pg/g fat weight) black-headed gull (84.2 pg/g fat weight), and common tern (47.1 pg/g fat weight). However, concentrations of DDTs and/or HCHs were great in some migratory species, such as little tern (mean DDT 6,200 ng/g fat weight) and black-necked grebe (HCHs 475 ng/g fat weight). This suggested that contamination of PCDD/Fs and PCBs in resident gulls are due to intake of locally contaminated fish near the NRE. Elevated OC pesticide levels in migratory birds indicated that these birds have been exposed to DDTs and HCHs during their migration in Southeast Asian countries where chlorinated pesticides are still used. 2,3,7,8-tetrachlorodibenzo-p-dioxin equivalent (TEQs) were calculated using the toxic equivalency factors (TEFs) reported by World Health Organization in 1998. Four of the 2,3,7,8-substituted congeners of PCDD/Fs contributed over 90 % of the TEQs.

Choi, Jae-Won; Kageyama, Takae; Matsuda, Muneaki; Kawano, Masahide; Min, Byung-Yoon; and Wakimoto, Tadaaki. 1998. PCDDs, PCDFs, and PCBs in avian species from Nakdong River estuary in Korea. Organohalogen Compounds. 39:43-46.
Rec #: 371
KEYWORDS: organochlorines; PCDDs; PCDFs; PCBs; toxicology; TEQs; Korea; Asia; fat; black-tailed gull; *Larus crassirostris;* black-headed gull; *Larus ridibundus*; bar-tailed godwit; *Limosa lapponica*; common gull; *Larus canus*; black-necked grebe; *Podiceps nigricollis*; common tern; *Sterna hirundo*; great knot; *Calidris tenuirostris*; greenshank; *Tringa nebularia*; herring gull; *Larus argentatus*; sanderling; *Crocethia alba*; little tern; *Sterna albifrons*
ABSTRACT: The PCDD/F and PCB levels were determined in avian species collected from the Nakdong River estuary, Korea. There were no significant differences in PCDD/DF levels between the resident and migratory species. The PCB levels in residents (black-tailed gull, BTG) ranged from 6700-27,000 ng/g (fat weight). Although the PCB concentrations of the herring gulls were the highest among the migratory species, the concentrations in the herring gull was lower than in the resident BTG. 4-6 Chlorinated PCDD/F congeners were persistent, and OCDD, OCDF, and non-2,3,7,8-substituted congeners were at most in small quantities detected in most of the species analyzed. In most of the species, PCDF levels were higher than PCDDs. Combustion processes and PCB impurities were likely to be partly responsible for the elevation of PCDFs in the resident and migratory birds. There were no critical levels of PCDD/Fs observed to generate acute toxicity in wild birds.

Christopher, S. J.; Vander Pol, S. S.; Pugh, R. S.; Day, R. D.; and Becker, P. R. 2002. Determination of mercury in the eggs of common murres (*Uria aalge*) for the seabird tissue

Annotated Bibliography (Cont.).

archival and monitoring project. J. Anal. At. Spectrom. 17(8):780-785.
Rec #: 497
KEYWORDS: metals; Hg; egg; common murre; *Uria aalge*; thick-billed murre; *Uria lomvia*; AK; USA; North America; Gulf of Alaska; Bering Sea; spatial variations
ABSTRACT: An analytical method using isotope dilution cold vapor inductively coupled plasma mass spectrometry (ID-CV-ICPMS) was developed for the detection of total Hg in the eggs of seabirds. Components including error magnification, verification of method accuracy and assignment of analytical uncertainty are presented in the context of collecting Hg data for single sample aliquots. 51 Egg samples collected from common murre (Uria aalge) colonies on Little Diomede and Saint George Islands in the Bering Sea and East Amatuli and Saint Lazaria Islands in the Gulf of Alaska yielded Hg mass fraction values ranging from approximately 0.010 $\mu g\ g^{-1}$ to 0.360 $\mu g\ g^{-1}$. Relative expanded uncertainties for the individual detections ranged from 1.2 % to 4.4 %. A one-way analysis of variance including pairwise comparisons across the colonies showed that Hg levels in eggs collected from the Gulf of Alaska colonies were significantly higher than their counterparts in the Bering Sea. Hg data from each colony were normally distributed, suggesting a ubiquitous regional deposition of Hg and corresponding incorporation into local food webs.

Clark, T. P. and Norstrom, R. J. 1986. Seasonal changes in lipid and diet: Effect on contaminant loading assessment from herring gull egg residue levels. IAGLR-86 Program. International Association for Great Lakes Research 29th Conference, May 26-29, 1986. 29 pp.
Rec #: 222
KEYWORDS: bioaccumulation; Great Lakes; North America; herring gull; *Larus argentatus*; egg; organochlorines; DDTs; mirex; dieldrin
NOTES: ABSTRACT ONLY - NOT AVAILABLE
ABSTRACT: A large data base exists on organochlorine residue levels in Herring Gulls (*Larus argentatus*) in the Great Lakes. A mathematical model of bioaccumulation has been developed to analyse trends in loadings to the Great Lakes from these data. In this paper the authors present an analysis of the effect of lipid levels and several proposed feeding regimes on contaminant levels in eggs and adult females using the bioaccumulation model, and compare the results to field studies. A scenario in which aquatic feeding in February and March was replaced by a non-aquatic food source resulted in the following reduction in egg contaminant levels averaged over a 3 years period: mirex, 32 %, dieldrin, 53 %, DDD, 70 %. Non-aquatic diet in June-July resulted in the following reductions: mirex, 10 %; dieldrin, 4 %; DDD, 0 %.

Clark, T. P.; Norstrom, R. J.; Fox, G. A.; and Won, H. T. 1987. Dynamics of organochlorine compounds in herring gulls (*Larus argentatus*): A two-compartment model and data for ten compounds. Environ. Toxicol. Chem. 6(7):547-559.
Rec #: 227
KEYWORDS: organochlorines; DDTs; HCB; chlordanes; HCHs; OCS; dieldrin; mirex; blood; fat; pharmacokinetics; herring gull; *Larus argentatus*; Newfoundland; Canada; Great Lakes; North America; dosing study; body burden
NOTES: SUPPLEMENTAL
ABSTRACT: A two-compartment open model using plasma and seasonally variable lipid compartments was developed and validated for several organochlorines in herring gulls (*Larus argentatus*). Plasma clearance rate constants k'_{pc}, L multiplied by kg^{-1} multiplied by d^{-1}, plasma:

Annotated Bibliography (Cont.).

whole-body lipid partition coefficients K_{pf} and compartment sizes for lipid and plasma were obtained for juvenile gulls injected i.p. with a mixture of p,p'-DDD, p,p'-DDE, hexachlorobenzene, oxychlordane, gamma -hexachlorocyclohexane, trans-chlordane, octachlorostyrene, dieldrin, mirex and photomirex. Concentrations in plasma were determined at 11 time points during the 239-d study, and whole-body contaminant burdens and lipid weights were determined at 3 time points. Mean K_{pf} for p,p'-DDD and p,p'-DDE (0.0038 plus or minus 0.0002) was different from that of the other organochlorines (0.0058 plus or minus 0.0005, n = 7).

Clausen, Jørgen and Berg, Ole. 1975. The content of polychlorinated hydrocarbons in Arctic ecosystems. Pure. Appl. Chem. 42:223-232.
Rec #: 525
KEYWORDS: organochlorines; PCBs; DDTs; HCHs; fat; Greenland; North America; *Somateria mollissima;* common eider; *Somateria spectabilis;* king eider; *Histrionicus histrionicus;* harlequin duck; *Clangula hyemalis;* long-tailed duck; *Calidris maritima;* purple sandpiper; *Uria lomvia;* thick-billed murre; *Phalacrocorax carbo*; cormorant; *Lagopus mutus;* rock ptarmigan; *Corvus corax;* common raven; biomagnification
ABSTRACT: The content of polychlorinated hydrocarbons was detected in fat tissue of Arctic birds and mammals. Among the birds the highest amounts of DDE [72-55-9] and polychlorinated biphenyls (PCB) were found in the cormorant and raven and among the mammals the polar bear contained highest levels of PCB. The mammals all contained relatively low amounts of DDE compared with the birds. Female Eskimos had a PCB content of fat tissue equal to that in the fat tissue of mammals, but lower than that in birds.

Cleemann, M; Riget, F; Paulsen, G. B; and Dietz, R. 2000. Organochlorines in Greenland glaucous gulls (*Larus hyperboreus*) and Icelandic gulls (*Larus glaucoides*). Sci. Total Environ. 245:117-130.
Rec #: 67
KEYWORDS: organochlorines; PCBs; HCHs; DDTs; HCB; chlordanes; Greenland; North America; glaucous gull; *Larus hyperboreus*; Iceland gull; *Larus glaucoides*; liver; age variations; gender differences; bioaccumulation; spatial variations
ABSTRACT: Glaucous gulls (*Larus hyperboreus*) and Icelandic gulls (*Larus glaucoides*) were sampled in 1994 from four different areas in Greenland, three on the west coast and one on the east coast. Livers of 93 glaucous gulls and seven Icelandic gulls were analysed for polychlorinated biphenyls (PCBs, IUPAC numbers 28, 31, 52, 101, 105, 118, 138, 153, 156 and 180), DDTs (p,p'-DDE, p,p'-DDD, p,p'-DDT), hexachlorocyclohexanes (α-,β - and γ-HCH), hexachlorobenzene (HCB) and trans-nonachlor (TNC). The overall geometric means of the concentrations found in glaucous gull liver were for ΣPCBs 388 (range 20-5557), for ΣDDTs 363 (17-8604), ΣHCHs 7.4 (1-53), HCB 47 (4-594) and trans-nonachlor 19 (3-187) µg kg^{-1} wet weight, respectively. The geometric means of concentrations in Icelandic gull liver were for ΣPCBs 112 (24Ż435), for ΣDDTs 95 (25Ż298), ΣHCHs 2.9 (1.4Ż5.2), HCB 22 (8Ż58) and trans-nonachlor 5.1 (2.4Ż8.6) µg kg^{-1} wet weight, respectively. Significantly (P=0.05) higher concentrations of PCBs, DDTs and HCHs were found in glaucous gulls at Ittoqqortoormiit at the east coast than in gulls from Qeqertarsuaq at the west coast of Greenland. This tendency was also seen for HCB and trans-nonachlor, but the differences were not statistically significant (P=0.05).

Annotated Bibliography (Cont.).

A decreasing trend in organochlorine concentrations followed the East Greenland Current, flowing from north to south down the east coast and to the north on the west coast. Gulls taken from the most northerly sampling area of the west coast, however, showed slightly higher concentrations than those from the central west coast. There appeared to be a tendency for higher concentrations to be found in males than females, and in adults compared to young glaucous gulls, but the differences were not statistically significant (P=0.05). The concentration ranges found in gulls from Greenland were similar to those reported previously for gulls from northern Norway and Russia. A principal component analysis revealed no obvious link between the presence of higher chlorinated PCBs and higher PCB concentrations in glaucous gulls. Significantly higher proportions of higher chlorinated PCBs were found in glaucous gulls than in Icelandic gulls, and in adult glaucous gulls compared to young gulls of 1-2 calendar years. As no such difference was found between female and male gulls it seems that PCBs of all degrees of chlorination may be passed equally well from mother to offspring.

Cooke, A. S. 1979. Egg shell characteristics of gannets *Sula bassana*, shags *Phalacrocorax aristotelis* and great black-backed gulls *Larus marinus* exposed to DDE and other environmental pollutants. Environ. Pollut. 19(1):47-65.
Rec #: 390
KEYWORDS: organochlorines; DDTs; dieldrin; PCBs; egg; eggshell thickness; shag; *Phalacrocorax aristotelis*; gannet; *Sula bassana*; great black-backed gull; *Larus marinus*; Great Britain; Europe; effects
ABSTRACT: Structure, strength and chemical composition of eggshells of gannets, shags, and great black-backed gulls were studied in relation to egg residues of DDE (I) [72-55-9]. For each species, eggs that were highly contaminated with I tended to have thin shells. For the gannet, shell thinning was due to loss of the vaterite cover. For the shag, cover loss was again considerable, but the true shell was also reduced in thickness. Gull shells do not have a cover, and for the great black-backed gull most of the shell thinning was due to a reduced palisade layer. The different types of structural defects observed in the shells of these and other species resulted from a single biochemical lesion such as enzyme inhibition in shell gland tissue. As the thickness of the shell decreased, the shell strength for the gannet and shag did not decrease as markedly as for the gull and other species. Partial loss of cover had little effect on strength compared with loss of true shell. Had the cover been entirely missing, strength would have been seriously affected.

Coulson, J. C.; Deans, I. R.; Potts, G. R.; Robinson, J.; and Crabtree, A. N. 1972. Changes in organochlorine contamination of the marine environment of Eastern Britain monitored by shag eggs. Nature. 236(5348):454-456.
Rec #: 395
KEYWORDS: organochlorines; dieldrin; DDTs; shag; *Phalacrocorax aristotelis*; egg; Great Britain; Europe; temporal trends
ABSTRACT: The concentrations of dieldrin (I) [60-57-1] and 1,1-dichloro-2,2-bis(p-chlorophenyl)-ethylene (p,p'-DDE) [72-55-9] in shag eggs at eastern Britain in years between 1964 and 1971 were measured, and the maximum concentration of I and p,p'-DDE occurred in 1966 and 1967-8, respectfully. Between 1966 and 1971 the average concentration of I decreased by .sim.66 %. A significant decline of p,p'-DDE concentration occurred by 1969. Between 1968 and 1971 the mean concentration of p,p'-DDE decreased by .sim.47 %.

Annotated Bibliography (Cont.).

Custer, T. W.; Custer, C. M.; Hines, R. K.; Stromborg, K. L.; Allen, P. D.; Melancon, M. J.; and Henshel, D. S. 2001. Organochlorine contaminants and biomarker response in double-crested cormorants nesting in Green Bay and Lake Michigan, Wisconsin, USA. Arch. Environ. Contam. Toxicol. 40(1):89-100.
Rec #: 398
KEYWORDS: double-crested cormorant; *Phalacrocorax auritus*; organochlorines; DDTs; PCBs; PCDDs; chlordanes; HCB; mirex; toxaphene; toxicology; EROD; TEQs; egg; carcass; liver; Great Lakes; WI; MN; SD; USA; North America; effects; eggshell thickness; temporal trends; age variations; bioaccumulation; spatial variations; biomagnification
ABSTRACT: Double-crested cormorant (*Phalacrocorax auritus*) eggs at pipping and sibling 10-day-old chicks were collected from 2 colonies in Green Bay, WI, 1 colony in Lake Michigan, WI and reference colonies in South Dakota and Minnesota. Egg contents and chicks were analyzed for organochlorine contaminants including polychlorinated biphenyl (PCB) congeners. Livers of embryos and chicks were assayed for hepatic microsomal ethoxyresorufin-O-dealkylase (EROD) activity. Eggshell thickness and the physical dimensions of embryo brains were measured. Concentrations of organochlorines, including p,p'-DDE (p,p'-dichlorodiphenyldichloroethylene), PCBs, and PCB congeners were generally an order of magnitude higher in eggs and chicks from Wisconsin than from reference locations. Total PCBs averaged 10-13 µg/g wet weight in eggs from 3 Wisconsin colonies compared to 0.9 µg/g PCBs from reference locations. Double-crested cormorant chicks accumulated on average 33-66 µg PCBs/day and 7-12 µg p,p'-DDE/day in the Wisconsin colonies compared to 0 µg PCBs/day and 1 µg p,p'-DDE/day in the reference colonies. At pipping, EROD activity in the livers of cormorant embryos was significantly higher in the Wisconsin colonies and significantly correlated with PCBs and the toxic equivalent (TEQs) of aryl hydrocarbon-active PCB congeners relative to 2,3,7,8-tetrachlorodibenzo-p-dioxin. However, in 10-day-old chicks, EROD activity was not consistently different among colonies and was not correlated with PCBs or TEQs. A significant negative relationship between embryo brain asymmetry and the size of the egg suggested that physical constraint might be an important factor influencing the response of this bioindicator. Thinner eggshells in 2 colonies located near Door County, Wisconsin suggested that historic p,p'-DDE residues associated with orchards are still an important source of p,p'-DDE in the local environment.

Custer, Thomas W.; Custer, Christine M.; Hines, Randy K.; Gutreuter, Steve; Stromborg, Kenneth L.; Allen, P. David; and Melancon, Mark J. 1999. Organochlorine contaminants and reproductive success of double-crested cormorants from Green Bay, Wisconsin, USA. Environ. Toxicol. Chem. 18(6):1209-1217.
Rec #: 399
KEYWORDS: organochlorines; PCBs; DDTs; dieldrin; chlordanes; mirex; HCB; eggshell thickness; toxicology; EROD; egg; carcass; double-crested cormorant; *Phalacrocorax auritus*; reproductive success; Great Lakes; WI; USA; North America; reproductive success; effects; abnormalities
ABSTRACT: In 1994 and 1995, nesting success of double-crested cormorants (*Phalacrocorax auritus*) was measured at Cat Island, in southern Green Bay, Lake Michigan, Wisconsin, USA. Sample eggs at pipping and unhatched eggs were collected and analyzed for organochlorines (including total polychlorinated biphenyls [PCBs] and DDE), hepatic microsomal

Annotated Bibliography (Cont.).

ethoxyresorufin-O-dealkylase (EROD) activity in embryos, and eggshell thickness. Of 1,570 eggs laid, 32 % did not hatch and 0.4 % had deformed embryos. Of 632 chicks monitored from hatching to 12 days of age, 9 % were missing or found dead; no deformities were observed. The PCB concentrations in sample eggs from clutches with deformed embryos (mean = 10.2 µg/g wet weight) and dead embryos (11.4 µg/g) were not significantly higher than concentrations in sample eggs from nests where all eggs hatched (12.1 µg/g). A logistic regression of hatching success vs. DDE, dieldrin, and PCB concentrations in sibling eggs identified DDE and not dieldrin or PCBs as a significant risk factor. A logistic regression of hatching success vs. DDE and eggshell thickness implicated DDE and not eggshell thickness as a significant risk factor. Even though the insecticide DDT was banned in the early 1970s. We suggest that DDE concentrations in double-crested cormorant eggs in Green Bay are still having an effect on reproduction in this species.

Custer, Thomas W.; Custer, Christine. M.; and Stromborg, Kenneth L. 1997. Distribution of organochlorine contaminants in double-crested cormorant eggs and sibling embryos. Environ. Toxicol. Chem. 16(8):1646-1649.
Rec #: 27
KEYWORDS: organochlorines; PCBs; DDTs; chlordanes; dieldrin; carcass; egg; embryo; bioaccumulation; double-crested cormorant; *Phalacrocorax auritus*; WI; USA; North America; Great Lakes
ABSTRACT: Double-crested cormorant (*Phalacrocorax auritus*) fresh eggs and sibling embryos at pipping were collected from a polychlorinated biphenyl (PCB)-contaminated colony in Green Bay, Wisconsin, USA. Egg contents were analyzed for organochlorine (OC) contaminants, including 15 arylhydrocarbon-active PCB congeners. In order to determine the significance of tissue removal on the subsequent estimate of contaminant burden, embryos were decapitated and the heads, yolk sac, liver, fecal sac (allantois), and carcass remainder were analyzed separately. The distribution of contaminant concentration in the embryos was yolk sac > liver > carcass > head > fecal sac. The distribution of contaminant mass in the embryos was yolk sac > carcass > liver > head > fecal sac. For example, mass of total PCBs (TPCB) was yolk sac = 58 %, carcass = 31 %, liver = 5 %, head = 3 %, and fecal sac = 1 %. Eighteen additional OCs, including 13 PCB congeners, had distribution patterns similar to that of TPCB concentration and mass. Excluding the head of the embryo from the chemical analysis overestimated TPCB concentrations by 15 % (16 vs 14 µg/g). In contrast, excluding the liver from the chemical analysis underestimated TPCB concentration by only 4 % (13.5 vs 14 µg/g). Mean concentrations of OCs were not significantly different between fresh eggs and sibling embryos.

Daelemans, F. F.; Mehlum, F.; and Schepens, P. J. C. 1992. Polychlorinated biphenyls in two species of Arctic seabirds from the Svalbard area. Bull. Environ. Contam. Toxicol. 48(6):828-834.
Rec #: 76
KEYWORDS: Svalbard; Norway; Europe; biomagnification; glaucous gull; *Larus hyperboreus*; organochlorines; PCBs; toxicology; AHH; EROD; liver; black guillemot; *Cepphus grylle*; glaucous gull; *Larus hyperboreus*
ABSTRACT: Polychlorinated hydrocarbons and more in particular polychlorinated biphenyls

Annotated Bibliography (Cont.).

(PCBs) seem to be one of the most dangerous pollutants of the Arctic environment. The archipelago of Svalbard in the European Arctic is an area with only minor local industrial activities, and the anthropogenic pollution recorded in the area is assumed to be mainly of foreign origin. The wildlife in Svalbard is rich in numbers and large populations of seabirds and marine mammals constitute major parts of the fauna. In the Glaucous Gull, *Larus hyperboreus*, especially high levels of PCBs have been recorded. This species partly acts as a predator and during the breeding season it is preying upon other seabird's eggs and chicks, and has similar ecological function as birds of prey in other places. Because of bioaccumulation of organochlorines one would suspect to find high levels of these substances in this species, which is the main avian predator in the Svalbard area.

Dale, I. M.; Baxter, M. S.; Bogan, J. A.; and Bourne, W. R. P. 1973. Mercury in seabirds. Mar. Pollut. Bull. 4(5):77-75.
Rec #: 538
KEYWORDS: metals; Hg; liver; Great Britain; Europe; common murre; *Uria aalge;* red-breasted merganser; *Mergus serrator*; common eider; *Somateria mollissima*; biomagnification
ABSTRACT: Concentrations of mercury in the liver of birds around the British coast vary from very low levels of 0.7 ppm in a young guillemot to 122 ppm in a redbreasted merganser. Eiders which feed on mussels known to accumulate mercury also have high concentrations of this pollutant.

Davis, Jay A.; Fry, D. Michael; and Wilson, Barry W. 1997. Hepatic ethoxyresorufin-O-deethylase activity and inducibility in wild populations of double-crested cormorants (*Phalacrocorax auritus*). Environ. Toxicol. Chem. 16(7):1441-1449.
Rec #: 444
KEYWORDS: double-crested cormorant; *Phalacrocorax auritus*; liver; CA; OR; USA; North America; toxicology; EROD; organochlorines; PCDDs; spatial variations
NOTES: SUPPLEMENTAL; FIGURES ONLY
ABSTRACT: Microplate fluorometric techniques were used to measure ethoxyresorufin O-deethylase (EROD) activity in hepatic microsomes and primary hepatocyte cultures from individual wild double-crested cormorant (*Phalacrocorax auritus*) embryos. Embryos were collected in 1993 and 1994 from Humboldt Bay and San Francisco Bay (CA, USA) and a reference site in coastal Oregon (USA). Median microsomal EROD activities in embryos collected from San Francisco Bay (in both 1993 and 1994) and from Humboldt Bay (1994) were four- to eightfold higher than the reference site median (Kruskal-Wallis). This degree of induction suggests that cormorant embryos in the two California locations were exposed to concentrations of dioxin-like compounds that are at the threshold for toxic effects in this species. Substantial variation in the EROD response in cultured hepatocytes was observed between individuals, populations, and the two bird species tested (cormorants and chickens [*Gallus gallus*]). Although most of the cormorant individuals displayed a consistent dose-response profile, a few individuals were uninducible, showing no appreciable increase over basal activity with increasing dose of inducer. Composite dose-response curves for two cormorant colonies appeared to be divergent in spite of small sample sizes, indicating that inducibility can also vary at the population level. These observations suggest that considerable variability in pollutant metabolism and sensitivity associated with single enzyme systems may exist within wild populations and species.

Annotated Bibliography (Cont.).

Day, Robert H.; Murphy, Stephen M.; Wiens, John A.; Hayward, Gregory D.; Harner, E. James; and Smith, Louise N. 1997. Effects of the Exxon Valdez Oil Spill on habitat use by birds in Prince William Sound, Alaska. Ecol. Appl. 7(2):593-613.
Rec #: 250
KEYWORDS: AK; USA; North America; oil; effects; common loon; *Gavia immer;* horned grebe; *Podiceps auritus*; red-necked grebe; *Podiceps grisegena*; fork-tailed storm-petrel; *Oceanodroma furcata*; double-crested cormorant; *Phalacrocorax auritus*; pelagic cormorant; *Phalacrocorax pelagicus*; red-faced cormorant; *Phalacrocorax urile*; great blue heron; *Ardea herodias*; Canada goose; *Branta canadensis*; green-winged teal; *Anas crecca*; mallard; *Anas platyrhynchos*; American wigeon; *Anas americana*; harlequin duck; *Histrionicus histrionicus*; long-tailed duck; *Clangula hyemalis*; black scoter; *Melanitta nigra*; surf scoter; *Melanitta perspicillata*; white-winged scoter; *Melanitta fusca*; common goldeneye; *Bucephala clangula;* Barrow's goldeneye; *Bucephala islandica*; bufflehead; *Bucephala albeola*; common merganser; *Mergus merganser;* red-breasted merganser; *Mergus serrator*; bald eagle; *Haliaeetus leucocephalus*; black oystercatcher; *Haematopus bachmani;* wandering tattler; *Heteroscelus incanus*; spotted sandpiper; *Actitis* macularia; red-necked phalarope; *Phalaropus lobatus*; pomarine jaeger; *Stercorarius pomarinus*; common gull; *Larus canus*; herring gull; *Larus argentatus*; glaucous-winged gull; *Larus glaucescens*; black-legged kittiwake; *Rissa tridactyla*; Arctic tern; *Sterna paradisaea*; common murre; *Uria aalge*; pigeon guillemot; *Cepphus columba;* marbled murrelet; *Brachyramphus marmoratus*; tufted puffin; *Fratercula cirrhata*; belted kingfisher; *Ceryle alcyon*; Steller's jay; *Cyanocitta stelleri*; black-billed magpie; *Pica pica;* northwestern crow; *Corvus caurinus*; common raven; *Corvus corax*
NOTES: SUPPLEMENTAL
ABSTRACT: Oil spills may affect species through direct effects on population size and structure and direct and indirect (toxicological) effects on reproduction. Spill effects on the habitats these organisms occupy have received less attention, but they are no less important. For 2.5 years following the Exxon Valdez oil spill in Prince William Sound, Alaska, we studied the use of oil-affected habitats by 42 species of marine-oriented birds. On 11 survey cruises, we surveyed bays that had received different levels of initial oiling. We related the abundance of individual species in the bays to the oiling gradient, using regression models that included habitat measures to control for variations among the sites in features other than oiling level. We defined a spill-induced impact as a statistically significant relationship between the abundance of a species and values along the oiling gradient, after accounting for the effects of variations in habitat features. We used among-year comparisons of regressions between oiling levels and abundance, controlled for season, to assess recovery. We concluded that recovery from a spill-induced impact had occurred when we no longer could detect a significant relationship between a species' abundance and oiling levels. Overall, 23 (55 %) of the 42 species exhibited no initial negative impacts on their use of oil-affected habitats. Of the 19 species that did exhibit negative impacts, 13 (68 %) showed evidence of recovery within 2.5 years (the final survey in 1991). Six species (Horned Grebe, Red-necked Grebe, Barrow's Goldeneye, Bufflehead, Mew Gull, and Northwestern Crow) showed no clear evidence of recovery by our final survey. The proportion of species recorded on individual surveys that exhibited negative impacts at that time declined over the study, from 54 % on the first survey after the spill in 1989 to 10 % in late 1991. A principal components analysis revealed extensive ecological overlap between species that were negatively impacted in their use of oil-affected habitats and those that were not. The six species

Annotated Bibliography (Cont.).

that had not recovered by late 1991 tended to be intertidal feeders and residents, but these traits also characterized some species that did not exhibit initial impacts and some species that subsequently recovered from impacts. We detected no obvious ecological differences between species that suffered spill impacts on habitat use and those that apparently were not affected, or between impacted species that later recovered in their use of habitats and species that had not yet recovered. These results indicate that the Exxon Valdez oil spill had clear initial negative impacts on habitat use by nearly half of the species examined, suggesting substantial initial effects on habitat suitability for these species. These impacts persisted for <2.5 years for most affected species. This rate of recovery in habitat use parallels the rapid recovery (usually <2 years) documented for other oil-affected communities (e.g., intertidal invertebrates, fishes, and birds) that have been studied in Alaska and elsewhere.

Day, Robert H.; Murphy, Stephen M.; Wiens, John A.; Hayward, Gregory D.; Harner, E. James; and Smith, Louise N. 1995. Use of oil-affected habitats by birds after the Exxon Valdez oil spill. In: Wells, P.G.; Butler, J.N.; Hughes, J.S. (Eds.). Exxon Valdez Oil Spill: Fate and Effects in Alaskan Waters. Philadelphia, PA: American Society for Testing and Materials. 726-761.
Rec #: 247
KEYWORDS: AK; USA; North America; oil; effects; common loon; *Gavia immer;* yellow-billed loon; *Gavia adamsii*; horned grebe; *Podiceps auritus*; red-necked grebe; *Podiceps grisegena*; northern fulmar; *Fulmarus glacialis*; sooty shearwater; *Puffinus griseus*; fork-tailed storm-petrel; *Oceanodroma furcata*; double-crested cormorant; *Phalacrocorax auritus*; pelagic cormorant; *Phalacrocorax pelagicus*; red-faced cormorant; *Phalacrocorax urile*; great blue heron; *Ardea herodias*; Canada goose; *Branta canadensis*; green-winged teal; *Anas crecca*; mallard; *Anas platyrhynchos*; American wigeon; *Anas americana*; ring-necked duck; *Aythya collaris*; harlequin duck; *Histrionicus histrionicus*; long-tailed duck; *Clangula hyemalis*; black scoter; *Melanitta nigra*; surf scoter; *Melanitta perspicillata*; white-winged scoter; *Melanitta fusca*; common goldeneye; *Bucephala clangula;* bufflehead; *Bucephala albeola*; Barrow's goldeneye; *Bucephala islandica*; common merganser; *Mergus merganser;* red-breasted merganser; *Mergus serrator*; ruddy duck; *Oxyura jamaicensis*; bald eagle; *Haliaeetus leucocephalus*; sharp-shinned hawk; *Accipiter striatus*; black oystercatcher; *Haematopus bachmani;* wandering tattler; *Heteroscelus incanus*; spotted sandpiper; *Actitis macularia*; least sandpiper; *Calidris minutilla*; red-necked phalarope; *Phalaropus lobatus*; pomarine jaeger; *Stercorarius pomarinus*; common gull; *Larus canus*; herring gull; *Larus argentatus*; Iceland gull; *Larus glaucoides*; glaucous-winged gull; *Larus glaucescens*; glaucous gull; *Larus hyperboreus*; black-legged kittiwake; *Rissa tridactyla*; Arctic tern; *Sterna paradisaea*; common murre; *Uria aalge*; pigeon guillemot; *Cepphus columba;* marbled murrelet; *Brachyramphus marmoratus*; ancient murrelet; *Synthliboramphus antiquus*; rhinoceros auklet; *Cerorhinca monocerata*; tufted puffin; *Fratercula cirrhata*; horned puffin; *Fratercula corniculata*; belted kingfisher; *Ceryle alcyon*; Steller's jay; *Cyanocitta stelleri*; black-billed magpie; *Pica pica;* northwestern crow; *Corvus caurinus*; common raven; *Corvus corax*
NOTES: SUPPLEMENTAL
ABSTRACT: This study investigated the effects of the Exxon Valdez oil spill on the use of oil-affected habitats by birds during 1989-1991. We measured densities of birds in bays that had been subjected to various levels of oiling from the spill during survey cruises that were conducted throughout the year in Prince William Sound (PWS) and during summer along the Kenai Peninsula. Overall, 23 of 42 (55 %) species in PWS and 22 of 34 (65 %) species on the

Annotated Bibliography (Cont.).

Kenai showed no evidence of oiling impacts on their use of habitats. Most species that did show initial negative impacts had recovered by late summer 1991 when our study concluded, although 6 of the 19 species initially impacted in PWS and 6 of the 12 species initially impacted along the Kenai did not exhibit clear signs of recovery by this time. A Principal Components Analysis of species examined from PWS revealed extensive overlap in ecological attributes among species that were and were not negatively impacted in their use of oil-affected habitats. Species that did not show clear evidence of recovery tended to be intertidal feeders and residents of PWS, but other ecologically similar species evidenced either no initial impacts or rapid recovery. These similarities suggest that the prognosis is good for the species for which we were unable to document recovery in habitat use. Our findings, together with the rapid rates of recovery in habitat features reported in other studies, suggested that impacts of the Exxon Valdez oil spill on avian use of oil-affected habitats generally were not persistent.

De Voogt, Pim; Van Raat, Patrick; Rozemeijer, Marcellino; and Green, Nick. 1996. Methylsulfonyl PCBs in cormorant chicks from the Netherlands. Organohalogen Compounds. 28:517-521.
Rec #: 445
KEYWORDS: organochlorines; PCBs; double-crested cormorant; *Phalacrocorax auritus*; fat; Netherlands; Europe
ABSTRACT: Methylsulfonyl-polychlorinated biphenyls (MeSO $_2$-PCBs) and PCBs were investigated in adipose tissue of cormorant chicks from 2 colonies [Biesbosch (BB), Brede Water (BW)] in the Netherlands. Levels of total PCBs (sum of 36 congeners) in BW and BB were 11 and 58 μg/g (lipid weight), respectively. In fish regurgitated by adult birds the sum of 6 congeners was 6 and 30 μg/g in BW and BB, respectively. PCB congener patterns reveal that lower chlorinated PCBs were reduced in eggs and hatchlings as compared to regurgitated food. The 3- and 4-MeSO2-metabolites (I and II) of PCB 49, 87, 101, 141, and 149 and the 4-MeSO2 of PCB 132 were analyzed in the adipose tissue from cormorant chicks. Total sum of MeSO2-PCB concentration (11 congeners) amounted to 30 and 70 ng/g in BW and BB, respectively. The most abundant metabolite was the II of PCB 149. The total concentration of I and II of PCB 149 was between 20 and 50 % of their parent compound PCB 149. For PCB 149 the concentration of the II predominated its I isomer. For PCB 87 and 101 para- and meta-isomer abundance was equal, whereas for PCB 49 and 141 the meta-isomer predominated. The ratio of total MeSO$_2$-PCBs to total PCBs was <0.3 %.

Debacker, V.; Eppe, G.; Massart, A.-C.; Xhrouet, C.; Jauniaux, T.; Huart, P.; Hauteclair, P.; Bouquegneau, J.-M.; and De Pauw, E. 2003. Polychlorinated dibenzo-*p*-dioxins and dibenzofurans in livers of an Atlantic seabird, the common guillemot Uria aalge: Influence of the general body condition. Organohalogen Compounds. 64:443-446.
Rec #: 492
KEYWORDS: organochlorines; PCDDs; PCDFs; toxicology; TEQs; liver; common murre; *Uria aalge*; Belgium; Brittany; Europe; effects; oil; spatial variations
ABSTRACT: The tissue distribution of polychlorinated dibenzo-p-dioxins and dibenzofurans in livers of the Atlantic seabird *Uria aalge* was studied considering the robustness of animals. Male guillemots washed ashore along the Belgian coast and after an oil spill off the Brittany coasts were examined. Increased levels of both PCDD and PCDF congeners (3- and 5-fold, respectively) were detected in the livers of the Belgian guillemots compared to their Brittany

Annotated Bibliography (Cont.).

counterparts. Increased levels of PCDD and PCDF were found parallel to increasing cachectic conditions. Levels of dioxin congeners were higher in severely debilitated birds.

Debacker, V.; Holsbeek, L.; Tapia, G.; Gobert, S.; Joiris, C. R.; Jauniaux, T.; Coignoul, F.; and Bouquegneau, J.-M. 1997. Ecotoxicological and pathological studies of common guillemots *Uria aalge* beached on the Belgian coast during six successive wintering periods (1989-90 to 1994-95). Dis. Aquat. Org. 29(3):159-168.
Rec #: 39
KEYWORDS: effects; organochlorines; PCBs; metals; Cu; Zn; Fe; Cd; Hg; organometallics; organic Hg; muscle; liver; kidney; *Uria aalge*; common murre; Belgium; Europe; biomarkers; seasonal variations; histopathology
ABSTRACT: During 6 successive wintering periods, 727 common guillemots *U. aalge* were recovered from Belgian beaches. One-third of the birds were already dead; the rest passed through rehabilitation centers where they eventually died. All birds were monitored for general condition (body mass, fat reserves), eventual status of oiling and pathological changes (cachexia, acute hemorrhagic gastroenteropathy); 339 birds were sampled for trace metals (total and organic Hg, Cu, Zn, Fe, Cd) and PCB (polychlorinated biphenyl) analysis. Oiling is still a major cause of death for wintering pelagic seabirds: half of the birds showed signs of external or internal oiling, probably a still greater number of oiled birds never reach the shores. Although a low body mass can be considered a normal winter condition for wintering guillemots, pathological results showed that three-quarters of the studied animals were in a state of cachexia with emaciated pectoral muscle and lowered muscle lipid content. Elevated levels of Cu, Zn, Hg, and PCBs were linked to the state of cachexia and may well represent an additional stress factor leading to the debilitation and death of part of the wintering guillemot population.

Debacker, V.; Jauniaux, T.; Coignoul, F.; and Bouquegneau, J.-M. 2000. Heavy metals contamination and body condition of wintering guillemots (*Uria aalge*) at the Belgian coast from 1993 to 1998. Environ. Res. 84(3):310-317.
Rec #: 342
KEYWORDS: metals; Zn; Cu; Fe; Cd; Belgium; Europe; common murre; *Uria aalge*; liver; kidney; muscle; effects; toxicology
ABSTRACT: A sample of 166 common guillemots (*Uria aalge*) recovered from Belgian beaches during five wintering seasons, from 1993-1994 to 1997-1998, were examined. At necropsy, postmortem examination including body mass, fat reserves, presence or not of intestinal contents, eventual status of oiling, and pathological changes (cachexia, acute hemorrhagic gastroenteropathy (GEAH)) was attributed to each individual. Mild to severe cachexia, a pathology characterized by moderate to severe atrophy of the pectoral muscle as well as reduced amounts or absence of subcutaneous and/or abdominal fat, was observed for most specimens (85.8 %). Heavy metal analyses (Cu, Zn, Fe, Cd, Ni, Cr, and Pb) of the tissues (typically liver, kidney, and pectoral muscle) were performed, and total lipids were determined (liver and pectoral muscle). The guillemots collected at the Belgian coast exhibited higher Cu and Zn concentrations compared to individuals collected in more preserved areas of the North Sea such as the northern colonies. A general decrease of their total body mass as well as liver, kidney, and pectoral muscle mass was associated to increasing cachexia severity. Moreover, significantly increasing heavy metal levels (Cu and Zn) in the tissues as well as depleted muscle lipid contents were observed parallel to increasing cachexia severity. On the contrary the organs'

Annotated Bibliography (Cont.).

total metal burden barely correlates to this status. These observations tend to indicate a general redistribution of heavy metals within the organs as a result of prolonged starvation and protein catabolism (cachectic status). Such a redistribution could well be an additional stress to birds already experiencing stressful conditions (starvation, oiling).

Debacker, V.; Rutten, A.; Jauniaux, T.; Daemers, C.; and Bouquegneau, J-M. 2001. Combined effects of experimental heavy-metal contamination (Cu, Zn, and CH3Hg) and starvation on quail's body condition: parallelism with a wild common guillemot (*Uria aalge*) population found stranded at the Belgian coast. Biological Trace Element Research. 82(1-3):87-107.
Rec #: 498
KEYWORDS: metals; Cu; Zn; organometallics; organic Hg; Japanese quail; *Coturnix coturnix;* common murre; *Uria aalge*; Belgium; Europe; liver; kidney; muscle; dosing study; effects; toxicology
NOTES: FIGURES ONLY
ABSTRACT: Combined effects of heavy-metal contamination (Cu, Zn, and CH3Hg) and starvation were tested on common quails (*Coturnix coturnix japonica*) and used as a model for comparison with a wild common guillemot (*Uria aalge*) population found stranded at the Belgian coast. Appropriate heavy-metal levels were given to the quails to obtain concentrations similar to those found in the seabird's tissues. The contaminated animals were then starved for 4 d to simulate the evident malnutrition symptoms observed at the guillemot's level. In such conditions, food intake and total body weight are shown to decrease in contaminated individuals with simultaneous significant hepatic and renal increase of the heavy-metal concentrations. Like guillemots, higher heavy-metal levels were observed in those contaminated quails that had also developed a cachectic status characterized by a general atrophy of their pectoral muscle and complete absence of s.c. and(or) abdominal fat depots. Although likely the result of a general protein catabolism during starvation, it is suggested that these higher metal levels could as well enhance a general muscle wasting process (cachectic status).

Debacker, V.; Schiettecatte, L.-S.; Jauniaux, T.; and Bouquegneau, J.-M. 2001. Influence of age, sex and body condition on zinc, copper, cadmium and metallothioneins in common guillemots (*Uria aalge*) stranded at the Belgian coast. Mar. Environ. Res. 52(5):427-444.
Rec #: 499
KEYWORDS: metals; Zn; Cu; Cd; metallothionein; common murre; *Uria aalge*; liver; kidney; Belgium; Europe; effects; toxicology; gender differences; age variations; bioaccumulation
ABSTRACT: The common guillemots, *Uria aalge*, found stranded at the Belgian coast, display high levels of Cu in both liver and kidneys. The condition index of the animals, defined as the ratio of liver to kidneys mass, influences both the metal concentration and its binding to metallothioneins (MT): the lower the condition index, the more emaciated the animals, and the higher the total Cu concentration and the concentration of Cu bound to MT. In less robust individuals, our results suggest that Cu could displace Zn from MT, rendering the Zn ions available to induce a new MT synthesis. Sex-related effects also emerged as significantly higher hepatic MT as well as Cu- and Zn-MT concentrations were found in emaciated male guillemots compared to females. In both organs, Cd concentrations remained low and typically demonstrated an age-dependent renal accumulation, with no noticeable effect of the condition index. As a whole, these results suggest that, for guillemots found stranded at the Belgian coast, Cu binding to hepatic and renal MT could function as a protective mechanism, rendering the

Annotated Bibliography (Cont.).

metal ions unavailable to exert any cytotoxic activity.

Debacker, Virginie. 2000. Common guillemot *Uria aalge* stranding at the Belgian coast: An ecotoxicological evaluation. Liege, Belgium: Universite De L'Etat a Liege, Faculty of Sciences, Oceanology. 179 pp.
Rec #: 493
KEYWORDS: common murre; *Uria aalge*; France; Belgium; North Sea; Europe; metals; Fe; Hg; Zn; Cu; Cd; organometallics; organic Hg; oil; liver; kidney; muscle; toxicology; metallothionein; oil; effects; biomarkers; age variations; bioaccumulation; gender differences; spatial variations
ABSTRACT: The aim of the present work is to determine the potential relationships existing between the common guillemot's health status and its contaminants loading (Cu, Zn, Fe and Cd) through:
- the study of the heavy metal levels in the tissues;
- the study of the heavy metal speciation using metallothionein analysis;
- the investigation of the potential relationship existing between the heavy metal levels in the tissues and the body condition.

Chapter 1 presents the studied species *(Uria aalge)* in its North Sea habitat. Heavy metals and their impact on seabirds are also presented in this chapter with a special interest given to studies relating heavy metal contamination to the bird's body condition.

The results are then presented in different chapters:
Chapter 2: <u>Ecotoxicological and pathological studies of common guillemots *Uria aalge* beached on the Belgian coast during six successive wintering periods (1989-90 to 1994-95).</u>
In this study, guillemots collected directly on the beaches and those which had been treated in rehabilitation centres were both analyzed. However, it soon appeared that, in addition to a general lack of information regarding the latter individuals, they were also characterized by significantly higher contamination levels. For these reasons, these birds were discarded from the sample and priority was given to those guillemots directly collected on the beaches.
In those beached birds, severe emaciation (cachectic status) was identified as one of the major lesion and linked to higher heavy metal levels (Cu, Zn and total Hg). At that stage, cachexia was evaluated as present or absent without intermediate status.
Compared to guillemots collected in more preserved area of the North Sea, those collected at the Belgian coast displayed higher heavy metal concentrations in their tissues (Cu, Zn and total Hg).

Chapter 3: <u>Heavy metals contamination and body condition of wintering guillemots *Uria aalge* at the Belgian coast from 1993 to 1998.</u>
In this study, the cachectic status is detailed depending on its severity (from non cachectic:'-' to low: '+1', moderate: '+2' and severely cachectic: '+3'). In addition, the general body condition of the guillemots is also evaluated using a condition index (liver to kidney weight ratio) which is clearly linked to the cachectic status: the lower the condition index, the more cachectic the bird. Heavy metal levels (Cu and Zn) in the tissues were shown to significantly increase with increasing cachexia severity while the organs' total metal burdens remained unchanged. A general redistribution of heavy metals within the organs as a result of prolonged starvation and protein catabolism is proposed.

Annotated Bibliography (Cont.).

Chapter 4: <u>Body condition and heavy metals: comparison between guillemots *(Uria aalge)* found stranded at the Belgian coast and those caught in the Erika's oil spill (Britanny, December 1999).</u>
Comparing those two samples permitted:
- to examine the heavy metal contents in the tissues of healthy and unhealthy guillemots. Results also pointed towards a probable general redistribution of heavy metals within the organs as a result of protein catabolism.
- to confirm that the southern North Sea, as a wintering ground for seabirds, including the guillemots, is likely to be more polluted: significantly higher heavy metal levels were systematically found in guillemots collected at the Belgium coast compared to their Brittany counterparts, even when comparing both guillemots to more robust individuals (condition index *>2.5)*.

Chapter 5: <u>Combined effects of experimental heavy metals contamination (Cu, Zn and CH_3Hg) and starvation on quails' body condition: parallelism with a wild common guillemots *(Uria aalge)* population found stranded at the Belgian coast.</u>
This experimental study using common quails was set up to evaluate the cause-effect relationship existing between the high heavy metal levels observed in the guillemots' tissues and their emaciation status: could these high levels cause emaciation or, on the contrary be the result of that emaciation process as previously suggested?

Chapter 6: <u>Influence of age, sex and body condition on Zn, Cu, Cd and metallothioneins in common guillemots *(Uria aalge)* stranded at the Belgian coast.</u>
Heavy metal speciation was examined through the study of metallothioneins and metalloprotein known to play a key role in metal detoxification. Analysis was conducted on individuals for which the condition index ranged from low moderate (>2) values, which also presented a wide range of total heavy metals in their tissues. The condition index appeared as significantly affecting the hepatic and renal metals distribution on the protein, with higher levels bound to the metallothioneins in extremely emaciated guillemots.

Delbeke, K.; Joiris, C.; and Decadt, G. 1984. Mercury contamination of the Belgian avifauna 1970-1981. Environ. Pollut. (B. Chem. Phys.). 7(3):205-221.
Rec #: 506
KEYWORDS: metals; Hg; Belgium; Europe; liver; kidney; heart; muscle; little grebe; *Tachybaptus ruficollis*; kingfisher; *Alcedo atthis*; northern fulmar; *Fulmarus glacialis*; great black-backed gull; *Larus marinus*; red-necked grebe; *Podiceps grisegena*; great crested grebe; *Podiceps cristatus*; common murre; *Uria aalge*; cormorant; *Phalacrocorax carbo;* grey heron; *Ardea cinerea*; effects; bioaccumulation; seasonal variations
ABSTRACT: Two hundred birds found dead in Belgium between 1970 and 1981, and belonging to 30 species, were analyzed for total Hg contamination. The contamination of aquatic birds ranged between 0.11 and 35.0 µg/g. For terrestrial birds, the extreme values were: not detectable and 14 µg/g. In both cases, differences in diet can explain the differences in contamination. The order of diets associated with increasing Hg contamination for aquatic birds was invertebrates, zooplankton and garbage, and fish; and for terrestrial birds this consisted of plants, invertebrates, mammals and birds. For raptors and owls, this effect of diet includes

Annotated Bibliography (Cont.).

geographical variations within species. A higher Hg contamination level in the winter and early spring was noted for 2 species of owls. For aquatic birds, the contamination of liver was higher than that of kidney, with ratios varying between 1.2 and 2.5. For terrestrial birds, the ratio was closer to 1. A few detections were also made for muscle and heart, giving respectively 0.25 and 0.6 of the liver contamination. Among the birds analyzed for their liver contamination, 15 % showed levels >3 µg/g and this level could have affected their reproduction; 3 % had levels >10 µg/g, and these birds could have died from Hg poisoning; and 6 % showed an abnormally high liver:kidney ratio, which could reflect an acute intoxication. There exists a striking parallelism between the levels of Hg and organochlorine residues (DDT) in birds of prey, suggesting the existence of common ecotoxicological mechanisms.

Delbeke, Katrien and Joiris, Claude. 1988. Accumulation mechanisms and geographical distribution of PCBs in the North Sea. Oceanis. 14(4):399-410.
Rec #: 242
KEYWORDS: organochlorines; PCBs; North Sea; Europe; carcass; common murre; *Uria aalge;* northern fulmar; *Fulmarus glacialis*; common gull; *Larus canus*; spatial variations; biomagnification
ABSTRACT: PCBs were determined in the main compartments of the North Sea ecosystems: suspended particulate matter, zooplankton, fish, sea birds and sediments. The PCB levels in particulate matter, net plankton and zooplankton from different geographical sites in the North Sea (51 degree -60 degree N) finally allow differently contaminated water masses to be distinguished: 1) high PCB levels in the Belgian coast zone decreasing with increasing distance from the coast (51 degree -53 degree N); 2) high PCB levels close to the Dogger Bank (53 degree -54 degree N) and 3) low PCB levels in the northern water mass (54 degree -60 degree N).

Denker, Eckhard. 1996. Seasonal and species specific differences of PCB burden and pattern in animals of the marine food web investigated by congener specific determination of all PCB congeners. Organohalogen Compounds. 29:82-87.
Rec #: 317
KEYWORDS: black-legged kittiwake; *Rissa tridactyla*; organochlorines; PCBs; North Sea; Germany; Europe; bioaccumulation
NOTES: FIGURES ONLY
ABSTRACT: Polychlorinated biphenyl (PCB) burden and pattern of sand eel (*Ammodytes tobianus*) and cod (*Gadus morhua*) caught in the North Sea and of kittiwakes (*Rissa tridactyla*), feeding on these fish were detected Differences in PCB burden and pattern existed between fish species and even between fish of one species of the same age class even in the spring and summer month, this means that randomly collected and investigated fish samples show only the status quo of the capture date and lead to wrong interpretations concerning the ecosystem. Complex relationships existed with respect to PCB burden and pattern in the water between biotic and abiotic parts of the ecosystem with processes to equilibrium partitioning as the most important factors. The predator kittiwake showed a higher chlorinated PCB mixture as the fish species, in this species as in other bird- and mammal-species uptake and metabolism of PCB are the dominating factors controlling PCB burden and pattern.

Denker, E.; Becker, P. H.; Beyerbach, M.; Büthe, A.; Heidmann, W. A.; and Staats de Yanes, G.

Annotated Bibliography (Cont.).

1994. Concentrations and metabolism of PCBs in eggs of waterbirds on the German North Sea coast. Bull. Environ. Contam. Toxicol. 52(2):220-225.
Rec #: 119
KEYWORDS: organochlorines; PCBs; metabolism; biomagnification; egg; Wadden Sea; Germany; Europe; herring gull; *Larus argentatus*; common tern; *Sterna hirundo*; oystercatcher; *Haematopus ostralegus*; sandwich tern; *Sterna sandvicensis*; shelduck; *Tadorna tadorna*; common gull; *Larus canus*; common eider; *Somateria mollissima*; avocet; *Recurvirostra avosetta*; black-headed gull; *Larus ridibundus*; ringed plover; *Charadrius hiaticula*; redshank; *Tringa totanus*
ABSTRACT: The authors show the interspecific variation in PCB concentrations and in the degree of metabolism of the PCB mixtures in eggs of 11 waterbird species breeding on the highly polluted German Wadden Sea coast, where the PCB contamination of birds has been increasing during the last decade.

Dietz, R.; Nielsen, C. O.; Hansen, M. M.; and Hansen, C. T. 1990. Organic mercury in Greenland birds and mammals. Sci. Total Environ. 95:41-51.
Rec #: 292
KEYWORDS: metals; Hg; organometallics; organic Hg; Greenland; North America; little auk; *Alle alle*; black guillemot; *Cepphus grylle*; thick-billed murre; *Uria lomvia*; northern fulmar; *Fulmarus glacialis*; Iceland gull; *Larus glaucoides*; glaucous gull; *Larus hyperboreus*; black-legged kittiwake; *Rissa tridactyla*; common eider; *Somateria mollissima*; king eider; *Somateria spectabilis*; pomarine jaeger; *Stercorarius pomarinus*; muscle; liver; kidney
ABSTRACT: Muscle, liver and kidney samples of 20 species of birds, seals, whales and polar bear were analyzed for total and organic mercury. Organic mercury concentrations varied considerably between individuals. A general tendency towards age accumulation was found, together with log-linear correlations between organic mercury concentrations in the three tissues. The major part of the muscle mercury was organic (maximum concentration found was 1235 micrograms kg-1 wet weight). This also applied to liver of birds, while in mammal liver organic mercury concentrations approached a level of 2000 micrograms kg-1 wet weight, which was not exceeded even when the total mercury concentration was greater than 100,000 micrograms kg-1 wet weight The percentage of organic mercury in relation to total mercury in kidney of seals and whales was 10-20 % (maximum 982 micrograms organic mercury kg-1 wet weight), while in polar bear it was less than 6 % (maximum 217 micrograms kg-1 wet weight). For the monitoring of local food in the Arctic, the simpler and less expensive analysis of total mercury suffices when testing muscle, whereas liver and kidney should be tested for organic mercury as well.

Dietz, R.; Riget, F.; and Born, E. W. 2000. An assessment of selenium to mercury in Greenland marine animals. Sci. Total Environ. 245(1-3):15-24.
Rec #: 456
KEYWORDS: metals; Hg; Se; Greenland; North America; muscle; liver; kidney; black guillemot; *Cepphus grylle*; thick-billed murre; *Uria lomvia*; little auk; *Alle alle*; common eider; *Somateria mollissima*; king eider; *Somateria spectabilis*; long-tailed duck; *Clangula hyemalis*; red-breasted merganser; *Mergus serrator*; Iceland gull; *Larus glaucoides*; glaucous gull; *Larus hyperboreus*; black-legged kittiwake; *Rissa tridactyla*; ivory gull; *Pagophila eburnea*; northern fulmar; *Fulmarus glacialis*; cormorant; *Phalacrocorax carbo*; pomarine jaeger; *Stercorarius pomarinus*

Annotated Bibliography (Cont.).

ABSTRACT: Information on mercury and selenium molar relation in muscle, liver and kidney tissue of Greenland marine animals is presented. In the majority of the samples selenium was present in a molar surplus to mercury. This was most clear in molluscs, crustaceans, fish and seabirds. A 1:1 molar ratio was found in tissues of marine mammals with high mercury concentrations (above approximately 10 nmol/g). This was most clearly demonstrated for liver and kidney tissue of polar bear and for ringed seal with high mercury concentration in the liver. These findings support previous results found in liver tissue of marine mammals, suggesting that Me mercury is detoxified by a chemical mechanism involving selenium. If the anthropogenic release of mercury to the environment increases in the future due to increasing energy demands, species such as polar bears and seals with high tissue mercury concentrations should be monitored to elucidate whether this protective mechanism can be maintained in target organs.

Dietz, R.; Riget, F.; and Johansen, P. 1996. Lead, cadmium, mercury and selenium in Greenland marine animals. Sci. Total Environ. 186:67-93.
Rec #: 9
KEYWORDS: Greenland; North America; metals; Pb; Cd; Hg; Se; liver; kidney; muscle; northern fulmar; *Fulmarus glacialis*; common eider; *Somateria mollissima*; king eider; *Somateria spectabilis*; Iceland gull; *Larus glaucoides*; glaucous gull; *Larus hyperboreus*; black-legged kittiwake; *Rissa tridactyla*; ivory gull; *Pagophila eburnea*; thick-billed murre; *Uria lomvia*; black guillemot; *Cepphus grylle*; little auk; *Alle alle*
ABSTRACT: Baseline concentrations of lead, cadmium, mercury and selenium are reported from different tissues in marine organisms from Greenland. Overall, lead levels in marine organisms from Greenland are low, whereas cadmium, mercury and selenium levels are high. Tissue differences are not very distinct for lead, whereas the opposite is the case for cadmium and mercury. Selenium shows an intermediate behaviour in this respect. In general, lead concentrations do not correlate with the age/size of animals, whereas cadmium, mercury and selenium increase with age/size of most species and tissues analysed. No clear conclusions can be drawn in relation to geographical differences in lead, mercury and selenium concentration in Greenland. In general, cadmium levels are higher in Northwest Greenland compared to southern areas. Local differences with increasing cadmium levels from inner fjords to the open sea in stationary species may be of the same order of magnitude as those observed over long distances in Greenland. There is no indication that lead and selenium levels increase in higher trophic levels, although this is clearly the case for cadmium and mercury. In almost all cases lead levels in marine organisms from Greenland are well below the Danish food standard limits, however, a substantial proportion of marine mammals and seabirds in Greenland have cadmium and mercury levels exceeding the Danish standard limits. No food standard limits are given for selenium in food, but in some cases human intake of selenium is estimated to be high.

Doi, Rikuo; Ohno, Hideki; and Harada, Masazumi. 1984. Mercury in feathers of wild birds from the mercury-polluted area along the shore of the Shiranui Sea, Japan. Sci. Total Environ. 40:155-167.
Rec #: 374
KEYWORDS: metals; Hg; feather; Japan; Asia; temporal trends; biomagnification; common gull; *Larus canus;* pelagic cormorant; *Phalacrocorax pelagicus;* Temminck's cormorant; *Phalacrocorax filamentosus*; Arctic loon; *Gavia arctica*; black-crowned night heron; *Nycticorax nycticorax*

Annotated Bibliography (Cont.).

ABSTRACT: Total Hg content in the feathers of 95 stuffed, wild birds collected all over the shore of the Shiranui Sea was measured. They were collected over 25 years, from 1955 to 1980. They showed relatively high Hg levels till the late 1970's. A strong correlation between feeding habits and Hg content was observed. Five groups in order of diminishing Hg content were: fish-eating sea birds (mean 7.1 ppm, n = 14), omnivorous water fowl (5.5 ppm, n = 17), predatory birds (3.6 ppm, n = 16), omnivorous terrestrial birds (1.5 ppm, n = 31), and herbivorous water fowl (0.9 ppm, n = 17). Stuffed, wild birds may be a good index of past environmental pollution.

Donaldson, G. M.; Braune, B. M.; Gaston, A. J.; and Noble, D. G. 1997. Organochlorine and heavy metal residues in breast muscle of known-age thick-milled murres (*Uria lomvia*) from the Canadian Arctic. Arch. Environ. Contam. Toxicol. 33(4):430-435.
Rec #: 262
KEYWORDS: thick-billed murre; *Uria lomvia*; Newfoundland; Canada; North America; muscle; organochlorines; DDTs; benzenes; HCHs; chlordanes; mirex; dieldrin; PCBs; metals; Se; Cd; Hg; Pb; age variations; bioaccumulation
ABSTRACT: Thick-billed murres (*Uria lomvia*) originating from breeding colonies in the Canadian Arctic were collected on their wintering grounds off the coast of Newfoundland. Murres had been previously banded such that the age of each bird could be determined upon collection. This allowed us to explore the possible relationships between age and contaminant levels in the thick-billed murre. Samples of breast muscle were analyzed for organochlorines (chlorobenzenes, hexachlorocyclohexanes, DDTs, chlordanes, mirex, dieldrin, and PCBs) and metals (selenium, cadmium, mercury, and lead). Levels of both organochlorine and metal residues were sufficiently low so that toxic effects were unlikely. First-year birds contained lower levels of DDTs, mirex, dieldrin, and PCBs compared with older birds, reflecting lower levels of contamination of these compounds in food chains at breeding colonies located at higher latitudes. Higher levels of chemical residues in older birds may reflect greater direct input of those organochlorines into the wintering grounds via the highly contaminated St. Lawrence River. Levels of chlorobenzenes, hexachlorocyclohexanes, and chlordanes, which reflect atmospheric deposition, were not detected at higher levels in older birds. Of the metals, only cadmium was detected at higher levels in older birds.

Elliott, J. E.; Noble, D. G.; Norstrom, R. J.; and Whitehead, P. E. 1989. Organochlorine contaminants in seabird eggs from the Pacific coast of Canada, 1971-1986. Environ. Monit. Assess. 12(1):67-82.
Rec #: 345
KEYWORDS: organochlorines; DDTs; dieldrin; chlordanes; mirex; HCB; HCHs; PCBs; egg; rhinoceros auklet; *Cerorhinca monocerata;* fork-tailed storm-petrel; *Oceanodroma furcata;* Leach's storm-petrel; *Oceanodroma leucorhoa;* double-crested cormorant; *Phalacrocorax auritus;* pelagic cormorant; *Phalacrocorax pelagicus;* ancient murrelet; *Synthliboramphus antiquus;* glaucous-winged gull; *Larus glaucescens;* peregrine falcon; *Falco peregrinus;* British Columbia; Canada; North America; temporal trends; spatial variations
ABSTRACT: Eggs were collected from 7 seabird species at colonies on the British Columbia coast, Canada, from 1983 to 1986 and analyzed for organochlorine contaminants. Total polychlorinated biphenyl levels (wet weight) were highest in *Phalacrocorax auritus* from the Fraser Estuary (2.91 mg/kg) and the Strait of Georgia (3.79 mg/kg). Highest DDE levels were in Oceanodroma furcata from the Queen Charlotte Islands (1.68 mg/kg). Organochlorine levels

Annotated Bibliography (Cont.).

were generally lower in eggs from the mid 1980s than in those collected in the early 1970s. Organochlorine levels in Pacific alcids and hydrobatids foraging in offshore locations were compared to those in the same or ecologically similar species from the Canadian Atlantic coast. DDT- and HCH-related compounds were higher in Pacific populations while levels of dieldrin, oxychlordane, and HCB were generally lower. With the exception of β-HCH, levels of all measured organochlorines were lower in cormorants breeding in the Fraser River Estuary than in cormorants from the St. Lawrence River Estuary on the Atlantic coast.

Elliott, J. E.; Noble, D. G.; Norstrom, R. J.; Whitehead, P. E.; Simon, M.; Pearce, P. A.; and Peakall, D. B. 1992. Patterns and trends of organic contaminants in Canadian seabird eggs, 1968-90. In: Walker, Colin H.; Livingstone, David R. Persistent Pollutants in Marine Ecosystems. New York: Pergamon Press. 181-194.
Rec #: 346
KEYWORDS: egg; organochlorines; DDTs; PCBs; dieldrin; chlordanes; HCB; HCHs; PCDDs; Newfoundland; British Columbia; New Brunswick; Canada; North America; temporal trends; rhinoceros auklet; *Cerorhinca monocerata;* Leach's storm-petrel*; Oceanodroma leucorhoa;* double-crested cormorant; *Phalacrocorax auritus;* glaucous-winged gull; *Larus glaucescens*; herring gull; *Larus argentatus*; ivory gull; *Pagophila eburnea*; Atlantic puffin*; Fratercula arctica*; thick-billed murre; *Uria lomvia*; northern fulmar; *Fulmarus glacialis*
NOTES: FIGURES ONLY
ABSTRACT: A discussion is given of the monitoring of marine environment organic pollution using seabirds (a Canadian seabird monitoring project). The organochlorines in ivory. gull eggs from the western Canadian Arctic showed that the mean residue levels remained the same or increased for all compounds, except DDT. Other species showed declined organochlorine in eggs and tissues during the same period.

Elliott, J. E. and Norstrom, R. J. 1985. Specimen Banking and Monitoring of Contaminants in Great Lakes Wildlife. Programs and Abstracts of the 28th Conference on Great Lakes Research University of Wisconsin-Milwaukee. 37 pp.
Rec #: 167
KEYWORDS: egg; herring gull; *Larus argentatus*; Canada; Great Lakes; North America; temporal trends
NOTES: ABSTRACT ONLY - NOT AVAILABLE
ABSTRACT: The Canadian Wildlife Service (CWS) has investigated the impact of toxic chemicals on Great Lakes wildlife, mainly colonial water birds, since 1970. Many samples from initial studies were fortuitously stored, deep frozen, for possible future analysis. Specimen banking is now a routine part of CWS toxic chemical monitoring and surveillance. Both intact specimens and aliquots of homogenized material are stored in the bank. The bulk of the samples are kept at -40 degree C with a smaller amount at -80 degree C. Investigations indicate that persistent organochlorine compounds remain reasonably stable at these temperatures for at least 5 years. Banking of samples permits retrodetermination of previously unidentified compounds as well as reanalysis using improved methodology. Recently, a series of Herring Gull (*Larus argentatus*) eggs from Scotch Bonnet Island, Lake Ontario, 1971-1982, were reanalyzed to obtain data on temporal trends.

Elliott, J. E.; Scheuhammer, A. M.; Leighton, F. A.; and Pearce, P. A. 1992. Heavy metal and

Annotated Bibliography (Cont.).

metallothionein concentrations in Atlantic Canadian seabirds. Arch. Environ. Contam. Toxicol. 22(1):63-73.
Rec #: 73
KEYWORDS: metals; Pb; Hg; Cd; Se; Zn; Cu; Mn; Cr; Mg; Fe; P; Ca; Na; metallothionein; Quebec; Newfoundland; New Brunswick; Bay of Fundy; Canada; North America; bioaccumulation; liver; kidney; bone; Leach's storm-petrel; *Oceanodroma leucorhoa;* double-crested cormorant; *Phalacrocorax auritus;* Atlantic puffin; *Fratercula arctica*; herring gull; *Larus argentatus*; spatial variations; biomagnification
ABSTRACT: Seabird tissues, collected during the 1988 breeding season from colonies on the Atlantic coast of Canada, were analyzed for toxic metals--Cd, Hg and Pb--and 18 other trace elements. Levels of most essential trace elements appear to be closely regulated in seabird tissues; values were in good agreement with those previously reported in the published literature. Liver-Se concentrations in Leach's storm-petrels (*Oceanodroma leucorhoa*) 77.6 + 7.49 µg/g dry weight) were much higher than values normally reported for freelying birds and mammals. Cd levels varied greatly among individuals, but were always higher in kidney than in liver. Highest mean Cd concentrations (183 + 65 µg/g dry weight) were in kidneys of the planktivorous Leach's storm-petrels from the Gulf of St. Lawrence. Cd and metallothionein (MT) concentrations were positively correlated in kidneys of Leach's storm-petrels, Atlantic puffin (*Fratercula arctica*) and herring gull (*Larus argentatus*). Concentrations of total Hg varied greatly among species and individuals, but were consistently higher in liver than in kidney.

Elliott, J. E.; Whitehead, P. E.; Martin, P. A.; Bellward, G. D.; and Norstrom, R. J. 1996. Persistent pulp mill pollutants in wildlife. In: Servos, Mark R., Munkittrick, Kelly R., Carey, John H., Van Der Kraak, Glen J. (Eds), Environmental Fate and Effects of Pulp and Paper Mill Effluents. Delray Beach, FL: St. Lucie Press. 297-314.
Rec #: 168
KEYWORDS: review; pulp mill; effects; biomagnification; reproductive success; biomarkers; toxicology; EROD; CYP1A; double-crested cormorant; *Phalacrocorax auritus*; great blue heron; *Ardea herodias*; herring gull; *Larus argentatus*; Great Lakes; British Columbia; Nova Scotia; New Brunswick; Quebec; Ontario; Canada; North America; organochlorines; PCDDs; PCDFs; egg; liver; muscle
ABSTRACT: Published and unpublished data on levels and effects of pulp mill contaminants in wildlife are reviewed. Polychlorinated dioxins and furans were detected at high concentrations in eggs of various fish-eating bird species from the Strait of Georgia on the west coast of British Columbia. Highest concentrations were near pulp mills: for example TCDD, PnCDD and HxCDD in cormorant (*Phalacrocorax* spp.) eggs were as high as 100, 275 and 950 ng kg^{-1} wet weight, respectively, near a pulp mill in 1986. In contrast, levels of these contaminants in cormorants collected from colonies near pulp mills on the Canadian Atlantic coast were typically <15 ng kg^{-1}. Polychlorinated dioxin and furan levels were also elevated in tissues of fish-eating waterfowl wintering in the Strait of Georgia, but lower in non-piscivorous waterfowl. Whales and porpoises sampled from the Strait of Georgia had lower levels than piscivorous waterbirds. Episodes of poor breeding success during the 1980s in a colony of great blue herons (*Ardea herodias*) near a bleached kraft pulp mill in the Strait of Georgia were associated with sublethal effects on embryos, including edema, reduced body weight and EROD induction. Sublethal responses including CYP1A induction and porphyria were linked to pulp mill contaminant

Annotated Bibliography (Cont.).

exposure in eagles and cormorants in the Strait of Georgia and to herring gulls breeding near a mill in Quebec.

Elliott, John E.; Martin, Pamela A.; and Whitehead, Philip E. 1997. Organochlorine contaminants in seabird eggs from the Queen Charlotte Islands. Occas. Pap. Can. Wildl. Serv. 93:137-148.
Rec #: 180
KEYWORDS: organochlorines; PCBs; DDTs; dieldrin; chlordanes; mirex; HCB; HCHs; egg; ancient murrelet; *Synthliboramphus antiquus*; Cassin's auklet; *Ptychoramphus aleuticus*; rhinoceros auklet; *Cerorhinca monocerata*; fork-tailed storm petrel; *Oceanodroma furcata*; Leach's storm-petrel; *Oceanodroma leucorhoa*; North America; British Columbia; Canada; temporal trends; spatial variations
ABSTRACT: Fresh eggs of three alcid species, ancient murrelets *Synthliboramphus antiquus*, Cassin's auklets *Ptychoramphus aleuticus*, and rhinoceros auklets *Cerorhinca monocerata*, and two hydrobatid species, fork-tailed storm petrels *Oceanodroma furcata* and Leach's storm-petrels *O. leucorhoa*, were collected between 1987 and 1991 in the Queen Charlotte Islands and at three other coastal colonies in British Columbia to measure residue levels of organochlorine pesticides and polychlorinated biphenyls. There were no consistent differences between contaminant levels in eggs from colonies in the Queen Charlotte Islands and those from other coastal locations, but interspecific differences were found. Contaminant levels in eggs from colonies of seabird species in the industrialized areas of coastal British Columbia, such as the Strait of Georgia and the Fraser estuary, showed significant declines during the decade following the ban on the use of most organochlorine pesticides and restrictions on the manufacture and use of PCBs, implemented during the early 1970s. Changes were far less apparent in eggs of birds from remote offshore colonies, such as the Queen Charlotte Islands. Trends in contaminant levels were assessed both temporally and geographically in this study.

Elliott, John E. and Scheuhammer, Anton M. 1997. Heavy metal and metallothionein concentrations in seabirds from the Pacific Coast of Canada. Mar. Pollut. Bull. 34(10):794-801.
Rec #: 153
KEYWORDS: metals; Ca; Mn; Fe; Mg; P; Cu; Zn; Se; Pb; Hg; metallothionein; liver; kidney; bone; British Columbia; Canada; North America; Leach's storm-petrel; *Oceanodroma leucorhoa*; rhinoceros auklet; *Cerorhinca monocerata;* Cassin's auklet; *Ptychoramphus aleuticus;* ancient murrelet; *Synthliboramphus antiquus;* fork-tailed storm-petrel; *Oceanodroma furcata*; spatial variations
ABSTRACT: Seabird tissues, collected during the 1990 breeding season from colonies on the Pacific coast of Canada, were analyzed for Cd, Hg, Pb and 19 other trace elements. Metallothionein (Mt) was measured in kidneys of 3 species. Ranges of essential trace metal concentrations were generally narrow, consistent with homeostatic control of these elements in seabird tissues. Cd concentrations were always higher in kidney than in liver. Highest mean Cd concentrations (306±78 µg/g dry weight) were in kidneys of planktivorous Leach's storm-petrels (*Oceanodroma leucorhoa*) from Hippa Island in the Queen Charlotte archipelago. Cd concentrations in kidneys of both Leach's storm-petrels and rhinoceros auklets (*Cerorhinca monocerata*) were significantly greater at northern colonies compared to those further south. Cd and Mt concentrations were positively correlated in kidneys across the 3 species for which measurements were made (rhinoceros auklet, Cassin's auklet (*Ptychoramphus aleuticus*), and

Annotated Bibliography (Cont.).

ancient murrelet (*Synthliboramphus antiquus*)) with an overall r=0.82, p<0.001. Hg accumulation was not sufficiently great to be of toxicological concern in any of the 5 species. Highest mean Hg concentrations (6.37 µg/g) were in livers of Leach's storm-petrels from Cleland Island on the west coast of Vancouver Island, and were significantly greater than in birds from further north on Hippa Island. Concentrations of hepatic Hg and Se were not correlated in the 3 species (rhinoceros auklet, Cassin's auklet and ancient murrelet) for which Se was measured. Pb concentrations were consistently greatest in bone, with highest mean concentrations in fork-tailed storm-petrels (*Oceanodroma furcata*)(6.2 µg/g dry weight).

Engwall, Magnus; Brunström, Björn; and Jakobsson, Eva. 1994. Ethoxyresorufin *O*-deethylase (EROD) and aryl hydrocarbon hydroxylase (AHH)-inducing potency and lethality of chlorinated naphthalenes in chicken (*Gallus domesticus*) and eider duck (*Somateria mollissima*) embryos. Arch. Toxicol. 68(1):37-42.
Rec #: 481
KEYWORDS: chicken; *Gallus domesticus*; common eider; *Somateria mollissima;* Baltic Sea; Sweden; Europe; organochlorines; PCNs; liver; toxicology; EROD; AHH; dosing study; effects
NOTES: SUPPLEMENTAL
ABSTRACT: The 7-ethoxyresorufin *O*-deethylase (EROD)- and aryl hydrocarbon hydroxylase (AHH)-inducing potencies and lethalities of a technical preparation of polychlorinated naphthalenes (PCNs) (Halowax 1014, approximate congener ratio: 20 % tetrachloronaphthalenes, 40 % pentachloronaphthalenes, 40 % hexachloronaphthalenes), a mixture of 50 % 1,2,3,5,6,7-hexachloronaphthalene and 50 % 1,2,3,4,6,7-hexachloronaphthalene (HxCN-mix), and 1,2,3,4,5,6,7-heptachloronaphthalene (HpCN) were studied in chicken (*Gallus domesticus*) and eider duck (*Somateria mollissima*) embryos. Mortality and hepatic EROD activity were determined on day 10 of incubation in chicken embryos exposed to various doses of the PCNs via the air-sacs of the eggs on day 7. The HxCN-mix and Halowax 1014 proved to have both embryolethal and EROD-inducing properties, while the HpCN had low EROD-inducing potency and embryolethality. ED_{50} values for EROD induction by the HxCN-mix and Halowax 1014 were estimated to be 0.06 mg/kg egg and 0.2 mg/kg egg, respectively. Fifty percent of the chicken embryos died (6/12) when given 3.0 mg/kg of the HxCN-mix while a similar dose of Halowax 1014 caused mortality in 4 out of 12 chicken embryos. The dose-response curve for EROD induction by Halowax 1014 exhibited a decline after the maximal level was reached. When Halowax 1014 (1.0 mg/kg egg) was coinjected with 3,3',4,4',5-pentachlorobiphenyl (PCB IUPAC #126) (0.1 µg/kg egg) no additive effects on EROD activity were found, but when the same dose of Halowax 1014 was coinjected with a dose of PCB #126, known to cause maximal induction (1.0 µg/kg egg), the resulting EROD activity was lower than that caused solely by 1.0 µg PCB #126/kg egg. These findings indicate that Halowax 1014 has both EROD-inducing and EROD-inhibiting properties. Mortality and EROD and AHH activities were determined on day 18 (chicken) or day 24 (eider) of incubation in embryos exposed to 1.0 mg/kg egg via the yolksac on day 4 (chicken) or day 5 (eider). The HxCN-mix and Halowax 1014 induced AHH and EROD in both chicken and eider, but the induction rates were higher in the eider embryos. The HxCN-mix and Halowax 1014 caused degenerative hepatic lesions and pericardial oedema in the chicken embryos but not in the eider embryos. The most toxic PCNs tested (the HxCN-mix and Halowax 1014) were approximately of the same EROD-inducing potency as previously found for the most toxic mono- *ortho*-chlorinated biphenyls (Brunström

Annotated Bibliography (Cont.).

1990), and 1000 times less toxic and potent as EROD inducers compared with PCB #126 (Brunström and Andersson 1988). HpCN was considerably less toxic and exhibited a low EROD-inducing potency. The chicken embryos were more sensitive to the hepatotoxic effects produced by Halowax 1014 and the HxCN-mix than the eider duck embryos, while the eider embryos were more responsive in terms of EROD and AHH induction. The two HxCNs studied usually make up approximately 1 % of the total quantity of PCNs present in Halowax 1014 [when determined with gas chromatography (flame ionization detection)]. Therefore, the relatively high toxic potency of Halowax 1014 cannot be explained by its content of the two HxCNs.

Falandysz, J. and Szefer, P. 1984. Chlorinated hydrocarbons in fish-eating birds wintering in the Gdansk Bay, 1981-82 and 1982-83. Mar. Pollut. Bull. 15(8):298-301.
Rec #: 537
KEYWORDS: organochlorines; HCB; HCHs; DDTs; PCBs; little auk; *Alle alle*; horned grebe; *Podiceps auritus*; red-necked grebe; *Podiceps grisegena*; great crested grebe; *Podiceps cristatus*; red-breasted merganser; *Mergus serrator*; common merganser; *Mergus merganser*; red-throated loon; *Gavia stellata*; Arctic loon; *Gavia arctica*; common murre; *Uria aalge;* black guillemot; *Cepphus grylle*; razorbill; *Alca torda*; Baltic Sea; Poland; Great Britain; Scotland; Finland; Switzerland; Europe; muscle; egg; liver; fat; temporal trends; biomagnification; gender differences; seasonal variations; spatial variations
ABSTRACT: Seventy-four samples of adipose fat of 10 species of fisheating birds commonly wintering in Gdask Bay and one of little auk, a rare species in this area, have been analysed for residues of HCB, BHC, DDT and PCBs. Auks (guillemot, black guillemot and razorbill), which probably stay a whole year in the Baltic area, contained in their fat twice the level of PCBs as mergansers, and 5–6 times as much as in divers and grebes—birds which stay in the sea mainly after the breeding season. DDT levels were somewhat higher but comparable in auks and grebes, and lower in mergansers and divers. BHC and HCB levels were 3–4 times and 5–10 times higher, respectively, in auks than in birds from other families studied. When comparing residue levels found in adipose fat of birds taken during the wintering seasons of 1981–1982 and 1982–1983 with those taken in 1975–1976 or 1980–1981, an initially rapid, but in recent years a rather slow decline in DDT level is detectable for grebes, mergansers and divers, and in the case of PCBs the residue levels fell slightly. In auks, the DDT residue level fell by half, but remains rather stable in recent years, while there are only minor fluctuations in the residue level of PCBs.

Falandysz, Jerzy. 1980. Chlorinated hydrocarbons in gulls from the Baltic south coast. Mar. Pollut. Bull. 11(3):75-80.
Rec #: 376
KEYWORDS: little gull; *Larus minutus*; black-headed gull; *Larus ridibundus*; common gull; *Larus canus*; herring gull; *Larus argentatus*; Baltic Sea; Europe; organochlorines; PCBs; DDTs; HCB; toxaphene; liver; muscle; fat; biomagnification
ABSTRACT: Polychlorinated terphenyls (PCT), polychlorinated biphenyls (PCB), DDT (I) [50-29-3], and hexachlorobenzene (II) [118-74-1] were detected in the tissues of Little Gull (*Larus minutus*), Black-headed Gull (*L. ridibundus*), Common Gull (*L. canus*), and Herring Gull (*L. argentatus*). The adult Herring Gull, were the most effected by chlorinated hydrocarbons; they contained in their livers II 0.36-2.2, I 50-320, and PCB 140-530 ppm. For the other gulls, the residue levels in their liver were II 0.005-0.17, I 0.15-8.3, and PCB 0.14-8.4 ppm,

Annotated Bibliography (Cont.).

respectfully. The presence of PCT in Black-headed Gulls and their absence in other gulls was related to the different feeding habits of the gull species.

Falandysz, Jerzy. 1992. Mercury content in liver, kidneys and pectoral muscles of guillemot (*Uria aalge* Pont.) wintering in the Gulf of Gdansk. Stud. Mater. Ocenaol. 62:5-11.
Rec #: 304
KEYWORDS: biomagnification; metals; Hg; common murre; *Uria aalge*; razorbill; *Alca torda*; horned grebe; *Podiceps auritus;* red-necked grebe; *Podiceps grisegena;* common goldeneye; *Bucephala clangula;* Poland; Europe; muscle; liver; kidney; spatial variations; temporal trends
ABSTRACT: Mercury content has been determined in tissues of 23 guillemots (*Uria aalge*) and 7 specimens of other marine bird species (razorbill, slavonian grebe, rednecked grebe and goldeneye) taken from fishing nets in a nearshore region of the Gulf of Gdansk in autumn 1983 and winter 1983-84. The mean level of mercury in pectoral muscles, liver and kidneys of male and female guillemots ranged between 0.12-0.15, 0.43-0.44 and 0.37-0.46 µg/g on a wet-weight basis, respectively, while for the remaining bird species the values noted were two to five times higher (liver and kidneys), on average.

Falandysz, Jerzy. 1986. Organochlorine compounds in tissues of golden eye, velvet scoter, eider and coot wintering in Gdansk Bay, 1975-1976. Bromatologia i Chemia Toksykologiczna. 19(1):55-60.
Rec #: 488
KEYWORDS: organochlorines; DDTs; PCBs; HCHs; HCB; common goldeneye; *Bucephala clangula*; white-winged scoter; *Melanitta fusca*; common eider; *Somateria mollissima*; coot; *Fulica atra;* Baltic Sea; Poland; Europe; liver; muscle; fat
ABSTRACT: The concentrations of HCB [118-74-1], total cyclohexane hexachloride isomers, sum of DDT and its metabolites, and PCB in the muscle, liver, and adipose tissue of diving ducks [golden eye (*Bucephla clangula*), velvet scoter (*Melanitta fusca*), and eider (*Somateria mollissima*)] and coots (*Fulica atra*) were severalfold higher in birds found dead on the coast of Gdansk Bay (Baltic Sea) than in living birds. The use of birds as bioindicators of environmental pollution is discussed.

Falandysz, Jerzy; Bergqvist, Per-Anders; and Rappe, Christoffer. 1998. Dioxins and furans (PCDD/Fs) in marine birds from the southern part of the Baltic Sea. Bromatologia i Chemia Toksykologiczna. 31(1):75-78.
Rec #: 372
KEYWORDS: organochlorines; PCDDs; PCDFs; toxicology; TEQs; biomagnification; fat; liver; muscle; Baltic Sea; Europe; white-tailed sea eagle; *Haliaeetus albicilla*; common gull; *Larus canus*; red-throated loon; *Gavia stellata*; razorbill; *Alca torda*; common murre; *Uria aalge*; biomagnification
ABSTRACT: PCDDs and PCDFs were detected in the adipose tissue, liver or breast muscle of a few marine birds, including mixed-feeders (common gull), fish-eaters (redthroated diver, razorbill, guillemot) and predator species (sea eagle) collected from the south coast of the Baltic Sea. High levels of 2,3,4,7,8-penta-CDF and 1,2,3,7,8-penta-CDDD as well as of numerous less toxic PCDD/Fs were found in all samples. A relation could be observed between the position of the test species in the food chain and body lipids PCDD/Fs concentrations, which were lower for the common gull than for a typical fish-eater and highest in the white-tailed sea eagle.

Annotated Bibliography (Cont.).

Falandysz, Jerzy and Szefer, Piotr. 1982. Chlorinated hydrocarbons in diving ducks wintering in Gdańsk Bay, Baltic Sea. Sci. Total Environ. 24(2):119-127.
Rec #: 489
KEYWORDS: organochlorines; HCB; HCHs; DDTs; PCBs; fat; Baltic Sea; Europe; tufted duck; *Aythya fuligula*; greater scaup; *Aythya marila*; common goldeneye; *Bucephala clangula*; long-tailed duck; *Clangula hyemalis;* black scoter; *Melanitta nigra*; white-winged scoter; *Melanitta fusca*; common eider; *Somateria mollissima*; gender differences
ABSTRACT: The levels of HCB [118-74-1], α-BHC [319-84-6], γ-BHC (I) [58-89-9], DDT [50-29-3] (plus analogs), and PCB were detected in adipose fat from species of diving ducks at their winter quarters in the southern Baltic. PCB, ΣDDT, and HCB were detected in all samples. PCB's were highest followed by ΣDDT and HCB. Residues of I were detected in only 4 of 129 samples examined; but, for all samples from the long-tailed duck, only levels of I were positive. Differences between HCB, ΣDDT and PCB residue levels between males and females of the scaup-duck were insignificant.

Fimreite, N.; Bjerk, J. E.; Kveseth, N.; and Brun, E. 1977. DDE and PCBs in eggs of Norwegian seabirds. Astarte. 10(1):15-20.
Rec #: 12
KEYWORDS: egg; effects; organochlorines; PCBs; DDTs; Norway; Europe; bioaccumulation; gannet; *Sula bassana*; herring gull; *Larus argentatus*; razorbill; *Alca torda*; common murre; *Uria aalge*; black-legged kittiwake; *Rissa tridactyla*; spatial variations
ABSTRACT: Concentrations of DDE and polychlorinated biphenyls (PCBs) were measured in 203 seabird eggs collected from 10 localities along the coast of Norway (58 degree 57'N-71 degree 05'N). The average concentrations of DDE were, on a wet weight basis, 2.05, 1.57, 1.20, 0.80, and 0.37 ppm in eggs of gannet (*Sula bassana*), herring gull (*Larus argentatus*), razorbill (*Alca torda*), guillemot (*Uria aalge*) and kittiwake (*Rissa tridactyla*), respectively. The corresponding concentrations of PCBs were 7.71, 8.49, 5.40, 2.19, and 2.87 ppm. Some geographical variation was found, but no south-north gradient. The possible biological effects are discussed.

Fimreite, Norvald and Bjerk, John Erik. 1979. Residues of DDE and PCBs in Norwegian seabird fledglings, compared to those in their eggs. Astarte. 12:49-51.
Rec #: 524
KEYWORDS: organochlorines; PCBs; Norway; Europe; egg; carcass; spatial variations; bioaccumulation; *Uria aalge;* common murre; *Rissa tridactyla;* black-legged kittiwake; *Larus argentatus;* herring gull
ABSTRACT: Residues of DDE and PCBs were recorded in fledglings of various seabird species from northern Norway. When corresponding levels in eggs were subtracted from those in fledglings after adjustment for weight increase (eggs - fledglings), the results were largely negative, indicating a decline of these compounds. This probably reflects a low degree of local contamination

Fimreite, Norvald; Brun, Einar; Frøslie, Arne; Frederichsen, Per; and Gundersen, Nil. 1974. Mercury in eggs of Norwegian seabirds. Astarte. 7(2):71-75.
Rec #: 531
KEYWORDS: metals; Hg; organometallics; organic Hg; egg; Norway; Europe; herring gull;

Annotated Bibliography (Cont.).

Larus argentatus; common murre; *Uria aalge*; razorbill; *Alca torda*; black-legged kittiwake; *Rissa tridactyla;* gannet; *Sula bassana*; spatial variations

ABSTRACT: The sampled material comprised 157 eggs of seabirds collected from nine localities along the coast of Norway. The mercury levels in Herring Gull eggs from Lyngør and Utsira, south of 59°N, averaged 0.34 ppm. From northern localities 67°30'-71°N, eggs of Herring Gull (*Larus argentatus*), Guillemot (*Uria aalge*), and Razorbill (*Alca torda*) contained less than 0.10 ppm, eggs of Kittiwake (*Rissa tridactyla*) less than 0.15 ppm, and Gannet *(Sula bassana)* 0.58 ppm, all average values. More than 90 per cent of the mercury was in methyl form.

Fisk, A. T.; Moisey, J.; Hobson, K. A.; Karnovsky, N. J.; and Norstrom, R. J. 2001. Chlordane components and metabolites in seven species of Arctic seabirds from the Northwater Polynya: Relationships with stable isotopes of nitrogen and enantiomeric fractions of chiral components. Environ. Pollut. 113:225-238.

Rec #: 341

KEYWORDS: biomagnification; stable isotopes; organochlorines; chlordanes; liver; fat; little auk; *Alle alle*; thick-billed murre; *Uria lomvia*; black guillemot; *Cepphus grylle*; black-legged kittiwake; *Rissa tridactyla*; ivory gull; *Pagophila eburnea*; glaucous gull; *Larus hyperboreus*; northern fulmar; *Fulmarus glacialis*; Greenland; Canada; North America; Northwater Polynya; Atlantic Ocean

ABSTRACT: The Northwater Polynya (NOW) is a large area of year-round open water found in the high Arctic between Ellesmere Island and Greenland. NOW has high biological productivity compared with other arctic marine areas, and supports large populations of several seabird species. Seven species of seabirds, dovekie (*Alle alle*, DOVE), thick-billed murre (*Uria lomvia*, TBMU), black guillemot (*Cepphus grylle*, BLGU), black-legged kittiwake (*Rissa tridactyla*, BLKI), ivory gull (*Pagophila eburnea*, IVGU), glaucous gull (*Larus hyperboreus*, GLGU) and northern fulmar (*Fulmarus glacialis*, NOFU) were collected in May and June 1998 to determine chlordane concentrations in liver and fat and to examine species differences, relationships with stable isotopes of nitrogen, and enantiomeric fractions (EFs) of chiral components. ΣCHLOR concentrations varied over an order of magnitude among species, from a low of 176±19 ng/g (lipid corrected) in TBMU liver to a high of 3190±656 ng/g (lipid corrected) in NOFU liver. Lipid-corrected concentrations of chlordane did not vary between sex for any species or between fat and liver except for the DOVE, that had fat concentrations that were significantly greater than the liver. $\delta^{15}N$ values described a significant percentage of the variability of concentrations for most chlordane components, although less than what has been reported for whole food chains. Slopes of $\delta^{15}N$ versus concentration of chlordane components and CHLOR were similar with the exception of those which were metabolized (*trans*-chlordane) or formed through biotransformation (oxychlordane). The relative proportions of chlordane components in seabirds were related to phylogeny; the procellariid (NOFU) had the greatest percentage of oxychlordane (>70 %), followed by the larids (BLKI, IVGU and GLGU; 40-50 %) and the alcids (DOVE and BLGU; 10-20 %). The exception was TBMU, an alcid, where oxychlordane made up >40 % of its chlordane. EFs of chiral components failed to predict concentration or trophic level, but did identify biotransformation differences between species and chlordane components. TBMU appeared to have a greater capacity to metabolize and eliminate chlordane, based on high proportions of oxychlordane, the highest EFs for oxychlordane and heptachlor epoxide, and a $\delta^{15}N$-ΣCHLOR value which was well below the relationships developed for all seabird species.

Annotated Bibliography (Cont.).

Focardi, S.; Fossi, C.; and Leonzio, C. 1985. Chlorinated hydrocarbons in some sea-birds. Oebalia. 11(1):127-140.
Rec #: 102
KEYWORDS: Danube Delta; Mediterranean Sea; Atlantic Ocean; Italy; Europe; black-headed gull; *Larus ridibundus;* slender-billed gull; *Larus genei;* herring gull; *Larus argentatus;* black-necked grebe; *Podiceps nigricollis;* little egret; *Egretta garzetta;* gull-billed tern; *Sterna nilotica;* black-winged stilt; *Himantopus himantopus;* avocet; *Recurvirostra avosetta;* coot; *Fulica atra;* Cory's shearwater; *Calonectris diomedea;* cormorant; *Phalacrocorax carbo;* pygmy cormorant; *Phalacrocorax pygmeus;* great white pelican; *Pelecanus onocrotalus;* Audouin's gull; *Larus audouinii;* common tern; *Sterna hirundo;* little tern; *Sterna albifrons*; organochlorines; PCBs; DDTs; egg; biomagnification; spatial variations
ABSTRACT: Chlorinated hydrocarbons have been tested in birds with different feeding habits, collected in the Mediterranean Sea, Danube Delta and in the East Atlantic Sea.

Focardi, S.; Fossi, C.; Leonzio, C.; Corsolini, S.; and Parra, O. 1996. Persistent organochlorine residues in fish and water birds from the Biobio River, Chile. Environ. Monit. Assess. 43(1):73-92.
Rec #: 36
KEYWORDS: organochlorines; PCBs; DDTs; HCB; HCHs; Chile; South America; bioaccumulation; spatial variations
ABSTRACT: Concentrations of polychlorinated biphenyls (PCBs), DDT and its metabolites, HCH isomers and hexachlorobenzene (HCB) were determined in fish and birds from different locations in the Biobio River basin (central Chile). Samples collected near the mouth of the river contained high concentrations of PCBs, reflecting the massive use of these xenobiotics in the urban and industrial areas of Concepcion and Talcauano. Samples collected in the central part of the basin contained very high concentrations of lindane that coincide with the widespread use of lindane-based pesticides (purified gamma -HCH) in this area. DDT was distributed homogeneously throughout the basin, except at Laguna Icalma, the source of the river in the Andes. Most PCB residues in fish and birds consisted of congeners between penta- and hepta-chlorobiphenyls. In fish, the predominant congeners were the pentachlorobiphenyl 23'44'5 (IUPAC number 118) and the hexachlorobiphenyl 22'344'55' (PCB-153); in birds 22'44'55'(PCB-180) prevailed.

Fog, Mette and Kraul, Inge. 1973. Levels of polychlorinated biphenyls (PCB) and organochlorine insecticides in eggs from eider (*Somateria mollissima*). Acta Vet. Scand. 14(2):350-352.
Rec #: 490
KEYWORDS: organochlorines; dieldrin; DDTs; PCBs; egg; common eider; *Somateria mollissima*; Denmark; Europe; temporal trends; eggshell thickness; effects
ABSTRACT: Residues of dieldrin [60-57-1] (mean = 0.39, 0.29, 0.24 mg/kg extracted fat), DDE [72-55-9] (mean = 2.12, 1.66, 1.37 mg/kg) and a 60 %-chlorinated PCB (mean = 35.0, 19.9, 14.0 mg/kg) were found in molluscivorous eider eggs taken from their natural environment in the years 1970, 1971 and 1972, respectively The gradual decrease in residue levels from 1970-1972 reflected a decrease in the level of persistent exposure to trace amounts of the pollutants, due to restriction in organochlorine insecticide use begun in 1970 in the area. Egg-shell thickness was apparently uneffected by even the greatest contamination (1970).

Annotated Bibliography (Cont.).

Fox, G. A. 1993. Temporal trends in biomarker responses of adult herring gulls from seven Great Lakes colonies, 1974-1991. OME 36th Conference of the International Association for Great Lakes Research, June 4-10, 1993. Program and Abstracts. 112.
Rec #: 203
KEYWORDS: North America; Great Lakes; herring gull; *Larus argentatus*; liver; biomarkers; toxicology; porphyrin; retinol; organochlorines; PCBs; DDTs; effects; temporal trends
NOTES: ABSTRACT ONLY - NOT AVAILABLE
ABSTRACT: We examined the temporal trends in biomarker responses known to be affected by PCBs and other persistent lipophilic contaminants for adult Herring Gulls (*Larus argentatus*) trapped on 7 Great Lakes colonies (4 in AOCs) and a reference colony on the Atlantic coast, 1974-1991. With the exception of Saginaw Bay and Middle Island (L. Erie), PCB and DDE levels in the livers of these birds declined at all sites. Retinyl palmitate stored in the liver increased at 5 of the Great Lakes locations and at the reference site, but remained relatively similar at Fighting Island (Detroit River) and in Saginaw Bay. Free retinol declined at all sites. Accumulations of highly carboxylated porphyrins in the liver have remained constant and low at the reference site but highly elevated and unchanged at Saginaw Bay. Whereas the relative thyroid mass of all Great Lakes gulls collected prior to 1985 were significantly greater than the reference site, goiter was not observed at any site in 1991. Further data and additional measures will be discussed in relation to contaminant trends at Great Lakes sites.

Fox, G. A.; Gilman, A. P.; Peakall, D. B.; and Anderka, F. W. 1978. Behavioural abnormalities of nesting Lake Ontario herring gulls. J. Wildl. Manage. 42(3):477-483.
Rec #: 138
KEYWORDS: reproductive success; abnormalities; herring gull; *Larus argentatus*; Great Lakes; New Brunswick; Canada; North America; effects
NOTES: SUPPLEMENTAL
ABSTRACT: Lake Ontario herring gulls (*Larus argentatus*) did not defend their nests in the normal manner and were less attentive (P <0 . 05) than those on Kent Island, New Brunswick. Incubating Lake Ontario gulls appeared to apply less heat to their eggs. Within Lake Ontario, a comparison of nest attentiveness and nest air temperatures between successful and unsuccessful nests revealed similar differences. The decreased nest defense of Lake Ontario gulls was sufficient to account for the high incidence of egg loss observed. The variability in nest air temperature was sufficient to increase embryonic mortality based on studies of other species. These abnormalities probably result from pollutant-induced endocrine dysfunction.

Fox, G. A.; Kennedy, S. W.; and Trudeau, S. 1993. Temporal and spatial variation in a battery of biomarkers in Great Lakes fish-eating birds in relation to known patterns of chemical contamination. Mar. Environ. Res. 35(1-2):230.
Rec #: 86
KEYWORDS: Great Lakes; North America; toxicology; temporal trends; spatial variations; herring gull; *Larus argentatus;* double-crested cormorant; *Phalacrocorax auritus*; abnormalities; blood; liver; organochlorines; PCBs; PCDDs
NOTES: NO TABLES
ABSTRACT: The Laurentian Great Lakes of North America constitute the largest and one of the most contaminated freshwater ecosystems in the world. We have used a number of biomarkers to measure contaminant-induced effects at the biochemical, organ, and individual

Annotated Bibliography (Cont.).

level in free-living populations of fish-eating birds breeding on these lakes. Spatial and temporal variation in hepatic levels of highly carboxylated porphyrins and retinoids, thyroid histology, and hormone levels in adult herring gulls, *Larus argentatus*, and the prevalence of birth defects in young double-crested cormorants, *Phalacrocorax auritus*, show a marked correspondence with relative levels of environmental contamination. Spatial differences in levels of free-plasma iodide are not consistent with thyroid histology and thyroxine levels. The increased prevalence of microfollicular hyperplasia and marked thyroxine depletion suggest that thyrotoxic factors are present in the food chain. The spatial and temporal similarity in severity of thyroxine and retinoid depletion is consistent with a common etiology, such as displacement from their shared carrier protein, transthyretin. Although it is both impossible to prove and biological improbable that a single component of the complex mixture of contaminants present is solely responsible for the effects observed, correlative evidence suggests that PCBs and related chemicals are major contributors to this toxicity. Similar effects have been observed in rats fed TCDD and some PCBs.

Fox, G. A.; Trudeau, S.; Won, H.; and Grasman, K. A. 1998. Monitoring the elimination of persistent toxic substances from the Great Lakes; Chemical and physiological evidence from adult herring gulls. Environ. Monit. Assess. 53(1):147-168.
Rec #: 365
KEYWORDS: organochlorines; PCBs; DDTs; dieldrin; mirex; toxicology; porphyrin; vitamin A; Great Lakes; Bay of Fundy; Canada; North America; herring gull; *Larus argentatus*; liver; gland; temporal trends; effects
ABSTRACT: To assess progress towards virtual elimination of PCBs, DDE, dieldrin and Mirex and their associated physiological effects, we compared their concentrations in pooled livers of adult herring gulls (*Larus argentatus*) repeatedly sampled at 8 Great Lakes colonies and a reference colony on the Atlantic coast between 1974 and 1993. We measured the relative thyroid mass and concentrations of highly carboxylated porphyrins and retinyl palmitate in the liver of each individual. PCBs, dieldrin and mirex declined in 7 of 8 colonies while DDE decreased in 6. The greatest decreases occurred pre-1985. PCBs and DDE did not decrease in gulls from Middle Island in western Lake Erie. Middle Island and Saginaw Bay had the highest concentrations of PCBs of 11 Great Lakes colonies in the 1990s. Thyroids of gulls from Great Lakes colonies were slightly enlarged but the degree of enlargement has decreased over time. In 1991, gulls from Great Lakes colonies had slight to moderately elevated concentrations of highly carboxylated porphyrins. In the early 1990s, hepatic stores of retinyl palmitate were very seriously depleted in gulls from the Detroit River, western basin of Lake Erie, and Lake Ontario, reflecting decreased availability and altered storage. We conclude that PCBs and/or other persistent toxic substances in the food of herring gulls have not been virtually eliminated.

Fox, Glen A.; Grasman, Keith A.; Hobson, Keith A.; Williams, Kim; Jeffrey, Deborah; and Hanbidge, Barbara. 2002. Contaminant residues in tissues of adult and prefledged herring gulls from the Great Lakes in relation to diet in the early 1990s. J. Great Lakes Res. 28(4):643-663.
Rec #: 356
KEYWORDS: bioaccumulation; biomagnification; egg; herring gull; *Larus argentatus*; kidney; liver; metals; Cd; Hg; Pb; Se; As; Cu; Fe; Zn; Mn; Co; Cr; Mo; Sn; Ni; V; Al; Ba; Tl; organochlorines; DDTs; dieldrin; mirex; chlordanes; benzenes; PCBs; PCDDs; PCDFs; toxicology; TEQs; Great Lakes; North America; Canada; spatial variations; temporal trends

Annotated Bibliography (Cont.).

ABSTRACT: In the early 1990s, herring gulls (*Larus argentatus*) were collected in 15 breeding colonies throughout the Great Lakes basin and in 2 reference colonies on Lake Winnipeg and the Bay of Fundy. Organochlorine and metal concentrations, and stable isotope ratios ($^{15}N/^{14}N$ and $^{13}C/^{12}C$) were measured in-their tissues, and the authors qualitatively assessed their diet. Breast muscle $\delta^{15}N$ suggested that adults fed on planctivorous or insectivorous fish at 6 colonies, on piscivorous fish at 4, and at a lower trophic level at the remaining 3. The concentrations of Co, Ni, Al, Cr, Sn, Fe, and Pb in kidneys of adults suggested anthropogenic enrichment in the Great Lakes basin. Concentrations of contaminants were highest most often in tissues of gulls from Lake Ontario and northern Lake Michigan colonies. Concentrations of Pb in adults from Hamilton Harbor and the Detroit River, and of Se in adults from the southern Lake Huron colony were similar to published toxicity thresholds. Tissue levels of Cd have increased, while those of Pb have decreased markedly since 1983. DDE, dieldrin, mirex, and ΣPCB concentrations in livers collected from 9 of these colonies revealed declines of 16 to 87 % at most locations since the early 1980s. 2,3,7,8-Tetrachlorodibenzo-p-dioxin equivalents were highest in adults from the offshore colonies in western Lake Erie and northern Lake Michigan, where gulls feed on piscivorous fish, and were driven by non-ortho PCBs. There was evidence of an unusually high bioavailability of organochlorines, especially dieldrin, near the northern Lake Michigan colony during the period of chick growth, and of an ongoing loading of mercury to eastern Lake Ontario. Tissues of adult gulls from colonies on Lakes Ontario, Erie, and Michigan best reflect local conditions whereas those from Lake Superior and northern Lake Huron reflect contaminants accumulated from time spent on the lower lakes.

Fox, Glen A.; Kennedy, Sean W.; Norstrom, Ross J.; and Wigfield, Donald C. 1988. Porphyria in herring gulls: A biochemical response to chemical contamination of Great Lakes food chains. Environ. Toxicol. Chem. 7(10):831-839.
Rec #: 223
KEYWORDS: liver; North America; Great Lakes; toxicology; porphyrin; effects; herring gull; *Larus argentatus*; organochlorines; dieldrin; mirex; chlordanes; benzenes; HCB; OCS; PCBs; DDTs; PCDDs; PCDFs; spatial variations
ABSTRACT: Concentrations of highly carboxylated porphyrins (HCPs) in the livers of adult herring gulls (*Larus argentatus*) from colonies throughout the Great Lakes were found to be markedly elevated in comparison with those in gulls from coastal areas and in seven other species of birds consuming diets uncontaminated with polyhalogenated aromatic hydrocarbons (PHAHs). The highest levels were found in gulls from lower Green Bay (Lake Michigan), Saginaw Bay (Lake Huron) and Lake Ontario. The authors suggest that the high levels of HCPs reflect PHAH-induced derangement of heme biosynthesis. Determination of HCPs offers promise as a specific and sensitive biological marker of PHAH-induced toxicity and as a measure of the toxicological significance of the chemical burden in gulls, terminal members of Great Lakes food chains.

Fox, Glen A.; Kennedy, Sean W.; Trudeau, Suzanne; Bishop, Christine A.; and Wayland, Mark. 1997. Hepatic porphyrin patterns in birds as a promising measure of effect and bioavailability of PCBs and other HAHs in water and sediments. Organohalogen Compounds. 33:366-370.
Rec #: 443
KEYWORDS: herring gull; *Larus argentatus*; double-crested cormorant; *Phalacrocorax auritus*; liver; Great Lakes; Canada; North America; organochlorines; PCBs; toxicology;

Annotated Bibliography (Cont.).

porphyrin; spatial variations
NOTES: NO TABLES
ABSTRACT: In susceptible bird species, chronic dietary exposure to polychlorinated biphenyls and other halogenated aromatic hydrocarbons results in the derangement of the heme biosynthetic pathway leading to the accumulation of uroporphyrin and highly carboxylated porphyrins (HCP). The observations cover prefledgling Herring Gulls (*Larus argentatus*) and Double-crested Cormorants (*Phalacrocorax auritus*) and prefledglings of the insectivorous Tree Swallow (*Tachycineta bicolor*). The birds were collected from the area of the Great Lakes and from reference areas in the prairie provinces and Atlantic Coast (Canada). On the basis of earlier surveys, uroporphyrin concentrations of >50 pmol/g and the presence of hexa- and penta-carboxylporphyrins are regarded as an indication of significant disruption of the heme biosynthetic pathway. Unlike Herring Gulls from the Atlantic coast, the concentration of uroporphyrins ranged from 11-166 pmol/g liver in adult Herring Gulls collected from Great Lakes colonies and a colony in Northern Lake Winnipeg in 1991. Uroporphyrin (<4-23.5 pmol/g) was the only HCP detected in 21- and 28-day old Double-crested cormorant chicks. Their was little variation in uroporphyrin concentrations within these collections. This suggests that this species is less sensitive to the porphyrinogenic effect chemicals than Herring Gulls. Tree Swallows feed on emergent insects and therefore provide a measure of sediment contamination, and the transfer of these contaminants to terrestrial food chains. For these Tree Swallows differences in the liver HCP content were found for swallows from the Wapiti River in Northern Alberta and those from the Great Lakes wetlands.

Frank, Adrian. 1986. In search of biomonitors for cadmium: Cadmium content of wild Swedish fauna during 1973-1976. Sci. Total Environ. 57:57-65.
Rec #: 486
KEYWORDS: metals; Cd; common eider; *Somateria mollissima*; black-headed gull; *Larus ridibundus*; herring gull; *Larus argentatus*; mute swan; *Cygnus olor;* whooper swan; *Cygnus cygnus*; green-winged teal; *Anas crecca;* mallard; *Anas platyrhynchos*; Canada goose*; Branta canadensis*; crane; *Grus grus*; common merganser; *Mergus merganser*; great-crested grebe; *Podiceps cristatus*; grey heron; *Ardea cinerea*; long-tailed duck; *Clangula hyemalis*; red-breasted merganser; *Mergus serrator*; kidney; liver; Sweden; Europe; biomagnification
ABSTRACT: Forty-five species of birds and 22 species of mammals of the terrestrial and aquatic fauna, herbivores as well as carnivores, were investigated during 1973-1976 for Cd accumulating properties to find biomonitors for Cd in the Swedish environment. Terrestrial herbivores, birds as well as mammals, demonstrate generally higher renal Cd levels than carnivores. The moose (*Alces alces*), roe deer (*Capreolus capreolus*) and hare (*Lepus europeus* and *Lepus timidus*) were suitable as biomonitors because of their common occurrence and uniform geographical distribution. The eider duck (*Somateria mollissima*) is suggested as a biomonitor of Cd for the aquatic environment. The accumulation rate of Cd in the kidneys is rapid with Cd levels 10 wk after hatching. Juvenile birds should be collected for monitoring purposes before leaving their feeding domains at the end of the summer.

Frank, Adrian. 1986. Lead fragments in tissues from wild birds: A cause of misleading analytical results. Sci. Total Environ. 54:275-281.
Rec #: 487
KEYWORDS: metals; Pb; common eider; *Somateria mollissima*; long-tailed duck; *Clangula*

Annotated Bibliography (Cont.).

hyemalis; liver; kidney; Sweden; Europe
NOTES: SUPPLEMENTAL
ABSTRACT: Seriously damaged eider ducks (*Somateria mollissima*) and long-tailed ducks (*Clangula hyemalis*) were shot in connection with an oil spill in 1974. Liver and kidney tissues were analyzed for environmental pollutants and lead analysis gave irreproducible results. By means of X-ray photographs, X-ray-dense particles could be observed in the tissues.
The foreign particles were extracted by dissolution of the organ tissues in Soluene-350 (Packard Instruments Co. Inc) and then washed with toluene. The insoluble particles consisted of lead and bone splinters of varying size. The form of the former ranged from irregular fragments to dust, and arose by disruption of lead pellets upon collision with bone tissue.
Birds shot with lead pellets should not be used for lead determination unless careful X-ray investigations are made prior to the chemical analysis. Determinations should be made on at least two different samples of the tissue examined.

Frank, Richard and Van Hove Holdrinet, Micheline. 1975. Residue of organochlorine compounds and mercury in birds' eggs from the Niagara Peninsula, Ontario. Arch. Environ. Contam. Toxicol. 3(2):205-218.
Rec #: 19
KEYWORDS: egg; biomagnification; organochlorines; DDTs; PCBs; dieldrin; metals; Hg; herring gull; *Larus argentatus*; black-crowned night-heron; *Nycticorax nycticorax*; black tern; *Chlidonias niger*; common tern; *Sterna hirundo;* Ontario; Canada; North America
ABSTRACT: Organochlorine insecticides, polychlorinated biphenyls (PCB), and total Hg were detected in eggs collected in 1971 from 20 species of birds from the Niagara Peninsula. Eggs from carnivorous species at the top of the aquatic food chain had the highest mean residues of total DDT (7.6-22.4 ppm), PCB (3.5-74.0 ppm), and total Hg (0.64-0.83 ppm). Eggs from herbivorous and insectivorous birds of both aquatic and terrestrial environments contained much lower residues. Eggs from some terrestrial carnivores (red-tailed hawk and great horned owl) also had relatively high residues, but levels were much lower than those found in eggs from aquatic-feeding carnivores. PBC residues were slightly lower in eggs among the terrestrial feeding species (0.05-2.0 ppm) than among the aquatic feeders (0.14-4.0 ppm) and tended to be lower in eggs from terrestrial species collected in rural rather than in city environs. Levels of total DDT were similar in both groups with eggs from terrestrial feeders containing mean residues between 0.15 and 2.64 ppm, and in those from aquatic feeders between 0.33 and 2.79 ppm.

Franson, J. Christian; Hollmén, Tuula; Poppenga, Robert H.; Hario, Martti; Kilpi, Mikael; and Smith, Milton R. 2000. Selected trace elements and organochlorines: Some findings in blood and eggs of nesting common eiders (*Somateria mollissima*) from Finland. Environ. Toxicol. Chem. 19(5):1340-1347.
Rec #: 313
KEYWORDS: common eider; *Somateria mollissima*; egg; blood; metals; Pb; Se; Hg; DDTs; organochlorines; Baltic Sea; Finland; Europe; effects; toxicology; ALAD
ABSTRACT: In 1997 and 1998, we collected blood samples from nesting adult female common eiders (*Somateria mollissima*) at five locations in the Baltic Sea near coastal Finland and analyzed them for lead, selenium, mercury, and arsenic. Eggs were collected from three locations in 1997 for analysis of selenium, mercury, arsenic, and 17 organochlorines (OCs). Mean blood

Annotated Bibliography (Cont.).

lead concentrations varied by location and year and ranged from 0.02 ppm (residues in blood on wet weight basis) to 0.12 ppm, although one bird had 14.2 ppm lead in its blood. Lead residues in the blood of eiders were positively correlated with the stage of incubation, and lead inhibited the activity of the enzyme delta-aminolevulinic acid dehydratase (ALAD) in the blood. Selenium concentrations in eider blood varied by location, with means of 1.26 to 2.86 ppm. Median residues of selenium and mercury in eider eggs were 0.55 and 0.10 ppm (residues in eggs on fresh weight basis), respectively, and concentrations of both selenium and mercury in eggs were correlated with those in blood. Median concentrations of p,p'-dichlorodiphenyldichloroethylene in eggs ranged from 13.1 to 29.6 ppb, but all other OCs were below detection limits. The residues of contaminants that we found in eggs were below concentrations generally considered to affect avian reproduction. The negative correlation of ALAD activity with blood lead concentrations is evidence of an adverse physiological effect of lead exposure in this population.

Franson, J. Christian; Petersen, Margaret R.; Meteyer, Carol U.; and Smith, Milton R. 1995. Lead poisoning of spectacled eiders (*Somateria fischeri*) and of a common eider (*Somateria mollissima*) in Alaska. J. Wildl. Dis. 31(2):268-271.
Rec #: 312
KEYWORDS: AK; USA; North America; effects; metals; Pb; blood; liver; spectacled eider; *Somateria fischeri*; common eider; *Somateria mollissima*
NOTES: NO TABLES
ABSTRACT: Lead poisoning was diagnosed in four spectacled eiders (*Somateria fischeri*) and one common eider (*Somateria mollissima*) found dead or moribund at the Yukon Delta National Wildlife Refuge, Alaska (USA) in 1992, 1993, and 1994. Ingested lead shot was found in the lower esophagus of one spectacled eider and in the gizzard of the common eider. Lead concentrations in the livers of the spectacled eiders were 26 to 38 ppm wet weight, and 52 ppm wet weight in the liver of the common eider. A blood sample collected from one of the spectacled eiders before it was euthanized had a lead concentration of 8.5 ppm wet weight. This is the first known report of lead poisoning in the spectacled eider, recently listed as a threatened species by the U.S. Fish and Wildlife Service.

Fry, D. Michael. 1995. Reproductive effects in birds exposed to pesticides and industrial chemicals. Environ. Health Perspect. 103(7):165-171.
Rec #: 2
KEYWORDS: review; organochlorines; PCBs; DDTs; dieldrin; HCHs; HCB; toxaphene; chlordanes; PCDDs; PCDFs; hydrocarbons; PAHs; effects
NOTES: NO TABLES
ABSTRACT: Environmental contamination by agricultural chemicals and industry waste disposal results in adverse effects on reproduction of exposed birds. The diversity of pollutants results in physiological effects at several levels, including direct effects on breeding adults as well as developmental effects on the embryos. The effects on embryos include mortality or reduced hatchability, failure of chicks to thrive (wasting syndrome), and teratological effects producing skeletal abnormalities and impaired differentiation of the reproductive and nervous systems through mechanisms of hormonal mimicking of estrogens. The range of chemical effects on adult birds covers acute mortality, sublethal stress, reduced fertility, suppression egg formation, eggshell thinning, and impaired incubation and chick rearing behaviors. The types of pollutants shown to cause reproductive effects include organochlorine pesticides and industrial

Annotated Bibliography (Cont.).

pollutants, organophosphate pesticides, petroleum hydrocarbons, heavy metals, and in a fewer number of reports, herbicides, and fungicides. o,p'-DDT, polychlorinated biphenyls (PCBs), and mixtures of organochlorines have been identified as environmental estrogens affecting populations of gulls breeding in polluted "hot spots" in southern California, the Great Lakes, and Puget Sound. Estrogenic organochlorines represent an important class of toxicants to birds because differentiation of the avian reproductive system is estrogen dependent.

Furness, R. W.; Muirhead, S. J.; and Woodburn, M. 1986. Using bird feathers to measure mercury in the environment: Relationships between mercury content and moult. Mar. Pollut. Bull. 17(1):27-30.
Rec #: 299
KEYWORDS: metals; Hg; feather; great skua; *Catharacta skua*; Atlantic petrel; *Pterodroma incerta*; soft-plumaged petrel; *Pterodroma mollis*; Kerguelen petrel; *Pterodroma brevirostris*; great shearwater; *Puffinus gravis*; black-legged kittiwake; *Rissa tridactyla*; northern fulmar; *Fulmarus glacialis*; Manx shearwater; *Puffinus puffinus*; Shetland; Great Britain; Europe; Atlantic Ocean
ABSTRACT: Feathers can be used to monitor mercury levels in marine, freshwater and terrestrial ecosystems. Previous studies have often failed to take into account the great differences in mercury levels between feathers of individual birds. Feathers replaced early in the moulting sequence have higher levels of mercury than those moulted later. The widely held idea that mercury levels in feathers reflect dietary intake at the time of feather growth is not supported by our data. We suggest that the amount of mercury stored in body tissues is the main factor determining levels in plumage. Although they are most often used, remiges and rectrices may not be the most suitable feathers if mercury levels in birds are to be examined: body feathers provide the most representative sample for estimating whole-bird mercury content.

Furness, R. W.; Thompson, D. R.; and Becker, P. H. 1995. Spatial and temporal variation in mercury contamination of seabirds in the North Sea. In: Franke, H-D. and Luening, K. (Eds.) The Challenge to Marine Biology in a Changing World. Hamburg (HFO): Biologische Anstalt Helgoland,. Helgoländer Meeresunters. 49(1-4):605-615.
Rec #: 58
KEYWORDS: feather; North Sea; Scotland; Germany; Europe; metals; Hg; organometallics; organic Hg; lesser black-backed gull; *Larus fuscus*; herring gull; *Larus argentatus;* common murre; *Uria aalge*; black-legged kittiwake; *Rissa tridactyla*; common tern; *Sterna hirundo*; spatial variations; temporal trends
ABSTRACT: Once the moult patterns have been taken into account, feather methylmercury levels can be used to accurately measure the mercury burdens of seabirds. We used body feathers from live seabirds and from museum collections to examine geographical and temporal patterns of mercury contamination in the North Sea. This approach identifies an increase in mercury concentrations in seabirds of the German North Sea coast during the last 100 years, especially high levels during the 1940s, and reduced contamination in the last few years. Comparisons among populations suggest that some increases in mercury levels are predominantly due to local pollution inputs, as on the German coast, while in other areas deposition from jet stream circulation of global contamination may be the major contributor. Mercury levels are far higher in seabirds from the German North Sea coast than in populations from the North and West North Sea or from most areas of the North Atlantic. We advocate the use of museum collections of

Annotated Bibliography (Cont.).

birds for studies of long-term changes in levels of mercury contamination.

Furness, Robert W.; Thompson, David R.; and Walsh, Paul M. 1990. Evidence from biological samples for historical changes in global metal pollution. In: Furness, Robert W. and Rainbow, Philip S. (Eds.). Heavy Metals in the Marine Environment. Boca Raton, FL: CRC Press. 219-225.
Rec #: 199
KEYWORDS: review; effects; metals; Hg; Cd; Pb; feather; black guillemot; *Cepphus grylle*; common murre; *Uria aalge*; thick-billed murre; *Uria lomvia*; Baltic Sea; Kattegat; Atlantic Ocean; Europe; temporal trends
NOTES: NO TABLES
ABSTRACT: A review is made of research work using historical biological material to assess time trends in metal levels in the environment either to detect instances of local pollution or to seek evidence for global trends. An evaluation is also made of the value of this approach in current and future studies, considering in particular the metals cadmium, mercury and lead. Particular to arctic seabirds, mercury concentration in feathers have increased from the pre-1900s to 1960s-70s.

Gabrielsen, Geir Wing and Henriksen, Espen O. 2001. Persistent organic pollutants in Arctic animals in the Barents Sea area and at Svalbard: Levels and effects. Memoirs of National Institute of Polar Research. 54:349-364.
Rec #: 432
KEYWORDS: review; glaucous gull; *Larus hyperboreus*; Barents Sea; Norway; Svalbard; Europe; toxicology; vitamin A; EROD; CYP1A; cytochrome P_{450}; porphyrin; organochlorines; PCDDs; PCBs; DDTs; mirex; liver; effects; temporal trends
NOTES: SUPPLEMENTAL
ABSTRACT: A review with many references is given. At Svalbard and in the Barents Sea area, high levels of persistent organic pollutants (POPs) have been found in glaucous gull (*Larus hyperboreus*), arctic fox (*Alopex lagopus*) and polar bear (*Ursus maritimus*). Studies of the possible toxic effects on the hormone-, vitamin-, enzyme-, immune- and reproduction system have been conducted during the last 5-10 years. Data obtained both from laboratory and field studies indicate that the present POP levels have an influence on biochemical-, physiological- and immunological parameters in glaucous gull and polar bear. In these 2 species, studies are currently being conducted to relate POP levels to biologic/toxic effects both on individuals and populations.

Gabrielsen, Geir Wing; Skaare, Janneche Utne; Polder, Anuschka; and Bakken, Vidar. 1995. Chlorinated hydrocarbons in glaucous gulls (*Larus hyperboreus*) in the southern part of Svalbard. Sci. Total Environ. 160/161:337-346.
Rec #: 41
KEYWORDS: glaucous gull; *Larus hyperboreus*; thick-billed murre; *Uria lomvia*; Norway; Europe; toxicology; organochlorines; HCHs; HCB; chlordanes; dieldrin; DDTs; PCBs; effects; biomagnification; egg; liver; brain; kidney; muscle
ABSTRACT: In 1989, a number of glaucous gulls (*Larus hyperboreus*) were found dead near a seabird cliff at south Svalbard. In an effort to elucidate the course of death, 12 individuals were sent to the Central Veterinary Institute for autopsy and analysis of chlorinated pesticides and

Annotated Bibliography (Cont.).

polychlorinated biphenyls (PCBs) in samples of liver, kidney, brain and muscle. Eggs of common and Brünnich's guillemots constitute important food for glaucous gulls during the nestling period, thus 13 eggs (3 eggs from common and 10 eggs from Brünnich's guillemots) from the same area were also analyzed for organochlorines. The autopsy of the birds did not reveal any specific cause of death. The following chlorinated pesticides were analyzed: hexachlorobenzene (HCB), hexachlorocyclohexanes (alpha -, beta -, gamma -HCH), chlordanes (oxychlordane, trans-nonachlor, heptachlor and heptachlor epoxide), the drin group (aldrin, dieldrin) and the DDT group (o,p', p,p'DDT, DDD and DDE). Relatively low hepatic concentrations of dieldrin were found (n.d.-0.17 mg/kg wet weight (w.w.), mean 0.04), capital sigma HCH (n.d.-0.14 mg/kg w.w., mean 0.02), HCB (0.03-1.01 mg/kg w.w., mean 0.51), oxychlordane (0.01-1.85 mg/kg w.w., mean 1.05), p,p'DDE (0.81-5.41 mg/kg w.w., mean 2.98), o,p'DDD (0-0.52 mg/kg w.w., mean 0.14) and capital sigma DDT (0.9-6.25 mg/kg w.w., mean 2.92). Levels of the other chlorinated pesticides were below the quantification level. Very high concentrations of PCBs were found. The capital sigma PCB (sum of wet weight concentrations of the 21 individual congeners, the IUPAC numbers 28, 52, 74, 99, 101, 110, 118, 114, 153, 105, 141, 138, 183, 128, 187, 156, 157, 180, 170, 194 and 209) in hepatic tissue was 0.8-32.3 mg/kg w.w. (mean 16.0), in brain 0.9-29.5 mg/kg w.w. (mean 14.8), in kidney 0.4-21.4 mg/kg w.w. (mean 9.7) and in muscle 0.5-6.0 mg/kg w.w. (mean 3.1). The concentrations of the individual PCB congeners are also given. Relatively low concentrations of all the organochlorines were found in guillemot eggs, mean concentrations of capital sigma DDT and PCBs were 0.25 and 0.47 mg/kg w.w., respectively. A toxicological evaluation of the PCB results is difficult; however, it cannot be excluded that PCBs might have contributed to the death of the gulls.

Giesy, John P. and Kannan, Kurunthachalam. 2001. Global distribution of perfluorooctane sulfonate in wildlife. Environ. Sci. Technol. 35(7):1339-1342.
Rec #: 364
KEYWORDS: PFOS; blood; egg; liver; Great Lakes; North America; Canada; Pacific Ocean; Arctic Ocean; herring gull; *Larus argentatus;* double-crested cormorant; *Phalacrocorax auritus*; spatial variations
ABSTRACT: The global distribution of perfluorooctane sulfonate (PFOS), a fluorinated organic contaminant. PFOS measured in the tissues of wildlife, including, fish, birds, and marine mammals is reported. Some of the species studied include bald eagles, polar bears, albatrosses, and various species of seals. Samples were collected from urbanized areas in North America, especially the Great Lakes region and coastal marine areas and rivers, and Europe. Samples were also collected from a number of more remote, less urbanized locations such as the Arctic and the North Pacific Oceans. The results demonstrated that PFOS is widespread in the environment. Concentrations of PFOS in animals from relatively more populated and industrialized regions, such as the North American Great Lakes, Baltic Sea, and Mediterranean Sea, were greater than those in animals from remote marine locations. Fish-eating, predatory animals such as mink and bald eagles contained concentrations of PFOS that were greater than the concentrations in their diets. This suggests that PFOS can bioaccumulate to higher trophic levels of the food chain. Currently available data indicate that the concentrations of PFOS in wildlife are less than those required to cause adverse effects in laboratory animals.

Gilbertson, Michael. 2001. Canadian wildlife service's herring gull monitoring program: importance to the development of environmental policy on organohalogens. Environ. Rev.

Annotated Bibliography (Cont.).

9(4):261-267.
Rec #: 425
KEYWORDS: review; herring gull; *Larus argentatus;* egg; disease; abnormalities; toxicology; effects; organochlorines; PCDDs; Great Lakes; Canada; North America
NOTES: SUPPLEMENTAL
ABSTRACT: A review and discussion. The biological context for the selection of the herring gull (*Larus argentatus*), from a long list of candidate species, as an indicator of the restoration of Great Lakes water quality is further elaborated. Embryo mortality and deformities in herring gull chicks led to the hypothesis that the Lake Ontario population was exhibiting chick edema disease. Subsequent observation of the suite of lesions associated with chick edema disease in herring gull chicks led to the hypothesis that the Great Lakes were contaminated with polychlorinated dibenzo-p-dioxins. Forensic toxicology, as a collaboration of environmental chemistry, biology and pathology, could form a rational basis for policy decisions about remedial actions to restore extirpated fish and wildlife populations and to protect public health.

Gilman, A. P.; Fox, G. A.; Peakall, D. B.; Teeple, S. M.; Carroll, T. R.; and Haymes, G. T. 1977. Reproductive parameters and egg contaminant levels of Great Lakes herring gulls. J. Wildl. Manage. 41(3):458-468.
Rec #: 149
KEYWORDS: herring gull; *Larus argentatus*; North America; Great Lakes; egg; organochlorines; DDTs; dieldrin; mirex; chlordanes; HCB; PCBs; metals; Hg; effects; reproductive success; spatial variations
ABSTRACT: Poor reproductive success and declines in colony size of herring gulls (*Larus argentatus*) have occurred in Lake Ontario at a time that dramatic increases in this species have been reported on the Atlantic seaboard. In 1975 herring gull productivity on Scotch Bonnet Island, Lake Ontario, was 0.15 chicks per pair of adults, one-tenth the productivity of colonies studies on Lakes Erie, Huron and Superior. Reduced nest site defense and decreases in eggs found, egg hatchability and chick survival were observed in the Lake Ontario colony. The major causes of egg failure were disappearance and embryonic death. Hatching success of Lake Ontario eggs by artificial incubation was 23-25 % compared to 53-79 % for eggs from other areas. Analysis of eggs from 9 gull colonies for organochlorine contaminants indicated that the pattern of relative contamination was: Lake Ontario > Michigan > Superior > Huron > Erie. Mirex levels were nearly 10 times higher in Lake Ontario than in the other lakes. Movements of herring gulls within the Great Lakes basin are offered as an explanation of variation in individual egg residues in each colony and the moderately high levels of chemical residues in some Lake Superior eggs.

Gilman, Andrew P.; Peakall, David B.; Hallett, Douglas J.; Fox, Glen A.; and Norstrom, Ross J. 1979. Herring gulls (*Larus argentatus*) as monitors of contamination in the Great Lakes. Animals As Monitors of Environmental Pollutants. Symposium on Pathobiology of Environmental Pollutants: Animal Models and Wildlife as Monitors, University of Connecticut, 1977. 280-289.
Rec #: 141
KEYWORDS: review; effects; biomagnification; egg; herring gull; *Larus argentatus*; Great Lakes; North America; organochlorines; DDTs; dieldrin; chlordanes; mirex; HCB; PCBs; metals; Hg; reproductive success; abnormalities

Annotated Bibliography (Cont.).

ABSTRACT: The herring gull (*L. argentatus*) is at the top of several food chains. Adults of this species are essentially year-round residents of the Great Lakes. The levels of contaminants found in them are higher than those in other Laridae in the Great Lakes. Their ability to bioaccumulate high loads of persistent contaminants allows investigators to identify compounds that would be difficult to determine in lower trophic levels. Chlorinated hydrocarbons and other persistent pollutants accumulate to high levels in gull tissues and are deposited into the eggs. Egg contaminant levels reflect the levels of lake contamination. High levels of PCB's, DDE, and mirex in Lake Ontario herring gulls and their association with early embryonic mortality, chick deformity, and aberrant adult behavior were examined.

Gochfeld, M. 1997. Spatial patterns in a bioindicator: Heavy metal and selenium concentration in eggs of herring gulls (*Larus argentatus*) in the New York Bight. Arch. Environ. Contam. Toxicol. 33(1):63-70.
Rec #: 278
KEYWORDS: metals; Se; Pb; Cd; Hg; Cr; Mn; egg; herring gull; *Larus argentatus*; NY; NJ; USA; North America; spatial variations; toxicology
ABSTRACT: Concentrations of selenium and five heavy metals (lead, cadmium, mercury, chromium, and manganese) in the eggs of herring gulls (*Larus argentatus*) were studied at six breeding colonies in the New York Bight to detect locational differences and to explore their use as a bioindicator of point source or nonpoint source pollution. The herring gull is widespread in North America, Europe, and Asia, and has urban-adapted counterparts in the southern hemisphere as well. We anticipated that the chromium contamination at Jersey City and high levels of manganese in industrial releases to the Passaic River would be reflected in the nearest colony (Shooter's Island), and that lead contamination from bridge remediation would be apparent in the Jamaica Bay colonies. There were significant locational differences in all metal levels, although the patterns were not the same for all metals. Shooter's Island in Newark Bay ranked first or second for five of the elements, but inexplicably had the lowest mercury level. Cadmium levels were highest at Canarsie Pol in Jamaica Bay, but mercury levels were highest at the relatively isolated Lavallette colony in northern Barnegat Bay. Chromium and manganese levels were indeed highest at Shooter's Island, but the lead levels in Jamaica Bay were only intermediate. We predicted that the essential trace elements, manganese, chromium, and selenium, which are known to be present at relatively high concentrations in various animal species, would have relatively low coefficients of variation, reflecting homeostatic mechanisms. This was confirmed. In conclusion, herring gull egg contents can be used to monitor metal concentrations at nearby colonies to indicate areas of concern for particular metals. They may confirm suspected associations or identify hitherto unsuspected problems.

Grand, James B.; Franson, J. Christian; Flint, Paul L.; and Petersen, Margaret R. 2002. Concentrations of trace elements in eggs and blood of spectacled and common eiders on the Yukon-Kuskokwim Delta, Alaska, USA. Environ. Toxicol. Chem. 21(8):1673-1678.
Rec #: 464
KEYWORDS: metals; As; Cd; Pb; Hg; Se; egg; blood; spectacled eider; *Somateria fischeri*; common eider; *Somateria mollissima*; AK; USA; North America; reproductive success; effects
ABSTRACT: We examined the relations among nesting success, egg viability, and blood and egg concentrations of As, Cd, Pb, Hg, and Se in a threatened population of spectacled eiders (*Somateria fischeri*) and a sympatric population of common eiders (*S. mollissima*) on the

Annotated Bibliography (Cont.).

Yukon-Kuskokwim Delta, Alaska, USA, during 1995 and 1996. During the early breeding season, males and females had mean Se concentrations in their blood of 19.2 µg/g and 12.8 µg/g wet weight, respectively Blood Se concentrations of females were correlated with egg concentrations During brood rearing, blood Se levels were higher in adult females than in ducklings. Blood concentrations of Pb in spectacled eider females were higher than in common eider females captured at hatching, but blood concentrations of Se were similar. Trace element concentrations were not related to nest success or egg viability. We submit that nest success and egg viability of spectacled eiders are not related to concentrations of the trace elements we measured. Because blood Se concentrations declined rapidly through the breeding season and were not related to nest success or egg viability, we suggest that spectacled eiders are exposed to high concentrations of Se during winter that pose little threat to this population.

Grasman, K. A.; Scanlon, P. F.; and Fox, G. A. 2000. Geographic variation in hematological variables in adult and prefledgling herring gulls (*Larus argentatus*) and possible associations with organochlorine exposure. Arch. Environ. Contam. Toxicol. 38(2):244-253.
Rec #: 329
KEYWORDS: herring gull; *Larus argentatus*; organochlorines; PCBs; DDTs; toxicology; TEQs; EROD; porphyrin; blood; liver; Great Lakes; Canada; North America
NOTES: FIGURES ONLY
ABSTRACT: The objectives of this study were (1) to describe variation in hematological values found in adult and prefledgling herring gulls (Larus argentatus) over a large geographic area, (2) to investigate relationships between hematological variables and other physiological indices, and (3) to examine potential associations between exposure to organochlorines and hematological variables. During 1991-93, we sampled 160 breeding adult gulls from 13 colonies and 101 4-week-old gulls from 11 colonies. All colonies were in the Great Lakes ecosystem, except for two colonies on Lake Winnipeg and the Atlantic coast. The hematological values measured in this study were similar to published values for herring gulls and related species. Significant intersite differences were found in hematological variables. Sex had little or no influence on leukocyte variables. Adults had lower total leukocyte counts and higher heterophil to lymphocyte ratios than chicks. PCV was lower in adult females than males. In adults, total leukocyte and total heterophil numbers were negatively associated with liver activity of ethoxyresorufin-O-deethylase (EROD) and concentrations of highly carboxylated porphyrins (HCPs), two biomarkers of organochlorine exposure. Total leukocyte and total heterophil numbers were positively associated with liver concentrations of DDE (1,1-dichloro-2,2-bis(p-chlorophenyl)ethylene), and total lymphocytes were associated positively with PCB (polychlorinated biphenyl) and HCP concentrations. The heterophil to lymphocyte ratio was negatively associated with liver EROD activity and HCPs. In chicks, there was a positive association between the heterophil to lymphocyte ratio and HG-TEQs (dioxin toxicity equivalents calculated using herring gull-specific equivalency factors). PCV was associated with some measures of contaminant exposure in adults and chicks. Additional research is needed to elucidate causal relationships between hematological indices and such factors as contaminants, disease, and nutrition.

Grasman, Keith. A.; Fox, Glen A.; Scanlon, Patrick F.; and Ludwig, James P. 1996. Organochlorine-associated immunosuppression in prefledgling caspian terns and herring gulls from the Great Lakes: An ecoepidemiological study. Environ. Health Perspect. 104(4):829-842.

Annotated Bibliography (Cont.).

Rec #: 1
KEYWORDS: biomarkers; toxicology; immunology; vitamin A; organochlorines; PCBs; DDTs; dieldrin; mirex; HCB; chlordanes; PCDDs; TEQs; Great Lakes; Canada; North America; egg; herring gull; *Larus argentatus;* Caspian tern; *Sterna caspia*
ABSTRACT: The objectives of study were to determine whether contaminant-associated immunosuppression occurs in prefledgling herring gulls and Caspian terns from the Great Lakes and to evaluate immunological biomarkers for monitoring health effects in wild birds. During 1992 to 1994, immunological responses and related variables were measured in prefledgling chicks at colonies distributed across a broad gradient of organochlorine contamination (primarily polychlorinated biphenyls), which was measured in eggs. The phytohemagglutinin skin test was used to assess T-lymphocyte function. In both species, there was a strong exposure-response relationship between organochlorines and suppressed T-cell-mediated immunity. Suppression was most severe (30-45 %) in colonies in Lake Ontario (1992) and Saginaw Bay (1992-1994) for both species and in western Lake Erie (1992) for herring gulls. Both species exhibited biologically significant differences among sites in anti-sheep red blood cells antibody titers, but consistent exposure-response relationships with organochlorines were not observed. In Caspian terns and, to a lesser degree, in herring gulls, there was an exposure-response relationship between organochlorines and reduced plasma retinol (vitamin A). In 1992, altered White blood cell numbers were associated with elevated organochlorine concentrations in Caspian terns but not herring gulls. The immunological and hematological biomarkers used in this study revealed contaminant-associated health effects in wild birds. An epidemiological analysis strongly supported the hypothesis that suppression of T-cell-mediated immunity was associated with high perinatal exposure to persistent organochlorine contaminants.

Greichus, Y. A.; Worman, J. J.; Pearson, M. A.; and Call, D. J. 1974. Analyses of polychlorinated biphenyls in bird tissues and Aroclor standards with gas chromatography and mass spectrometry. Bull Environ Contam Toxicol. 11(2):113-120.
Rec #: 451
KEYWORDS: double-crested cormorant; *Phalacrocorax auritus*; white pelican; *Pelecanus erythrorhynchos*; SD; USA; North America; carcass; egg; fat; organochlorines; PCBs
NOTES: SUPPLEMENTAL
ABSTRACT: The polychlorinated biphenyls (PCB's) in Aroclor 1254 [11097-69-1] and Aroclor 1260 [11096-82-5] and in cormorant carcass and eggs and pelican fat were identified by gas-liquid chromatography and mass spectrometry, and the numbers of Cl's in the PCB's were detected by observing the appropriate m/e values of the mol. ions and were verified by the relative intensities of the peaks in the mol. ion clusters. The relative retention times (rrt) of the major peaks in the gas chromatograms of the cormorant and pelican samples were correlated with the rrt of the major peaks in the standards via their relative intensities and identical mass spectral fragmentation patterns. Multiple component peaks were verified by observations of the correct mol. weight and mol. ion clusters for each PCB. Thus, all major PCB's occurring in cormorant carcass and eggs and pelican fat also occurred in Aroclors 1254 and (or) 1260. The pattern of PCB's found in the biological samples was not the same as those in Aroclors 1254 or 1260, and further changes in pattern were observed when comparing cormorant carcass and eggs.

Greichus, Yvonne A.; Greichus, A.; and Emerick, R. J. 1973. Insecticides, polychlorinated biphenyls, and mercury in wild cormorants, pelicans, their eggs, food, and environment. Bull

Annotated Bibliography (Cont.).

Environ Contam Toxicol. 9(6):321-328.
Rec #: 452
KEYWORDS: SD; USA; North America; organochlorines; PCBs; dieldrin; DDTs; metals; Hg; double-crested cormorant; *Phalacrocorax auritus*; white pelican; *Pelecanus erythrorhynchos*; carcass; egg; muscle; liver; kidney; biomagnification
ABSTRACT: The concentration of organochlorine insecticide residues in and around Lake Poinsett, South Dakota, decreased in the order fish>bottom sediments>water. The levels of polychlorinated biphenyls (PCB) were greater in fish than in bottom sediments, and the PCB levels in both bottom sediments and fish were higher than those of insecticides. Mercury [7439-97-6] residues were not detected in water, bottom sediments, or fish except carp which also had the highest levels of insecticides and PCB residues. The bodies of adult cormorants and pelicans had 250 and 280 times greater insecticide levels, respectively, than fish and 60 and 30 times, respectively, more PCB. Levels of PCB and insecticides in nestling cormorants apparently reflected the levels in local fish as both the fish and young birds had more PCB residues than insecticides whereas the opposite was true with the adults. The insecticide and PCB levels in cormorant eggs reflected body levels but in pelican eggs they did not.

Greichus, Yvonne A. and Hannon, Michael R. 1973. Distribution and biochemical effects of DDT, DDD and DDE in penned double-crested cormorants. Toxicol. Appl. Pharmacol. 26(4):483-494.
Rec #: 416
KEYWORDS: brain; heart; liver; carcass; metabolism; organochlorines; DDTs; double-crested cormorant; *Phalacrocorax auritus*; toxicology; effects; vitamin A; carotene; CBC; dosing study; gender differences
ABSTRACT: Penned double-crested cormorants were fed 2, 5, and 10 ng of a combination of DDT (I) [50-29-3], DDD (II) [72-54-8], and DDE (III) [72-55-9] in their daily diet; birds stressed by a 1/2 decrease in food after the cessation of 9 weeks of treatment and birds that died of I toxicity showed a marked increase in brain and liver residues and a decrease in carcass residues. Higher brain residue levels were correlated with decreased body weight, and carcass lipid content. Brain concentrations were 24-85 ppm in birds that died of toxic effects and 0.4-29 ppm in survivors, so 30ppm indicated toxicity. Brain concentrations of I and metabolites in wild cormorants indicated no immediate danger of toxicity. A decrease in total liver vitamin A [11103-57-4], and liver and heart weights as percentage of total body weight was also observed after treatment. Liver weight was negatively correlated with brain concentrations of total I, II, and III.

Haffner, G. Douglas; Straughan, Cameron A.; Weseloh, D. V. Chip; and Lazar, Rodica. 1997. Levels of polychlorinated biphenyls, including coplanar congeners, and 2,3,7,8-T$_4$CDD toxic equivalents in double-crested cormorant and herring gull eggs from Lake Erie and Lake Ontario: A comparison between 1981 and 1992. J. Great Lakes Res. 23(1):52-60.
Rec #: 301
KEYWORDS: organochlorines; PCBs; TEQs; toxicology; temporal trends; egg; North America; Great Lakes; double-crested cormorant; *Phalacrocorax auritus;* herring gull; *Larus argentatus*
ABSTRACT: Eggs of double-crested cormorants (*Phalacrocorax auritus*) and herring gulls (*Larus argentatus*) collected from Lakes Erie and Ontario during 1981 and 1992 were analyzed

Annotated Bibliography (Cont.).

for PCB congener concentrations, including non-ortho congeners 77, 126, and 169. Total PCB in herring gulls, measured as Aroclor 1254/1260, was significantly lower in 1992, although differences in chemical concentrations in Lake Erie birds were not of the same magnitude as those observed in Lake Ontario. Changes in concentration of total PCB in cormorant populations were not significant in Lake Ontario, and only a small change was observed in Lake Erie cormorants. In 1981, cormorants and herring gull eggs were similarly contaminated in Lake Ontario and Lake Erie, but in 1992, Lake Erie eggs were significantly more contaminated than those from Lake Ontario. An examination of changes in concentrations of individual congeners suggested that in cormorants, the decrease in PCB concentrations was due primarily to the loss of low K_{ow} congeners. All congeners, however, contributed to the decline of PCBs in herring gulls. Toxic equivalents (TEQs) estimates revealed that congener 126 dominated the TEQs in both species in Lake Ontario and Lake Erie, and were highest in herring gulls. Although estimates of total TEQs in herring gulls in both lakes were lower in 1992, there was little change in TEQs in double-crested cormorants. These results support the conclusion that chemical accumulation patterns are regulated to some degree by both ecological and limnological processes.

Hallett, D. J.; Norstrom, R. J.; Onuska, F.; and Comba, M. 1978. Incidence of chlorinated benzenes and other lower molecular weight organochlorines in Lake Ontario herring gulls. In: Extended Abstracts of the International Symposium on the Analysis of Hydrocarbons and Halogenated Hydrocarbons, #40, University of Toronto, Ontario.
Rec #: 553
KEYWORDS: organochlorines; benzenes; HCB; Great Lakes; North America; herring gull; *Larus argentatus*; egg
NOTES: NOT ABLE TO OBTAIN; SEE #519 FOR INFO

Hallett, D. J.; Norstrom, R. J.; Onuska, F. I.; and Comba, M. E. 1982. Incidence of chlorinated benzenes and chlorinated ethylenes in Lake Ontario herring gulls. Chemosphere. 11(3):277-285.
Rec #: 519
KEYWORDS: herring gull; *Larus argentatus*; egg; North America; Great Lakes; organochlorines; PCBs; HCB; benzenes; ethylenes; spatial variations
ABSTRACT: Great Lakes Herring Gulls (*Larus argentatus*) and their eggs have proven to be useful integrators on a lakewide basis of high molecular weight, relatively involatile organochlorine pollutants such as PCBs. A search for relatively volatile organochlorine compounds by GC/MS also revealed the presence of tri- and tetrachloroethylene, and isomers of di-, tri-, tetra-, penta- and hexachlorobenzene in the body lipid of adult Herring Gulls from Lake Ontario. Analysis of pooled eggs from colonies throughout the Great Lakes in 1978 showed that pentachlorobenzene and hexachlorobenzene were ubiquitous contaminants at levels from 14–50 ng/kg and 90–350 ng/kg, respectively. Hexachlorobenzene levels were 2–3 times higher in Lake Ontario than the other lakes, whereas pentachlorobenzene was more evenly distributed geographically. Levels of 1,2,3,4-tetra-, 1,2,4,5-tetra- and 1,2,4-trichlorobenzene near the detection limit of 10–20 ng/kg were found in a few samples.

Hallett, D. J.; Shear, H.; Weseloh, D. V.; and Mineau, P. 1981. Surveillance of wildlife contaminants on the Great Lakes. Verh. Internat. Verein. Limnol. 21:1734-1740.
Rec #: 286
KEYWORDS: organochlorines; PCBs; egg; temporal trends; herring gull; *Larus argentatus*;

Annotated Bibliography (Cont.).

North America; Great Lakes
NOTES: FIGURES ONLY
ABSTRACT: The use of biological indicator species, such as herring gulls (*Larus argentatus*) which represent the top trophic level of the Great Lakes aquatic food web, has proven to be an effective, holistic approach to contaminant monitoring. Complementary programs acquiring data from predatory fish, and data from sediment are now allowing proper interpretation of the apparent dynamic trends in organochlorine residues within the food chain. Together they should provide an early warning of serious problems related to contaminants which could jeopardize the balance of the Great Lakes ecosystem.

Hannon, Michael Robert. 1972. Distribution and physiological effects of DDT, DDD, and DDE in penned double-crested cormorants (*Phalacrocorax a. auritus*). Brookings, SD, USA: South Dakota State Univ. 52 pp.
Rec #: 411
KEYWORDS: double-crested cormorant; *Phalacrocorax auritus*; organochlorines; DDTs; CBC; metals; Ca; Na; P; feces; blood; brain; carcass; feather; SD; USA; North America; gender differences; dosing study; effects; toxicology; vitamin A; carotene
ABSTRACT: Levels, distribution and excretion of DDT, DDD and DDE were determined in tissues of penned cormorants administered doses of 2, 5 and 10 mg of total DDT and metabolites (40 percent DDT, 30 percent DDD and 30 percent DDE) daily in their diet, approximately 5, 12 and 25 ppm. Analysis of residues was by electron capture gas chromatography.
Carcass retention of total experimental dose was estimated at 32 to 54 percent. Analysis of excreta indicated that 19 to 29 percent of the daily dose was lost via this media. Male cormorants retained greater proportions of total experimental dose than females, while females that survived retained more than those that died during the experiment.
Average DDE, DDD and DDT concentration increased with higher dosage in all sample types analyzed. DDE was detected at the highest concentrations in all samples, followed by DDD, then DDT. DDE was in greater and DDT lesser proportion than present in treatment.
Birds subjected to a period of two weeks on reduced diet after cessation of insecticide treatment had increased brain and decreased carcass residue concentrations. Increased brain and decreased carcass residues were also associated with death.
All birds diagnosed on the basis of tremors and convulsions as dying of insecticide toxicosis were females, suggesting a greater susceptibility of young female cormorants to DDT toxicosis. Residue levels differed markedly between, females that died on treatment and those that survived. Brain DDD concentrations showed the clearest separation between survivors and dead. Ranges were 1 to 24 ppm in survivors and 24 to 84 ppm in those that died.
Carcass lipid content declined with reduced diet and death, while brain lipid varied little with diet, death or sex. Higher brain residue levels were significantly correlated with decreased body weight ($P < .01$) and decreased carcass lipid content ($P < .05$).
Liver vitamin A and carotene levels appeared to be little affected by administration of DDT, DDD or DDE.
Analysis of variance showed significant treatment differences in liver ($P < .05$) and heart ($P < .05$) weights, but not in brain or spleen weights. Stress appeared to have the greatest effect on organ weights. A significant and negative correlation was found between liver weight and brain levels of DDE, DDD and DDT ($P < .05$).
Statistical analysis of blood chemistry data failed to show significant treatment effects on

Annotated Bibliography (Cont.).

parameters measured (calcium, sodium, phosphorus, blood urea nitrogen, total protein, hematocrit and hemoglobin).
Residue levels determined in wild adult cormorants were not diagnostic of immanent DDT toxicosis. However, concentrations present pose a potential danger, as mobilization of insecticide...laden carcass lipids have been shown to raise brain residues. Mobilization of the carcass residues determined in wild cormorants (10 ppm) could raise brain residues to a level diagnostic of DDT toxicosis in penned birds.

Hario, Martti; Hirvi, Juha-Pekka; Hollmén, Tuula; and Rudbäck, Eeva. 2004. Organochlorine concentrations in diseased vs. healthy gull chicks from the northern Baltic. Environ. Pollut. 127(3):411-423.
Rec #: 477
KEYWORDS: organochlorines; HCHs; chlordanes; HCB; DDTs; PCBs; toxicology; TEQs; lesser black-backed gull; *Larus fuscus*; herring gull; *Larus argentatus*; liver; Finland; Baltic Sea; Europe; disease; effects; reproductive success; biomagnification
ABSTRACT: The population decline of the nominate lesser black-backed gull *Larus fuscus fuscus* in the Gulf of Finland (northern Baltic) is caused by an exceedingly high chick mortality due to diseases. The chick diseases include degeneration in various internal organs (primarily liver), inflammations (mainly intestinal), and sepsis, the final cause of death. The hypothesis of starvation causing intestinal inflammations (leading to sepsis) was tested by attempting to reproduce lesions in apparently healthy herring gull L. argentatus chicks in captivity. The herring gull chicks were provided a similar low food-intake frequency as observed for the diseased chicks in the wild. However, empty alimentary tract per se did not induce the intestinal inflammations and therefore, inflammations seem to be innate or caused by other environmental factors in the diseased lesser black-backed chicks. They had very high concentrations of PCB in their liver; but the concentrations were not significantly higher than those of the healthy herring gull chicks, indicating a common exposure area for both species (i.e. the Baltic Sea). When compared to NOEL and LOEL values for TEQs in bird eggs our TEQ levels clearly exceed most or all of the values associated with effects. Compared with published data on fish-eating waterbirds, the DDE concentrations in the diseased lesser black-backed chicks were well above the levels previously correlated with decreased reproduction, while the residues in apparently healthy herring gulls were below those levels. The DDE/PCB ratio in lesser black-backs was significantly elevated, indicating an increased exposure to DDTs as compared with most other Baltic and circumpolar seabirds. The possible exposure areas of DDT in relation to differential migration habits of the two gull species are discussed. Elevated DDE/PCB ratio correlates with a high rate of chick diseases in the endangered nominate lesser black-backed gull.

Harris, Megan L.; Wilson, Laurie K.; Norstrom, Ross J.; and Elliott, John E. 2003. Egg concentrations of polychlorinated dibenzo-*p*-dioxins and dibenzofurans in double-crested (*Phalacrocorax auritus*) and pelagic (*P. pelagicus*) cormorants from the Strait of Georgia, Canada, 1973-1998. Environ. Sci. Technol. 37(5):822-831.
Rec #: 350
KEYWORDS: double-crested cormorant; *Phalacrocorax auritus*; pelagic cormorant; *Phalacrocorax pelagicus*; British Columbia; Canada; North America; PCDDs; PCDFs; PCBs; toxicology; EROD; TEQs; egg; temporal trends
ABSTRACT: Eggs of double-crested and pelagic cormorants were collected between 1973 and

Annotated Bibliography (Cont.).

1998 from colonies in the Strait of Georgia, BC, Canada, and assayed for concentrations of polychlorinated dibenzo-*p*-dioxins (PCDDs), dibenzofurans (PCDFs), and non-ortho- and mono-ortho-biphenyls (PCBs). Double-crested cormorant eggs contained (on average) up to 433 ng kg^{-1} wet weight 1,2,3,6,7,8-HxCDD, 151 ng kg^{-1} 1,2,3,7,8-PnCDD, and 74 ng kg^{-1} 2,3,7,8-TCDD, whereas pelagic cormorant eggs contained up to 300, 99, and 28 ng kg^{-1} wet weight of these respective congeners. The dominant non-ortho-PCB was CB-126, which ranged as high as 2263 ng kg^{-1} in double-crested cormorant eggs. Concentrations of PCDDs and PCDFs fell dramatically in the early 1990s, following both severe restrictions on the use of chlorophenolic wood preservatives and antisapstains and a switch from molecular chlorine bleaching to alternative bleaching technologies at pulp mills in the region. Concentrations of PCBs did not show similar marked declines over time. On the basis of total TEQs $\geq$148 ng kg and previously published documentation of effects in siblings of the cormorant eggs analyzed here, double-crested cormorant young may have exhibited significantly elevated EROD activity and/or brain asymmetries at all colonies from 1973 to 1989 and even at some colonies during the 1990s. Pelagic cormorant eggs collected from a few colonies in 1988-1989 also contained total TEQs greater than the threshold value estimated for double-crested cormorants.

Headley, Alistair D. 1996. Heavy metal concentrations in peat profiles from the high Arctic. Sci. Total Environ. 177(1-3):105-111.
Rec #: 437
KEYWORDS: Norway; Europe; black-legged kittiwake; *Rissa tridactyla*; glaucous gull; *Larus hyperboreus*; feces; metals; Fe; Mn; Pb; Cu; Zn; Ni
ABSTRACT: The concentration and possible sources of heavy metal input to a high Arctic mire in Kongsfjord, West Spitsbergen were examined; levels of lead, copper and zinc are highest in the surface humus closest to the carboniferous limestone cliff upon which seabirds nest. The concentrations of iron, manganese, nickel, copper, zinc and lead in two cores of peat increase markedly in the upper 5 cm of consolidated peat, with the highest concentrations found in the uppermost centimeter of peat. There has, therefore, been an increase in heavy metal loading to this high Arctic mire within the last 100 years. This cannot be attributed to the direct input of heavy metals from precipitation or coal dust from the local mines as the concentrations of the same elements in these materials are three orders of magnitude lower. It is only in the vicinity of seabird colonies that this pattern is shown and the concentrations (μg g^{-1} dry weight) of lead, copper and zinc in fecal samples of kittiwake and glaucous gull are in the range 17-32, 35-65 and 63-260, respectively. The concentrations of these heavy metals in the feces are close to those found in the surface layers of humus closest to the cliff. This indicates that the seabirds are acting as a vector for the movement of heavy metals between the marine and terrestrial ecosystems.

Hebert, C. E.; Norstrom, R. J.; Simon, M.; Braune, B. M.; Weseloh, D. V.; and Macdonald, C. R. 1994. Temporal trends and sources of PCDDs and PCDFs in the Great Lakes: Herring gull egg monitoring, 1981-1991. Environ. Sci. Technol. 28(7):1268-1277.
Rec #: 210
KEYWORDS: egg; Great Lakes; North America; organochlorines; PCDDs; PCDFs; herring gull; *Larus argentatus*; spatial variations; temporal trends; biomagnification
ABSTRACT: Levels of individual polychlorinated dibenzodioxin (PCDD) congeners were measured in pooled herring gull (*Larus argentatus*) eggs collected from colonies in the Great

Annotated Bibliography (Cont.).

Lakes and the St. Lawrence River between 1981 and 1991. Polychlorinated dibenzofurans (PCDFs) were quantified from 1984 to 1991. 2,3,7,8-TCDD, 1,2,3,7,8-PnCDD, 1,2,3,6,7,8-HxCDD, and 2,3,4,7,8-PnCDF were detectable in all samples;1,2,3,4,6,7,8-HpCDD, OCDD, 2,3,7,8-TCDF, 1,2,3,4,7,8-HxDF, and 1,2,3,6,7,8-HxDF were frequently detected. Eggs from Saginaw Bay, Lake Huron, had the highest PCDD/PCDF levels. Levels of TCDD, PnCDD, and HxCDD declined in most colonies between 1981 and 1984. There were no obvious temporal trends after 1984. Using multivariate analyses, colonies were separated into two classes based upon differences in egg bioaccumulation patterns. PCDD and PCDF patterns in a variety of potential sources were compared to these two herring gull classes. Patterns of PCDD accumulation were similar in herring gull eggs, lake trout, and walleye emphasizing the similarity of these species as regional indicators of PCDD/PCDF contamination.

Hebert, C. E.; Shutt, J. L.; and Norstrom, R. J. 1997. Dietary changes cause temporal fluctuations in polychlorinated biphenyl levels in herring gull eggs from Lake Ontario. Environ. Sci. Technol. 31(4):1012-1017.
Rec #: 231
KEYWORDS: North America; Great Lakes; organochlorines; PCBs; temporal trends; stable isotopes; egg; herring gull; *Larus argentatus;* biomagnification
NOTES: FIGURES ONLY
ABSTRACT: After adjusting Lake Ontario herring gull (*Larus argentatus*) egg PCB concentrations for the influence of time, an analysis was conducted to explain the remaining variation in annual egg PCB concentrations. In years with cold winters and/or high alewife (*Alosa pseudoharengus*) abundance, egg PCB concentrations were greater than predicted. PCB levels were also greater than predicted in years when alewife condition was low. Increasing the proportion of alewives in the gull's diet may lead to increased PCB levels in eggs. Stable isotope analysis $^{15}N/^{14}N$ of herring gull eggs provided evidence supporting this hypothesis. Consumption of alewives by gulls (as influenced by gull metabolism, alewife abundance/condition, and alewife overwinter mortality) and alewife population characteristics (growth rates and age distribution) may be the keys to explaining fluctuations in Lake Ontario herring gull egg PCB levels.

Hebert, Craig E.; Norstrom, Ross J.; and Weseloh, D. V. Chip. 1999. A quarter century of environmental surveillance: The Canadian Wildlife Service's Great Lakes Herring Gull Monitoring Program. Environ. Rev. 7(4):147-166.
Rec #: 426
KEYWORDS: herring gull; *Larus argentatus*; Great Lakes; Canada; North America; egg; temporal trends; organochlorines; mirex; benzenes; PBDEs; PCDDs; PCBs; DDTs; hydrocarbons; PAHs; toxicology; effects; spatial variations
NOTES: FIGURES ONLY
ABSTRACT: The Great Lakes Herring Gull Monitoring Program has annually provided information concerning levels of environmental contaminants in herring gull eggs since 1974, making it one of the longest running biomonitoring programs in the world. The program was initiated in response to observations of poor reproductive success in colonial waterbirds on the Great Lakes. Initial studies examined the role of halogenated hydrocarbons (HAHs) in causing this reproductive dysfunction. By the late 1970s, reproductive success in herring gulls had improved greatly and emphasis was placed on developing more sensitive indicators to measure

Annotated Bibliography (Cont.).

the subtle effects associated with HAH exposure. Geographical and temporal trends in Great Lakes contamination were also elucidated. Analysis of herring gull tissues led to the identification of HAHs (Mirex, Photomirex, polynuclear aromatic hydrocarbons, chlorobenzenes, dioxins) previously undetected in Great Lakes upper trophic level biota. Data collected as part of this program have improved our understanding of contaminant sources and fate in the Great Lakes and have provided us with a means to assess our progress in controlling contaminant inputs. The extensive nature of this dataset has allowed detailed examination of the factors that regulate contaminant levels in this species. Most monitoring programs rely on less extensive datasets for the interpretation of environmental trends and may benefit from the mechanisms identified here. Research has also identified other stressors, e.g., dietary deficiencies, that may affect the success of Great Lakes herring gull populations. Ongoing monitoring of this species will continue to provide new insights into the dynamic Great Lakes ecosystem

Hebert, Craig E.; Norstrom, Ross J.; Zhu, Jiping; and Macdonald, Colin R. 1999. Historical changes in PCB patterns in Lake Ontario and Green Bay, Lake Michigan, 1971 to 1982, from herring gull egg monitoring data. J. Great Lakes Res. 25(1):220-233.
Rec #: 331
KEYWORDS: herring gull; *Larus argentatus*; egg; organochlorines; PCBs; Great Lakes; North America; temporal trends; spatial variations
ABSTRACT: Patterns of PCB congener bioaccumulation were examined in archived herring gull (Larus argentatus) eggs collected from Big Sister Island in Green Bay, Lake Michigan, and Scotch Bonnet Island in Lake Ontario from 1971 to 1982 as part of the Canadian Wildlife Service's Great Lakes Herring Gull Monitoring Program. Concentrations of 97 PCB congeners were measured. From 1971 to 1982, ecological half-lives of most congeners, particularly the tri- through hexachlorobiphenyls, were greater in eggs from Green Bay than Lake Ontario. Comparing sum PCB levels in eggs collected in 1971 and 1982, concentrations declined 80 % at Scotch Bonnet Island and 74 % at Big Sister Island. PCB congener patterns were different in eggs from the two colonies. Principal components analysis showed that inter-site differences in congener patterns became more apparent after 1976. This indicated that regional PCB sources were the most influential in determining patterns of biologically-available PCBs during the 1971 to 1982 period in these two lakes, via recycling of historical PCBs from sediments or gradually decreasing loading. Trend analysis of selected congeners specific to Aroclors 1242, 1254, and 1260 revealed that the rapid decline of less chlorinated congeners, observed from 1971 to 1976 in Lake Ontario, was explained by a decrease in loading of Aroclor 1242 to the lake. At both colonies, ecological half life of the congeners was significantly ($p < 0.001$) correlated with log K_{ow} and with -log HLC. Changes in PCB composition, after 1976 in Lake Ontario and from 1971 to 1982 in Green Bay, could be explained by differences in the physical behavior of individual congeners affecting removal by volatilization and sedimentation.

Heidmann, W. A.; Beyerbach, M.; Böckelmann, W.; Büthe, A.; Knüwer, H.; Peterat, B.; and Rüssel-Sinn, H. A. 1987. Chlorierte Kohlenwasserstoffe und Schwermetalle in tot an der deutschen Nordseeküste aufgefundensen Seevögeln. Die Vogelwarte. 34:126-133.
Rec #: 554
KEYWORDS: organochlorines; PCBs; HCB; DDTs; dieldrin; metals; Hg; Pb; Cd; liver; black-legged kittiwake; *Rissa tridactyla*; common eider; *Somateria mollissima*; common murre; *Uria*

Annotated Bibliography (Cont.).

aalge; shelduck; *Tadorna tadorna*; herring gull; *Larus argentatus*; common gull; *Larus canus*; black-headed gull; *Larus ridibundus*; oystercatcher; *Haematopus ostralegus*; red-throated loon; *Gavia stellata*; North Sea; Germany; Europe; biomagnification

ABSTRACT: Residues of chlorinated hydrocarbons and heavy metals in 149 livers of nine species of birds living on sea and coast were determined. These birds were found dead on the North Sea coast mostly in 1980 to 1984 in the wintertime. Almost all birds were bad nourished. Relatively low residues of ΣDDT, HCB, and Dieldrin were found. However, the PCB residues were highest. The burden towards chlorinated hydrocarbons of the species examined corresponds with their trophic position. The situation of the residues of heavy metals is different: differences in nutrition cannot serve as an unequivocal explanation of the residues found. Possible reasons of the different burden are discussed.

Heinz, Gary H. 1998. Contaminant effects on Great Lakes fish-eating birds: A Population Perspective. In: Kendall, Ronald J., Richard L. Dickerson, John P. Giesy, William P. Suk (Eds.). Principles and Processes for Evaluating Endocrine Disruption in Wildlife, Proceedings From Principles and Processes for Evaluating Endocrine Disruption in Wildlife; March 1996; Kiawah Island, S. C. Pensacola, FL: Society of Environmental Toxicology and Chemistry (SETAC). 515 p. 141-153.
Rec #: 400
KEYWORDS: review; Great Lakes; North America; effects; reproductive success; eggshell thickness; abnormalities; EDCs; organochlorines; DDTs; double-crested cormorant; *Phalacrocorax auritus*; ring-billed gull; *Larus delawarensis*; bald eagle; *Haliaeetus leucocephalus*; herring gull; *Larus argentatus*; black-crowned night-heron; *Nycticorax nycticorax*; common tern; *Sterna hirundo*; Caspian tern; *Sterna caspia*; Forster's tern; *Sterna forsteri*
NOTES: SUPPLEMENTAL
ABSTRACT: A review and discussion with many references. Endocrine effects may have contributed to declines in fish-eating bird populations in the Great Lakes, but the greatest harm probably was caused by DDE-induced eggshell thinning. Following the ban on DDT in 1972, DDE levels in the Great Lakes declined, eggshells of birds began to get thicker, and reproductive success improved. Populations of double-crested cormorants (*Phalacrocorax auritus*) and ring-billed gulls (*Larus delawarensis*) have increased dramatically since the bans on DDT and other organochlorine pesticides. Bald eagles (*Haliaeetus leucocephalus*) still may not be reproducing at a completely normal rate along the shores of the Great Lakes, but success is much improved. Other species, such as herring gulls (*Larus argentatus*) and black-crowned night-herons (*Nycticorax nycticorax*), seem to be having improved reproductive success, but data on Great Lakes-wide population changes are incomplete. Reproductive success of common terns (*Sterna hirundo*), Caspian terns (*Sterna caspia*), and Forster's terns (*Sterna forsteri*) seems to have improved in recent years, but again, data on population changes are not very complete, and these birds face many habitat-related problems as well as contaminant problems. Although contaminants are still producing toxic effects, and these effects may include endocrine dysfunction, fish-eating birds in the Great Lakes seem largely to be weathering these effects, at least as far as populations are concerned.

HELCOM. 1996. Third Periodic Assessment of the State of the Marine Environment of the Baltic Sea, 1989-93. Helsinki : HELCOM. No. 64 A & B:28 + 262 pp.

Annotated Bibliography (Cont.).

Rec #: 280
KEYWORDS: oil; organochlorines; PBDEs; PCDDs; PCDFs; common murre; *Uria aalge*; Baltic Sea; Europe; temporal trends
NOTES: SUPPLEMENTAL; FIGURES ONLY
ABSTRACT: In the framework of the Convention on the Protection of the Marine Environment of the Baltic Sea Area (Helsinki, 1974, revised 1992), the state of the Baltic Sea is regularly assessed in about 5-year intervals. Hundreds of experts from a multitude of disciplines participate in this assessment process. The outcome is a basically scientific background document covering most of the topics (meteorology, hydrology, hydrography, contaminants and inputs, fish stocks and diseases, nature conservation and biodiversity, special problems, and new monitoring aspects) related to the state of the Baltic Sea. The respective situation, trends and tendencies are highlighted. The multidisciplinary approach attempts to secure a balanced view on the different problems. As typical for any science, a compilation of today's results does not provide 'final answers' even when the Baltic Sea belongs to one of the longest and most intensively studied sea areas of the world. The assessments are understood to be a time limited consensus which has been reached between scientists participating in long-term studies.

Henny, Charles J.; Blus, Lawrence J.; Thompson, Steven P.; and Wilson, Ulrich W. 1989. Environmental contaminants, human disturbance and nesting of double-crested cormorants in northwestern Washington. Colonial Waterbirds. 12(2):198-206.
Rec #: 189
KEYWORDS: double-crested cormorant; *Phalacrocorax auritus*; North America; USA; WA; effects; reproductive success; eggshell thickness; organochlorines; DDTs; PCBs; metals; Hg; Se; Al; Cd; Cu; Fe; Mn; Zn; Sb; As; egg; liver; kidney; muscle; spatial variations
ABSTRACT: Double-crested Cormorants (*Phalacrocorax auritus*) in extreme northwestern Washington produced few young (0.27/occupied nest) in 1984; the clutch size was generally small and eggs, if laid at all, were laid later than usual. Residues (geometric means, wet weight) of DDE (0.58 and 0.59 ppm) in eggs from Colville Island and Protection Island were lower than from other locations in the Pacific Northwest, while PCBs (2.19 and 1.37 ppm) were similar to those at most locations. Both contaminants in 1984 were below levels associated with reproductive problems. Eggs also contained concentrations of mercury (0.26 and 0.27 ppm) and selenium (0.31 and 0.28 ppm) below levels associated with reproductive problems. The distribution of nesting colonies in the study area changed dramatically since 1984. The cormorants were most likely responding to increased human disturbance in the San Juan Islands, coupled to additional protection and reduced human activity on Protection and Smith Islands.

Henny, Charles J.; Hill, Elwood F.; Hoffman, David J.; Spalding, Marilyn G.; and Grove, Robert A. 2002. Nineteenth century mercury: Hazard to wading birds and cormorants of the Carson River, Nevada. Ecotoxicology. 11(4):213-231.
Rec #: 397
KEYWORDS: metals; Hg; organometallics; organic Hg; stomach contents; liver; kidney; brain; blood; feather; egg; toxicology; histopathology; snowy egret; *Egretta thula*; black-crowned night-heron; *Nycticorax nycticorax*; double-crested cormorant; *Phalacrocorax auritus*; NV; USA; North America; age variations; effects
ABSTRACT: Contemporary mercury interest relates to atmospheric deposition, contaminated fish stocks and exposed fish-eating wildlife. The focus is on methylmercury (MeHg) even though

Annotated Bibliography (Cont.).

most contamination is of inorganic (IoHg) origin. However, IoHg is readily methylated in aquatic systems to become more hazardous to vertebrates. In response to a classic episode of historical (1859–1890) IoHg contamination, we studied fish-eating birds nesting along the lower Carson River, Nevada. Adult double-crested cormorants (*Phalacrocorax auritus*), snowy egrets (*Egretta thula*) and black-crowned night-herons (*Nycticorax nycticorax*) contained very high concentrations of total mercury (THg) in their livers (geo. means 134.8µg/g wet weight (ww), 43.7 and 13.5, respectively) and kidneys (69.4, 11.1 and 6.1, respectively). Apparently tolerance of these concentrations was possible due to a threshold-dependent demethylation coupled with sequestration of resultant IoHg. Demethylation and sequestration processes also appeared to have reduced the amount of MeHg redistributed to eggs. However, the relatively short time spent by adults in the contaminated area before egg laying was also a factor in lower than expected concentrations of mercury in eggs. Most eggs (100 % MeHg) had concentrations below 0.80µg/g ww, the putative threshold concentration where reproductive problems may be expected; there was no conclusive evidence of mercury-related depressed hatchability. After hatching, the young birds were fed diets by their parents averaging 0.36–1.18µgMeHg/g ww through fledging. During this four to six week period, accumulated mercury concentrations in the organs of the fledglings were much lower than found in adults, but evidence was detected of toxicity to their immune (spleen, thymus, bursa), detoxicating (liver, kidneys) and nervous systems. Several indications of oxidative stress were also noted in the fledglings and were most apparent in young cormorants containing highest concentrations of mercury. This stress was evidenced by increased thiobarbituric acid-reactive substances, low activities of enzymes related to glutathione metabolism and low levels of reduced thiols, plus an increase in the ratio of oxidized to reduced glutathione. At lower concentrations of mercury, as was found in young egrets, we observed elevated activities of protective hepatic enzymes, which could help reduce oxidative stress. Immune deficiencies and neurological impairment of fledglings may affect survivability when confronted with the stresses of learning to forage and the ability to complete their first migration.

Henny, Charles J.; Rudis, Deborah D.; Roffe, Thomas J.; and Robinson-Wilson, Everett. 1995. Contaminants and sea ducks in Alaska and the circumpolar region. Environ. Health Perspect. 103(4):41-49.
Rec #: 46
KEYWORDS: AK; USA; North America; effects; metals; Cd; Se; Cu; Hg; Fe; Mg; Mn; Zn; Al; kidney; liver; surf scoter; *Melanitta perspicillata*; white-winged scoter; *Melanitta fusca*; black scoter; *Melanitta nigra*; long-tailed duck; *Clangula hyemalis*; spectacled eider; *Somateria fischeri*; Steller's eider; *Polysticta stelleri*; histopathology; effects; disease
ABSTRACT: We review nesting sea duck population declines in Alaska during recent decades and explore the possibility that contaminants may be implicated Aerial surveys of the surf scoter (*Melanitta perspicillata*), white-winged scoter (*M. fusca*), black scoter (*M. nigra*), oldsquaw (*Clangula hyemalis*), spectacled eider (*Somateria fischeri*), and Steller's eider (*Polysticta stelleri*) show long-term breeding population declines, especially the latter three species. The spectacled eider was recently classified threatened under the Endangered Species Act. In addition, three other diving ducks, which commonly winter in coastal areas, have declined from unknown causes. Large die-offs of all three species of scoters during molt, a period of high energy demand, were documented in August 1990, 1991, and 1992 at coastal reefs in southeastern Alaska. There was no evidence of infectious diseases in those scoters. The die-offs may or may not be associated with the long-term declines. Many scoters had elevated renal concentrations of

Annotated Bibliography (Cont.).

cadmium (high of 375 µg/g dry weight [dw]). Effects of cadmium in sea ducks are not well understood. Selenium concentrations in livers of nesting white-winged scoters were high; however, the eggs they laid contained less selenium than expected based on relationships for freshwater bird species. Histological evaluation found a high prevalence of hepatocellular vacuolation (49 %), a degenerative change frequently associated with sublethal toxic insult. Cadmium and selenium mean liver concentrations were generally higher in those birds with more severe vacuolation; however, relationships were not statistically significant. We do not know if sea duck population declines are related to metals or other contaminants.

Henriksen, E. O.; Gabrielsen, G. W.; Skaare, J. U.; Skjegstad, N.; and Jenssen, B. M. 1998. Relationships between PCB levels, hepatic EROD activity and plasma retinol in glaucous gulls, *Larus hyperboreus*. Mar. Environ. Res. 46(1-5):45-49.
Rec #: 322
KEYWORDS: glaucous gull; *Larus hyperboreus*; liver; blood; toxicology; EROD; CYP1A; organochlorines; PCBs; retinol; Svalbard; Norway; Europe; effects
NOTES: FIGURES ONLY
ABSTRACT: Fifteen adult glaucous gulls, *Larus hyperboreus*, were captured in the vicinity of Ny-Ĺlesund, Svalbard. The birds were kept captured and fed a diet of polar cod, *Boreogadus saida*, caught off the coast of Spitsbergen. After 40 days of captivity, the birds were killed. The presence of hepatic homologues to mammalian CYP1A and CYP2B proteins was demonstrated by Western blotting. High concentrations of PCBs were found in the liver (median ΣPCB111 µg g^{-1} lipid), but hepatic EROD activity was low and significantly negatively correlated with concentrations of PCB-congener numbers 28, 47, 66, 74, 105 and 187 in the liver. Plasma retinol did not correlate significantly with liver concentrations of PCBs, but a significant positive correlation was found between retinol concentration in plasma and hepatic EROD activity.

Henriksen, E. O.; Gabrielsen, G. W.; Trudeau, S.; Wolkers, J.; Sagerup, K.; and Skaare, J. U. 2000. Organochlorines and possible biochemical effects in glaucous gulls (*Larus hyperboreus*) from Bjørnøya, the Barents Sea. Arch. Environ. Contam. Toxicol. 38(2):234-243.
Rec #: 330
KEYWORDS: glaucous gull; *Larus hyperboreus*; organochlorines; PCBs; HCB; chlordanes; DDTs; mirex; stable isotopes; toxicology; porphyrin; retinol; EROD; cytochrome P_{450}; liver; Norway; Europe; effects
ABSTRACT: To study possible biochemical effects of organochlorine contaminants (OCs) in glaucous gulls (Larus hyperboreus), 40 adult individuals were collected from colonies on Bjørnøya in the Barents Sea. OCs (four pesticides and nine PCB congeners), microsomal 7-ethoxyresorufin O-deethylase (EROD) activity, microsomal testosterone hydroxylation, highly carboxylated porphyrins (HCPs), retinol, and retinyl palmitate were quantified in liver samples. The hepatic vitamin A stores in glaucous gulls were larger than in herring gulls (Larus argentatus) from other studies conducted in contaminated locations in North America. No significant relationships were found between liver retinoid concentrations and OC levels. The hepatic EROD activity was low compared to other studies on fish-eating birds and only marginally associated with PCB levels. Microsomal testosterone hydroxylase activity was only observed at the 6beta-position and could not be related to OC levels. The low P_{450}-associated enzyme activities in the glaucous gull suggests that they have a low capacity for metabolizing OCs, which may contribute to the high accumulation of OCs in this species. HCPs were only

Annotated Bibliography (Cont.).

elevated (138 pmol g^{-1}) in the sample with highest OC levels, whereas the remaining samples contained low levels of HCPs (<30 pmol g^{-1}). The weak association between EROD activity and PCB levels and the low level of HCPs suggest that these biochemical parameters were unaffected by OCs in most of the sampled gulls. Thus, the glaucous gull seems not to be particularly sensitive toward Ah-receptor mediated effects.

Henriksen, Espen O.; Brunström, Björn; Skaare, Janneche Utne; and Gabrielsen, Geir Wing. 1998. Bioassay-derived 2,3,7,8-tetrachlorodibenzo-p-dioxin equivalents and mono-ortho polychlorinated biphenyl concentrations in liver of glaucous gulls, *Larus hyperboreus*, from Svalbard. Organohalogen Compounds. 39:415-418.
Rec #: 323
KEYWORDS: glaucous gull; *Larus hyperboreus*; liver; Svalbard; Norway; Europe; toxicology; EROD; TEQs; PCBs; organochlorines; effects
ABSTRACT: Bioassay-derived 2,3,7,8-tetrachlorodibenzo-p-dioxin-equivalent (Bio-TEQ) in glaucous gull liver extracts were compared with the concentration of mono-ortho polychlorinated biphenyls (PCB) in the samples. The bioassay detection is based on 7-ethoxyresorufin O-deethylase (EROD) induction in cultured chick embryo livers. Glaucous gulls used were either captured and fed with polar cod for 24-41 days or found dead in the same area. The Bio-TEQ were 5-254 ng/g lipid (extractable lipids 3.33-5.83 %; wet weight Bio-TEQ 205-8450 ppt), with the 2 highest concentrations in birds found dead. PCB-118 was the major mono-ortho PCB. Based on TEQ, PCB-156 and PCB-105 were more important. The concentration of mono-ortho PCB in the gulls found dead were 3-13 times higher than the concentration in the gulls that were kept in captivity. Assuming that the induction effects of the individual PCB-congeners are additive, the TEQ concentration associated with mono-ortho PCB in the captive gulls was 2.8 ng/g lipid, which is 11 % of the Bio-TEQ. Non-ortho PCB contributed mainly to the Bio-TEQ, and this contribution is almost entirely due to PCB-126. Mono-ortho PCB explain 74 % of the variation in Bio-TEQ (based on Log10-transformed values). The coefficient of detection (r^2) for the correlation between Log(Bio-TEQ) and Log(PCB-153) was 0.76. In spite of high Bio-TEQ levels in the liver extracts, hepatic EROD activities in the same individuals were low (≤70 pmol/min mg protein).

Henriksen, Espen O.; Gabrielsen, Geir W.; and Skaare, Janneche Utne. 1998. Validation of the use of blood samples to assess tissue concentrations of organochlorines in glaucous gulls, *Larus hyperboreus*. Chemosphere. 37(13):2627-2643.
Rec #: 112
KEYWORDS: blood; organochlorines; HCB; HCHs; chlordanes; DDTs; mirex; PCBs; biomagnification; brain; liver; fat; Norway; Svalbard; Europe; glaucous gull; *Larus hyperboreus*
ABSTRACT: Fifteen adult glaucous gulls, *Larus hyperboreus*, were captured near Ny-Alesund, Svalbard. The birds were kept in captivity for 24 - 41 days and fed a diet of polar cod, *Boreogadus saida*. A range of organochlorines (OCs) were quantified in blood, brain, liver, and subcutaneous fat tissue. For more than 80 % of the quantified OCs, r^2 values >0.75 were found for the blood-liver concentration correlations. Repeated sampling revealed intra-individual temporal variability in blood OC concentrations. Much of the temporal variability in OC blood concentrations was associated with changes in nutritional condition.

Henriksen Espen O.; Gabrielsen, Geir Wing; and Skaare, Janneche Utne. 1996. Levels and

Annotated Bibliography (Cont.).

congener pattern of polychlorinated biphenyls in kittiwakes (*Rissa tridactyla*), in relation to mobilization of body-lipids associated with reproduction. Environ. Pollut. 92(1):27-37 .
Rec #: 315
KEYWORDS: brain; liver; fat; organochlorines; PCBs; seasonal variations; reproductive success; elimination; metabolism; black-legged kittiwake; *Rissa tridactyla*; Norway; Europe
ABSTRACT: Three groups of female kittiwakes (*Rissa tridactyla*) were collected in a North-Norwegian colony: (i) before breeding; (ii) immediately after egg-laying; and (iii) late in the chick-rearing period. Concentrations of 21 selected individual polychlorinated biphenyls (PCBs) were determined in liver, brain and fat tissues by capillary gas-chromatography (GC-ECD). This was done in order to investigate how the mobilization of lipids associated with breeding influences concentrations and compositions of PCBs in these tissues. The results indicate an average decrease in body mass from pre-breeding to late chick-rearing of almost 20 %. During this period, the mean concentration of PCBs in brain tissue approximately quadruples. This increase can be attributed to the redistribution of PCBs from utilized depot fat to metabolizing organs. A strong negative correlation was found between body mass and lipid weight PCB-concentrations in all three tissues. The relative amounts of different PCB-congeners were quite similar in different tissues, and showed only minor changes during the breeding period.

Henshel, Diane S.; Martin, J. William; Norstrom, Ross J.; Elliott, John; Cheng, Kimberly M.; and DeWitt, Jamie C. 1997. Morphometric brain abnormalities in double-crested cormorant chicks exposed to polychlorinated dibenzo-p-dioxins, dibenzofurans, and biphenyls. J. Great Lakes Res. 23(1):11-26.
Rec #: 99
KEYWORDS: organochlorines; PCBs; PCDDs; PCDFs; toxicology; TEQs; brain; Great Lakes; North America; double-crested cormorant; *Phalacrocorax auritus*; abnormalities; effects
NOTES: NO TABLES
ABSTRACT: We examined brains from double-crested cormorant (*Phalacrocorax auritus*) hatchlings from colonies known to be contaminated with low, moderate, and high levels of PCBs, as well as PCDDs and PCDFs, in order to evaluate whether such chemical exposure during embryonic development would result in a TCDD- and TEQ-related induction of gross brain asymmetry as described previously in similarly exposed great blue herons. Our results demonstrate that cormorant embryos exposed in ovo to elevated mixtures of PCBs, PCDDs, and PCDFs in the environment are likely to hatch with asymmetric brains, when the width (medial-lateral), angle (oblique to the medio-lateral axis), height (rostro-caudal), and depth (dorso-ventral) were measured both from dorsal and ventral perspectives. The degree and frequency of this asymmetry correlates to dose based on both TCDD alone and TEQs. This indicates that the TEQ-related brain asymmetry is not a species-specific response. The gross brain asymmetries are present in hatchlings and are readily quantifiable. Thus, they could potentially be used as a biomarker of the effects of TCDD-related compounds on neuromorphological development.

Herzke, Dorte; Gabrielsen, Geir Wing; Evenset, Anita; and Burkow, Ivan C. 2003. Polychlorinated camphenes (toxaphenes), polybrominated diphenylethers and other halogenated organic pollutants in glaucous gull (*Larus hyperboreus*) from Svalbard and Bjørnøya (Bear Island). Environ. Pollut. 121(2):293-300.
Rec #: 378
KEYWORDS: glaucous gull; *Larus hyperboreus*; liver; intestine; organochlorines; PCBs;

Annotated Bibliography (Cont.).

chlordanes; HCB; mirex; DDTs; PBDEs; toxaphene; Norway; Svalbard; Europe; gender differences

ABSTRACT: The levels of polychlorinated camphenes (toxaphenes) were studied in liver samples from 18 glaucous gulls (*Larus hyperboreus*) from Bjørnøya (74°N, 19°E) and four individuals from Longyearbyen (78°N, 15°E). Additional brominated flame retardants (BFRs), PCBs and chlorinated pesticides were studied in liver and intestinal contents of 15 of the glaucous gulls from Bjørnøya. Of the analyzed BFRs only 2,2',4,4'-tetra- and 2,2',4,4',5-pentabrominated diphenylethers (PBDE 47 and 99) could be detected. The concentrations were 2-25 ng/g ww. In addition, high resolution measurements with GC/HRMS revealed the existence of several, not quantified, PBDEs and polybrominated biphenyls (PBBs) congeners in the samples. B9-1679 and B8-1413 were the dominating toxaphenes with median concentrations of 8 and 15 ng/g ww. Concentrations of toxaphenes and PBDEs were ≤100-times lower than the concentrations of PCB and some of the pesticides. PCB and p,p'-DDE constituted 90 % of the contaminants found.

Hesse, Larry W.; Brown, Robert L.; and Heisinger, James F. 1975. Mercury contamination of birds from a polluted watershed. J. Wildl. Manage. 39(2):299-304.
Rec #: 448
KEYWORDS: metals; Hg; muscle; liver; kidney; SD; USA; North America; double-crested cormorant; *Phalacrocorax auritus*; biomagnification
NOTES: NO TABLES
ABSTRACT: Total Hg [7439-97-6] concentrations in the muscle, liver, and kidney were detected in 22 species of birds collected from Hg-contaminated western South Dakota watershed. Elevated Hg levels were detected in the fish-eating birds, especially in double-crested cormorants (*Phalacrocorax auritus*). Levels in nonfish-eating birds were lower, but the mean residues were significantly higher than the mean levels of control birds. In general, greater Hg accumulations were found in the livers of fish-eating birds and the kidneys of nonfish-eaters. A significant correlation was found for Hg residues in muscle-liver, muscle-kidney, and liver-kidney tissue combinations.

Hogstad, O.; Nygård, T.; Gätzschmann, P.; Lierhagen, S.; and Thingstad, P. G. 2003. Bird skins in museum collections: Are they suitable as indicators of environmental metal load after conservation procedures? Environ. Monit. Assess. 87(1):47-56.
Rec #: 480
KEYWORDS: goshawk; *Accipiter gentilis*; eagle owl; *Bubo bubo*; common pigeon; *Columba livia;* common eider; *Somateria mollissima*; feather; metals; Cd; Hg; Pb; Mg; Al; Mn; Fe; Cu; Zn; As; Norway; Europe
NOTES: SUPPLEMENTAL
ABSTRACT: To find out whether modern conservation treatments alter the level of metals in feathers, the authors analyzed the content of 10 metals in feathers before and after skins were washed with detergent and treated with Eulan U-33 (a commonly used preservative at museums). Feathers of 31 birds of Goshawk *Accipiter gentilis*, Eagle Owl *Bubo bubo*, Feral Pigeon *Columba livia* domest. and Common Eider *Somateria mollissima* were analyzed. The authors found that in most cases metals were partly washed out of the feathers, but the effects were related to species and type of feather. The value of bird skins as indicators of environmental

Annotated Bibliography (Cont.).

metal load is therefore affected by this treatment. It is recommended that the conservation techniques used at museums should be reconsidered if skins are intended for specimen banking for future reference in environmental monitoring schemes and research.

Hollmén, Tuula; Franson, J. Christian; Poppenga, Robert H.; Hario, Martti; and Kilpi, Mikael. 1998. Lead poisoning and trace elements in common eiders *Somateria mollissima* from Finland. Wildl. Biol. 4(4): 193-203.
Rec #: 511
KEYWORDS: metals; As; Cd; Cr; Cu; Fe; Pb; Mg; Mn; Hg; Mo; Se; Zn; common eider; *Somateria mollissima*; liver; blood; Finland; Europe; effects; abnormalities; toxicology; age variations; gender differences; bioaccumulation
ABSTRACT: We collected carcasses of 52 common eider *Somateria mollissima* adults and ducklings and blood samples from 11 nesting eider hens in the Gulf of Finland near Helsinki in 1994, 1995 and 1996. Samples of liver tissue were analysed for arsenic, cadmium, chromium, copper, iron, lead, magnesium, manganese, mercury, molybdenum, selenium and zinc. Blood was analysed for lead, mercury and selenium. Most of the 21 adults examined at necropsy were emaciated with empty gizzards, and no ingested shotgun pellets or other metal were found in any of the birds. Three adult females had a combination of lesions and tissue lead residues characteristic of lead poisoning. Two of these birds had acid-fast intranuclear inclusion bodies in renal epithelial cells and high concentrations of lead (73.4 and 73.3 ppm; all liver residues reported on dry weight basis) in their livers. The third was emaciated with a liver lead concentration of 47.9 ppm. An adult male had a liver lead concentration of 81.7 ppm, which is consistent with severe clinical poisoning. Two other adults, one male and one female, had liver lead concentrations of 14.2 and 8.03 ppm, respectively. Lead concentrations in the blood of hens ranged from 0.11 to 0.63 ppm wet weight. Selenium residues of ≥60 ppm were found in the livers of five adult males. Selenium concentrations in the blood of hens ranged from 1.18 to 3.39 ppm wet weight. Arsenic concentrations of 27.5-38.5 ppm were detected in the livers of four adult females. Detectable concentrations of selenium, mercury and molybdenum were found more frequently in the livers of adult males arriving on the breeding grounds than in incubating females, while the reverse was true for arsenic, lead and chromium. Mean concentrations of selenium, copper and molybdenum were higher in the livers of arriving males than in the livers of incubating hens, but hens had greater concentrations of iron and magnesium. Concentrations of trace elements were lower in the livers of ducklings than in the livers of adults

Holm, E.; Persson, B. R. R.; Hallstadius, L.; Aarkrog, A.; and Dahlgaard, H. 1983. Radiocesium and transuranium elements in the Greenland and Barents Seas. Oceanologica Acta. 6(4):457-462.
Rec #: 461
KEYWORDS: radionuclides; Greenland; North America; Barents Sea; Atlantic Ocean; Europe; Sweden; Norway; Svalbard; common eider; *Somateria mollissima*; black-legged kittiwake; *Rissa tridactyla*; thick-billed murre; *Uria lomvia*; northern fulmar; *Fulmarus glacialis*; carcass; biomagnification; spatial variations
ABSTRACT: The results of ^{137}Cs, ^{239,240}Pu, and ^{241}Am analyses in seawater, sediment, and biota from the Swedish Arctic expedition, Ymer-80, in the Barents Sea and Greenland Sea show that ^{137}Cs released from European nuclear fuel reprocessing facilities is effectively transferred by the Gulf stream to these latitudes. The transuranium elements ^{239,240}Pu and ^{241}Am present in the investigated areas originate mainly from local fallout, but a substantial mixing with Atlantic

Annotated Bibliography (Cont.).

waters containing these elements was observed The levels of ^{137}Cs found in biota like algae, seals, polar bears, and birds are quite moderate and do not differ much from those found in other areas, such as Greenland.

Holt, Gunnar; Frøslie, Arne; and Norheim, Gunnar. 1979. Mercury, DDE, and PCB in the avian fauna of Norway. Acta Vet. Scand. Suppl. 70:1-28.
Rec #: 526
KEYWORDS: metals; Hg; organochlorines; DDTs; PCBs; Norway; Europe; grey heron; *Ardea cinerea*; osprey; *Pandion haliaetus*; white-tailed sea eagle; *Haliaeetus albicilla*; eagle owl; *Bubo bubo*; effects
ABSTRACT: In general, contamination of Norwegian birds with Hg, DDE (I) [72-55-9], and polychlorinated biphenyls (PCB) appeared to be low. Highest levels were found in the grey heron, osprey, white-tailed eagle, and eagle owl. With the possible exception of the eagle owl, no adverse effects on the bird population of these contaminants were observed

Honda, Katsuhisa; Marcovecchio, Jorge Eduardo; Kan, Shinya; Tatsukawa, Ryo; and Ogi, Haruo. 1990. Metal concentrations in pelagic seabirds from the North Pacific Ocean. Arch. Environ. Contam. Toxicol. 19(5):704-711.
Rec #: 75
KEYWORDS: metals; Hg; Cd; Fe; Zn; Cu; Mn; muscle; liver; kidney; biomagnification; elimination; black-footed albatross; *Diomedea nigripes;* Laysan albatross; *Diomedea immutabilis*; northern fulmar; *Fulmarus glacialis*; sooty shearwater; *Puffinus griseus*; short-tailed shearwater*; Puffinus tenuirostris*; Swinhoe's storm petrel; *Oceanodroma monorhis*; Japanese cormorant; *Phalacrocorax capillatus*; glaucous winged gull*; Larus glaucescens;* black-legged kittiwake; *Rissa tridactyla*; thick-billed murre; *Uria lomvia*; common murre; *Uria aalge*; ancient murrelet*; Synthliboramphus antiquus*; Cassin's auklet*; Ptychoramphus aleuticus*; parakeet auklet; *Aethia psittacula*; crested auklet; *Aethia cristatella*; least auklet; *Aethia pusilla*; rhinoceros auklet; *Cerorhinca monocerata*; horned puffin; *Fratercula corniculata*; tufted puffin; *Fratercula cirrhata*; Pacific Ocean; Bering Sea; China Sea; Sea of Japan; Gulf of Alaska; AK; North America; Japan; China; Asia; spatial variations
ABSTRACT: Concentrations of four essential elements (Fe, Mn, Zn, and Cu) and two toxic metals (Cd and Hg) were measured in selected tissues of 19 pelagic seabird species collected in the North Pacific and neighboring waters. Essential metal concentrations were generally highest in the liver and less variable than toxic metals among species and also within each species. Fe concentrations in the muscle were higher in *Alcidae* than in the other families, whereas the opposite trend was found for Fe and Mn in the liver. Zn concentrations varied among species, depending on the Cd concentrations. Toxic metal concentrations were highest in the liver or kidney and varied widely among species, greatly depending on differences in the diet among species. Extraordinarily high Hg concentrations were found in Black-footed Albatrosses, *Diomedea nigripes*, exceeding 300 µg/g wet weight in some, and seemed to be due to constraints on the elimination of Hg. Also, some geographical differences in Cd and Hg concentrations of the seabirds were observed.

Hontelez, L. C. M. P.; Van den Dungen, H. M.; and Baars, A. J. 1992. Lead and cadmium in birds in the Netherlands: A preliminary survey. Arch. Environ. Contam. Toxicol. 23(4):453-456.

Annotated Bibliography (Cont.).

Rec #: 467
KEYWORDS: metals; Pb; Cd; kidney; liver; bone; buzzard; *Buteo buteo;* grey heron; *Ardea cinerea*; common eider; *Somateria mollissima*; Netherlands; Europe
ABSTRACT: Three birds species (*Buteo buteo*, *Ardea cinerea*, and *Somateria mollissima*) from the Netherlands were investigated for Pb and Cd concentrations in kidneys, livers, and tibiae. Common eiders contained a higher Cd load in liver and kidney than buzzards and grey herons. They had a higher Pb burden in bone than grey herons. The 3 birds species, all standing at the end of a different food chain, seem appropriate indicators for environmental contamination with heavy metals.

Hop, Haakon; Borgå, Katrine; Gabrielsen, Geir Wing; Kleivane, Lars; and Skaare, Janneche Utne. 2002. Food web magnification of persistent organic pollutants in poikilotherms and homeotherms from the Barents Sea. Environ. Sci. Technol. 36(12):2589-2597.
Rec #: 352
KEYWORDS: liver; organochlorines; HCB; HCHs; chlordanes; DDTs; PCBs; stable isotopes; biomagnification; Barents Sea; Norway; Europe; biomagnification; black guillemot; *Cepphus grylle;* glaucous gull*; Larus hyperboreus;* black-legged kittiwake; *Rissa tridactyla;* thick-billed murre; *Uria lomvia*
ABSTRACT: Food web magnification of persistent organic pollutants (POPs) was detectedfor the Barents Sea food web using $\delta^{15}N$ as a continuous variable for assessing trophic levels (TL). The food web investigated comprised zooplankton, ice fauna and fish (poikilotherms, TL 1.7-3.3), and seabirds and seals (homeotherms, TL 3.3-4.2), with zooplankton representing the lowest and glaucous gull the highest trophic level. Concentrations of lipophilic and persistent organochlorines were orders of magnitude higher in homeotherms than in poikilotherms. These compounds had significantly higher rates of increase per trophic level in homeotherms relative to poikilotherms, with the highest food web magnification factors (FWMFs) for cis-chlordane and p,p'-DDE. Some compounds, such as trans-nonachlor and HCB, had similar rates of increase throughout the food web, whereas compounds that are more readily eliminated (γ-HCH) showed no relationship with trophic level. It is preferable to calc. FWMFs with regard to thermal groups, because the different energy requirements and biotransformation abilities between poikilotherms and homeotherms may give different rates of contaminant increase with trophic level. When biomagnification is compared between ecosystems, FWMFs are preferable to single predator-prey biomagnification factors. FWMFs represent a trophic level increase of contaminants that is average for the food chain rather than an increase for a specific predator-prey relationship. The Barents Sea FWMFs were generally comparable to those detectedfor marine food webs with similar food chain lengths in the Canadian Arctic.

Howard, Edwin B.; Esra, Gerald N.; and Young, David. 1979. Acute foodborne pesticide toxicity in cormorants (*Phalacrocorax sp.*) and seagulls (*Larus californicus*). Animals As Monitors of Environmental Pollutants. Symposium on Pathobiology of Environmental Pollutants: Animal Models and Wildlife As Monitors, University of Connecticut, 1977. 290-296.
Rec #: 18
KEYWORDS: effects; biomagnification; organochlorines; DDTs; PCBs; liver; muscle; brain; California gull; *Larus californicus*; Brandt's cormorant; *Phalacrocorax penicillatus*; Guanay cormorant; *Phalacrocorax bougainvillii*
ABSTRACT: This report describes the acute neurological death of captive seagulls and

Annotated Bibliography (Cont.).

cormorants due to the ingestion of bottom-feeding fish that were contaminated with pesticides. The report also describes the presence of organochlorine compounds in marine mammals stranded on the Los Angeles coast. Data on pesticide concentration in bird and fish tissues are presented.

Hughes, Kimberley D.; Weseloh, D. Vaughn; and Braune, Birgit M. 1998. The ratio of DDE to PCB concentrations in Great Lakes herring gull eggs and its use in interpreting contaminants data. J. Great Lakes Res. 24(1):12-31.
Rec #: 211
KEYWORDS: organochlorines; PCBs; DDTs; egg; North America; Great Lakes; spatial variations; herring gull; *Larus argentatus*; bioaccumulation; metabolism
ABSTRACT: The ratio of DDE to PCB (DDE:PCB) concentrations was examined in herring gull (*Larus argentatus*) eggs collected from thirteen sites on the Great Lakes from 1979 to 1996. This ratio has been shown to have had a number of biological interpretations in the past and the data in this study have been used to test the validity of these interpretations. The findings suggest that the consistency of DDE:PCB over many years reflects the relative availability of DDT and PCB in different geographical areas and provides an indication of a bird's general foraging ecology. The ratio can also be used as a reflection of the relative rates of increase or decrease of DDE and PCBs in food over time. A significant increase was found in the ratio in eggs sampled from sites on the upper Great Lakes and Lake Ontario during the study period. This is attributed to PCB levels decreasing faster than DDE levels at these sites. At Lake Erie and its two connecting channels, a significant increase was detected in the ratio but at a rate 56 % or less than that found at other Great Lakes sites. This lower rate is attributed to DDE and PCB levels decreasing at high and equal rates. Similarities/differences in the fates of these contaminants among the Great Lakes would never have been realized upon an examination of individual contaminant levels alone. The ratio is valuable as a measure of the relative exposure of the two contaminants in nonmigratory birds such as herring gulls. The use of the ratio as an indication of contamination movement through the food chain could not be assessed; the ratio cannot be used as a reflection of the interspecific differences in the accumulation and metabolism of the two contaminants. Similarities in the patterns of these ratios shown by eggs collected from sites in close proximity or within the same lake reinforce the fact that herring gulls are important as monitors of regional contaminant conditions.

Hutton, M. 1981. Accumulation of heavy metals and selenium in three seabird species from the United Kingdom. Environ. Pollut. (A Ecol. Biol.). 26:129-145.
Rec #: 539
KEYWORDS: metals; Cd; Hg; Pb; Zn; Se; oystercatcher; *Haematopus ostralegus*; herring gull; *Larus argentatus*; great skua; *Catharacta skua*; biomagnification; age variations; bioaccumulation; gender differences; kidney; liver; metallothionein; toxicology
ABSTRACT: The levels of 3 toxic metals, Cd, Hg, and Pb, and 2 essential elements, Zn and Se, were measured in selected tissues of oystercatcher (*Haematopus ostralegus*), herring gull (*Larus argentatus*), and great skua (*Catharacta skua*). Considerable interspecific differences existed in the contents of toxic metals and these were related both to the feeding behavior of the sea birds and to the extent of habitat contamination. Cd levels in tissues of great skua and oystercatcher increased with age, and kidney values in female oystercatchers were higher than in males. A Cd-binding protein with properties similar to metallothionein was present in great skua kidney.

Annotated Bibliography (Cont.).

Positive correlations were obtained between Zn and Cd in kidney and between Se and Hg in kidney and liver. Thus, these associations may reflect antagonistic interactions between Zn on Cd toxicity and Se on Hg toxicity. The suitability of tissues selected for heavy metal monitoring programs in marine birds is discussed briefly.

Iwata, Hisato; Tanabe, Shinsuke; Iida, Tetsuji; Baba, Norihisa; Ludwig, James P.; and Tatsukawa, Ryo. 1998. Enantioselective accumulation of α-hexachlorocyclohexane in northern fur seals and double-crested cormorants: Effects of biological and ecological factors in the higher trophic levels. Environ. Sci. Technol. 32(15):2244-2249.
Rec #: 401
KEYWORDS: organochlorines; HCHs; double-crested cormorant; *Phalacrocorax auritus;* Great Lakes; North America; muscle; liver; bone; brain; skin; carcass; gender differences; age variations; bioaccumulation; metabolism; effects
ABSTRACT: Tissues of northern fur seals (*Callorhinus ursinus*) from the Pacific coast of Japan and double-crested cormorants (*Phalacrocorax auritus*) from the Great Lakes were analyzed in order to explore the enantioselective accumulation of α-hexachlorocyclohexane (HCH). The effects of biological and ecological factors such as species, tissue, sex, age, feeding habit, and habitat, which may be attributable to the differences in accumulation between enantiomers, were also investigated. The enantiomeric ratios (ERs) of (+)-/(-)-α-HCH in fat tissue of female fur seals, composed of different age groups, collected in 1986 (1.58 ± 0.25) exhibited greater values than those in abiotic and lower trophic levels previously reported. No age trend of ERs was found in female northern fur seals. There appeared to be a temporal transition of ERs in adult female northern fur seals collected in 1971-1988. Regression analysis showed a significant relationship between ERs and feeding habits ($p = 0.003$). Analysis of breast muscle of double-crested cormorants exhibited no sex difference in ERs. ERs (1.26 ± 0.13) in cormorants from Lake Michigan were significantly higher than those (1.01 ± 0.18) from Lake Superior ($p = 0.002$), suggesting the effects of factors such as feeding habit and habitat. Enantiomeric accumulation in the body of double-crested cormorants was tissue-specific. No age trend of ERs was seen in breast muscle of cormorants. The result implies that sexual maturity, aging and breeding activities are less effective for changing ERs. The ERs in higher trophic animals could be influenced by species-specific metabolism and transport process in the body as biological factors and by feeding habit and habitat as ecological factors.

Jansson, Bo; Asplund, Lillemor; and Olsson, Mats. 1987. Brominated flame retardants - Ubiquitous environmental pollutants? Chemosphere. 16(10-12):2343-2349.
Rec #: 505
KEYWORDS: PBDEs; PBBs; common murre; *Uria aalge*; thick-billed murre; *Uria lomvia*; white tailed sea eagle; *Haliaeetus albicilla*; muscle; Baltic Sea; North Sea; Arctic Ocean; Europe; spatial variations
ABSTRACT: Biological samples from marine environments were analyzed for brominated organic compounds. All samples contained a number of such substances and some of them were characterized as polybrominated biphenyls (PBB) and di-Ph ethers (PBDE). The chromatographic pattern of the 2 groups agreed well with the corresponding patterns from 2 com. flame retardants based on these types of substances. In seals, guillemots and white-tailed sea eagles from the Baltic region, the levels of PBB and PBDE were 2-5-fold higher than in the

Annotated Bibliography (Cont.).

animals from the North Sea and the Arctic Ocean. The presence of the compounds in animals from remote areas indicates a world-wide distribution and emphasizes the risk in the use of polyhalogenated hydrocarbons.

Jansson, Bo; Vaz, Reggie; Blomkvist, Gun; Jensen, Sören; and Olsson, Mats. 1979. Chlorinated terpenes and chlordane components found in fish, guillemot and seal from Swedish waters. Chemosphere. 8(4):181-190.
Rec #: 507
KEYWORDS: organochlorines; chlordanes; DDTs; PCBs; toxaphene; common murre; *Uria aalge*; egg; muscle; Baltic Sea; Sweden; Europe; biomagnification
ABSTRACT: Fish samples from the Baltic, the North Sea and lake Vattern were contaminated with chlorinated terpenes. These compounds were also present in grey seal (*Halichoerus gryphus*) and guillemot (*Uria aalge*) from the Baltic at the same concentration, approximately 10 mg/kg fat as in their prey, the herring (*Clupea harengus*). Chlordane [12789-03-6]-related compounds were also found in all samples. The most common substances were trans-nonachlor (I) [39765-80-5] and oxychlordane [26880-48-8], but differences could be observed between species. These organochlorine pesticides are not used in Sweden and the probable explanation for their presence in a Swedish lake is air transportation.

Jarman, Walter M.; Hobson, Keith A.; Sydeman, William J.; Bacon, Corinne E.; and McLaren, Elizabeth B. 1996. Influence of trophic position and feeding location on contaminant levels in the Gulf of the Farallones food web revealed by stable isotope analysis. Environ. Sci. Technol. 30(2):654-660.
Rec #: 336
KEYWORDS: organochlorines; DDTs; HCHs; chlordanes; HCB; PCBs; metals; Pb; Hg; Se; stable isotopes; biomagnification; egg; common murre; *Uria aalge*; Brandt's cormorant; *Phalacrocorax penicillatus*; rhinoceros auklet; *Cerorhinca monocerata*; pigeon guillemot; *Cepphus columba*; bioaccumulation; CA; USA; North America
ABSTRACT: In this study, we present the levels of organochlorine (DDT, HCH, chlordane, HCB, and PCBs) and metal (Pb, Hg, and Se) contaminants and their relationship to stable carbon and nitrogen isotope values in the Gulf of the Farallones marine food web. This food web consisted of two species of euphausiids (*Euphausia pacifica* and *Thysanoessa spinifera*), two fish species [short-bellied rockfish (*Sebastes jordani*) and anchovy (*Engraulis mordax*)], four bird species [common murre (*Uria aalge*), Brandt's cormorant (*Phalacrocorax penicillatus*), rhinoceros auklet (*Cerorhinca monocerata*), and pigeon guillemot (*Cepphus columba*)], and the northern sea lion (*Eumetopias jubatus*). We used a novel method of using egg albumen to determine stable isotope values. The values of 13C ranged from -20.1" in the euphausiids to -15.0" in the northern sea lion and were consistent with a pelagic/offshore vs benthic/inshore results found in other studies. Values of 15N in the Gulf of the Farallones food web ranged from 11.2" in the euphausiids to 19.8" in the northern sea lion and generally demonstrate an equivalence with trophic level. The levels of organochlorine compounds were lowest in the euphausiids [DDT 11, and PCB 4.5 µg/kg dry weight geometric mean (GM)] and highest in the northern sea lion blubber (DDT 9500 and PCB 3500 µg/kg dry weight GM). The highest levels of organochlorine compounds in the birds were in the common murre (DDT 8200 and PCB 5900 µg/kg dry weight GM). Levels of Pb, Hg, and Se ranged from 80 to 1000, from 100 to 19000, and from 1900 to 4100 µg/kg dry weight GM, respectively. All of the organochlorine

Annotated Bibliography (Cont.).

compounds and Hg were significantly correlated with 15N values in the food web. Lower values of 15N in egg albumen than in the muscle tissue from common murres reflect a switch in diet to a lower trophic position during the egg formation period. The high contaminant levels in the murre suggest a mobilization of stored lipids into the eggs.

Jarman, Walter M.; Sydeman, William J.; Hobson, Keith A.; and Bergqvist, Per-Anders. 1997. Relationship of polychlorinated dibenzo-p-dioxin and polychlorinated dibenzofuran levels to stable-nitrogen isotope abundance in marine birds and mammals in coastal California. Environ. Toxicol. Chem. 16(5):1010-1013.
Rec #: 418
KEYWORDS: stable isotopes; biomagnification; organochlorines; PCDDs; PCDFs; common murre; *Uria aalge*; Brandt's cormorant; *Phalacrocorax penicillatus*; rhinoceros auklet; *Cerorhinca monocerata*; pigeon guillemot; *Cepphus columba;* egg; CA; USA; North America
ABSTRACT: Levels of polychlorinated dibenzo-p-dioxins (PCDDs) and polychlorinated dibenzofurans (PCDFs) were determined in common murre (*Uria aalge*), Brandt's cormorant (*Phalacrocorax penicillatus*), rhinoceros auklet (*Cerorhinca monocerata*), and pigeon guillemot (*Cepphus columba*) eggs, and Steller sea lion (*Eumetopias jubatus*) blubber collected from the Gulf of the Farallones National Marine Sanctuary in 1993. In addition, the samples were analyzed for stable-nitrogen isotopes ($\delta 15N$). Of the PCDDs and PCDFs, the 2,3,7,8-TCDD (TCDD) and 2,3,7,8-TCDF (TCDF) congeners were the most prominent in the birds. The levels of TCDD in the eggs ranged from 0.2 to 6.6 ng/wet kg in the pigeon guillemot and Brandt's cormorant, respectfully. The TCDF ranged from 0.30 to 2.25 ng/kg in the pigeon guillemot and Brandt's cormorant eggs, respectfully. Other prominent PCDD and PCDF congeners detected in all bird species were 1,2,3,6,7,8-HxCDD, 2,3,4,7,8-PeCDF, 1,2,3,7,8-PeCDD, and 1,2,3,4,6,7,8-HpCDD. In the Steller sea lion, the most prominent congeners were 1,2,3,7,8-PeCDD at 3.2 ng/kg, 2,3,7,8-TCDD at 2.9 ng/kg, OCDF at 2.2 ng/kg, 1,2,3,6,7,8-HxCDD at 1.92 ng/kg, and 1,2,3,4,7,8-HxCDF at 1.3 ng/kg.

Jensen, S.; Johnels, A. G.; Olsson, M.; and Otterlind G. 1969. DDT and PCB in marine animals from Swedish waters. Nature. 224(5216):247-250.
Rec #: 510
KEYWORDS: organochlorines; PCBs; DDTs; common murre; *Uria aalge;* white tailed sea eagle; *Haliaeetus albicilla*; grey heron; *Ardea cinerea*; egg; muscle; brain; Sweden; Baltic Sea; Europe; spatial variations; biomagnification
ABSTRACT: A variety of animal tissues including those from mussel, fish, seals, and birds were analyzed for PCB, DDT, and DDT metabolites. Analyses were made by gas chromatography with an electron capture detector. The concentration of PCB and DDT was approximately 10 times greater in tissues of animals of the Baltic proper area, the Gulf of Bothnia, and the Archip elago of Stockholm than that for comparable species from the North Se a area and the Atlantic. Levels of PCB decreased from southern fish to northern fish and seal tissues. The concentration of DDT was greater in fish than in other marine animals and the concentration of DDT metabolites was highest in eagles and herons.

Jermyn-Gee, K.; Pekarik, C.; Havelka, T.; Barrett, G.; and Weseloh, D. V. 2004. An atlas of contaminants in the eggs of fish-eating colonial birds of the Great Lakes (1998-2001). Technical Report Series No. ???. Burlington, Ontario, Canada: Canadian Wildlife Service Ontario Region.

Annotated Bibliography (Cont.).

214 pp.
Rec #: 517
KEYWORDS: great black-backed gull; *Larus marinus*; herring gull; *Larus argentatus*; black-crowned night-heron; *Nycticorax nycticorax*; black tern; *Chlidonias niger*; Forster's tern; *Sterna forsteri*; organochlorines; benzenes; chlordanes; HCB; DDTs; PCBs; PCDDs; PCDFs; egg; Great Lakes; North America; temporal trends; spatial variations
ABSTRACT: During 1998-2001, Canadian Wildlife Service (Ontario Region) collected 1252 eggs from 32 sites. Five species of fish-eating colonial waterbirds were sampled: Herring Gull (*Larus argentatus*), Great Black-backed Gull (*Larus marinus*), Black-crowned Night-Heron (*Nycticorax nycticorax*), Black Tern (*Chlidonias niger*), Forster's Tern (*Sterna forsteri*). The purpose was to measure the levels of the following compounds: organochlorine pesticides, chlorinated benzenes, polychlorinated biphenyls, dioxins and furans, lipid and moisture. The data presented in this report were generated as part of a monitoring program started in 1970 to understand the temporal and spatial trends of environmental contaminant levels in Great Lakes wildlife. Since the 1970s the levels of most chlorinated hydrocarbons have decreased significantly at most colonies on the Great Lakes. The change-point regression analysis, which we have used since 1997, continues to show that most contaminant levels at most sites (72.4 %) are declining as fast as or faster now than they did in the past. This is particularly evident for dieldrin, HCB, HE and DDE. The rates of decline have slowed for some compound-site comparisons (21.9 %), particularly PCBs and mirex. Since the last atlas was published (Pekarik et al. 1998), levels and trends of a relatively new contaminant, brominated diphenyl ethers (BDEs), have been documented in Great Lakes Herring Gull eggs (Norstrom et al. 2002). At the time of this writing (early 2004), routine analysis for BDEs is just being incorporated into CWS protocols. Hence, the data of Norstrom et al. (2002) are not included here. BDE data for Herring Gull eggs will be included for 2004 onwards in the next atlas. The data from 1998-2001 are summarized in two volumes. Volume I contains contaminant data for all five species summarized by location as well as non-coplanar PCB data for all species. Volume II contains contaminant data for all five species summarized by compound. Both volumes contain maps of sample locations and the means and standard deviations or the pooled analysis values for organochlorine pesticides, chlorinated benzenes, non-ortho polychlorinated biphenyls, dioxins and furans, and percent lipid and moisture. Non-coplanar PCB data is presented only in volume I and is only summarized by location. Additionally, contaminant data for Black and Forster's Terns from 1996 have been added, since they were not included from Pekarik *et al.,* 1998a; b. Since the last atlas (Pekarik et al. 1998), several papers have been published or are in press from the Herring Gull database. These include: DiMaio et al. 1999, Hebert and Weseloh (2002), Weseloh et al. (In Press), Weseloh et al. (In Review).

Joiris, C. R. 1997. Ecotoxicology of stable pollutants - organochlorines and heavy metals - in seabirds and marine mammals. Bulletin De La Societe Royale Des Sciences De Liege. 66(1-3):51-59.
Rec #: 503
KEYWORDS: organochlorines; PCBs; metals; Cd; Hg; Fe; Zn; Cu; organometallics; organic Hg; liver; muscle; common murre; *Uria aalge;* black-headed gull; *Larus ridibundus*; North Sea; Belgium; Europe
NOTES: FIGURES ONLY

Annotated Bibliography (Cont.).

ABSTRACT: Organochlorine and heavy metal levels in guillemots (*Uria aalge*), harbor porpoises (*Phocoena phocoena*), and common dolphins (*Delphinus delphis*) collected or found dead in the North Sea area were reported.

Joiris, Claude R.; Tapia, German; and Holsbeek, Ludo. 1997. Increase of organochlorines and mercury levels in common guillemots *Uria aalge* during winter in the southern North Sea. Mar. Pollut. Bull. 34(12):1049-1057.
Rec #: 38
KEYWORDS: metals; Hg; organometallics; organic Hg; organochlorines; PCBs; DDTs; bioaccumulation; North Sea; Belgium; Europe; muscle; liver; kidney; common murre; *Uria aalge*; seasonal variations
ABSTRACT: Beached seabirds, mainly common guillemots *Uria aalge*, were collected on the Belgian coast during winter from 1990 to 1995. Concentrations of total and organic mercury, and of organochlorines (PCBs and pesticides) were determined in muscle, liver and kidney. They were high compared to summer data (up to one order of magnitude), and increased during winter. This increase is not due to changes of total body weight nor polar lipid content, and thus reflects an actual increase of the seabirds' contamination while wintering in the southern North Sea. The observed annual cycle can be understood by assuming differences in prey contamination: higher during winter in the southern North Sea ecosystem than during summer in the Atlantic water ecosystem.

Jones, A. M.; Jones, Yvonne; and Stewart, W. D. P. 1972. Mercury in marine organisms of the Tay region. Nature. 238(5360):164-165.
Rec #: 491
KEYWORDS: metals; Hg; liver; kidney; gland; muscle; heart; lung; common eider; *Somateria mollissima*; Scotland; Europe; spatial variations; biomagnification
ABSTRACT: The concentrations of mercury [7439-97-6] were detected in organisms collected from the Tay river region of Scotland. The concentrations varied according to the sampling area examined and were highest in organisms receiving direct discharge from the Tay (near the mouth). Total Hg was detected by flameless atomic absorption spectrophotometry. High levels of Hg were found in algae and molluscs at a region subject to the discharge of fresh water from the river while very low levels (<.001 µg Hg/g wet weight of tissue) were found in a sampling area in a fully marine environment. High levels of Hg were also present in a grey seal and an eider duck, with concentrations in the liver and kidney of both species. In the Tay estuary, Hg concentrations ranged from 0.067 to 6.26 µg/g wet weight of algae and from 0.036 to 0.639 µg/g wet weight in molluscs.

Jones, P. D.; Giesy, J. P.; Newsted, J. L.; Verbrugge, D. A; Ludwig, J. P.; Ludwig, M. E.; Auman, H. J.; Crawford, R.; Tillitt, D. E.; and Kubiak, T. J. 1994. Accumulation of 2,3,7,8-tetrachlorodibenzo-p-dixoin equivalents by double-crested cormorant (*Phalacrocorax auritus*, Pelicaniformes) chicks in the North American Great Lakes. Ecotoxicol. Environ. Saf. 27(2):192-209.
Rec #: 31
KEYWORDS: organochlorines; PCDDs; PCBs; egg; carcass; herring gull; *Larus argentatus*; double-crested cormorant; *Phalacrocorax auritus*; Great Lakes; North America; biomagnification; TEQs

Annotated Bibliography (Cont.).

ABSTRACT: Concentrations of polychlorinated biphenyls (PCBs), and 2,3,7,8-tetrachlordibenzo-p-dioxin equivalents (TCDD-EQ) were determined in eggs and chicks of double-crested cormorants (DCC) which were collected in 1989 from eight locations in the Laurentian Great Lakes. The mean biomagnification factor (BMF) from forage fish to eggs was found to be 31.3. Absolute and relative concentrations as well as rates of accumulation of total concentrations of PCBs and TCDD-EQ were measurable in all of the samples. The concentrations of both PCBs and TCDD-EQs decreased immediately upon hatching of chicks, due to growth dilution. Initial decreases in absolute masses of TCDD-EQ in chicks were also observed, which indicates that there can be significant elimination of these compounds during early development. The initial rates of accumulation by chicks were dependent only on the mass of fish consumed. After the chicks began thermoregulating, the rates of accumulation, expressed as a concentration, normalized to body weight, became greater. Rates of accumulation of both PCBs and TCDD-EQ were correlated with their respective concentrations in forage fish consumed by the chicks. The relative potency, expressed as the ratio of the concentration of TCDD-EQ to that of total PCBs was calculated to determine if there was significant trophic-level enrichment of the TCDD-EQs, relative to total concentrations of PCBs. A significant enrichment was observed at the more and less contaminated locations, but the degree of enrichment was greater at the less contaminated locations (26 vs 72 µg/g).

Järnberg, U.; Asplund, L.; de Wit, C.; Egebäck, A.-L.; Wideqvist, U.; and Jakobsson, E. 1997. Distribution of polychlorinated naphthalene congeners in environmental and source-related samples. Arch. Environ. Contam. Toxicol. 32(3):232-245.
Rec #: 502
KEYWORDS: organochlorines; PCBs; PCNs; egg; common murre; *Uria aalge;* white tailed sea eagle; *Haliaeetus albicilla*; Baltic Sea; Sweden; Europe; biomagnification
NOTES: FIGURES ONLY FOR *URIA AALGE*
ABSTRACT: Polychlorinated naphthalene (CN) congener profiles in environmental and source related samples were compared graphically and by principal component analysis. Samples investigated included biological, sediment, water, and air samples, tech. polychlorinated biphenyl (PCB) and polychlorinated naphthalene (PCN) formulations, as well as municipal waste incineration (MWI) fly ash and graphite electrode sludge. Biological samples showed a preferential enrichment of planar, 1,3,5,7-substituted tetra-, penta-, and hexachlorinated congeners and most of these samples showed profiles that displayed some similarity to those found in the tech. PCB formulations. Sediment samples representing diffuse pollution, i.e., sediment samples from remote sites, showed an elevated abundance of the planar hexa- and heptaCN congeners (1,2,3,4,6,7-/1,2,3,5,6,7- and 1,2,3,4,5,6,7-). The CN congener profile found in these sediment samples and the two air samples were more similar to the tech. PCB formulations than to the investigated MWI and graphite sludge samples. Samples from three PCB contaminated lakes displayed similar congener profiles as Aroclor 1242, 1254 and Clophen A40. Two sediment samples and a pike sample collected from the vicinity of a chloroalkali plant showed profiles that were closely related to the investigated samples displayed profiles similar to low or medium chlorinated tech. PCN (Halowax 1099, 1013, and 1014).

Järnberg, Ulf; Asplund, Lillemor; de Wit, Cynthia; Grafström, Anna Karin; Haglund, Peter; Jansson, Bo; Lexén, Karin; Strandell, Michael; Olsson, Mats; and Jonsson, Björn. 1993. Polychlorinated biphenyls and polychlorinated naphthalenes in Swedish sediment and biota:

Annotated Bibliography (Cont.).

Levels, patterns, and time trends. Environ. Sci. Technol. 27(7):1364-1374.
Rec #: 513
KEYWORDS: organochlorines; PCBs; PCNs; PCDDs; common murre; *Uria aalge;* egg; Sweden; Europe; temporal trends; biomagnification
ABSTRACT: Levels of non-ortho-polychlorinated biphenyls (PCBs), some mono-ortho-PCBs, and resolved peaks of tetra- to heptachloronaphthalenes are reported in biological and sediment samples from the Swedish environment. The results show that levels of individual PCB congeners, and especially PCB 126, may pose a greater threat to the environment than the TCDD and 2,3,7,8-tetrachlorofuran when expressed as tetrachlorodibenzodioxin toxic equivalent. Polychlorinated naphthalene is as widespread as PCBs, and at some locations the pollution situation indicates a specific source, leading to a bioaccumulating hexachloronaphthalene contributing considerably to the total TCDD-like toxicity. A time trend study of guillemot eggs from the Baltic Proper seems to indicate that levels of non-ortho-PCB and -PCN have decreased since the 1970s.

Kahle, Silke and Becker, Peter H. 1999. Bird blood as bioindicator for mercury in the environment. Chemosphere. 39(14):2451-2457.
Rec #: 368
KEYWORDS: common gull; *Larus canus;* blood; feather; metals; Hg; Wadden Sea; Germany; Europe
ABSTRACT: Mercury concentrations were studied in blood, down and feathers of Common Gull (*Larus canus L.*) to investigate the suitability of bird blood as a matrix for biomonitoring of mercury in the marine environment. Chicks were collected in 1996 on the Elbe river and the Jade Bay. Like the side feathers, blood indicated site differences in mercury contamination. Correlational analyses showed that mercury concentrations in blood are significantly related to levels in side feathers ($p < 0.001$; Pearson), but not to those in down ($p > 0.05$; Pearson). Therefore, blood can be considered as a suitable matrix to indicate the current mercury burden in wild birds.

Kallenborn, Roland; Evenset, Anita; Herzke, Dorte; Christensen, Guttorm; and Schlabach, Martin. 2001. Chlorobornanes in biota samples, related to a typical freshwater food web at Bjørnøya (Bear Island). Organohalogen Compounds. 52:387-391.
Rec #: 455
KEYWORDS: glaucous gull; *Larus hyperboreus*; Svalbard; Norway; Europe; gut; organochlorines; toxaphene; spatial variations
NOTES: FIGURES ONLY
ABSTRACT: Samples from selected species belonging to a typical freshwater food web were collected at Lake Ellasjøen and Lake Øyangen in Bjørnøya, Norway, and analyzed for their chlorobornane content. Gut samples from glaucous gull were also taken and analyzed, since guano input from seabirds into the freshwater system cannot be neglected as a possible source of persistent organic pollutants (POP). Although significant differences were found between the samples from the two lakes, the highest chlorobornane contamination cannot be considered as unusually high compared to published literature to date for Arctic marine top predators. The results identified glaucous gull guano as one of the major chlorobornane sources for the Lake Ellasjøen freshwater system. The distinct local chlorobornane level differences between the two lakes demonstrate the important influence of locally defined and restricted ecological,

Annotated Bibliography (Cont.).

geographical, and meteorological environmental factors on the overall contaminant levels.

Kallenborn, Roland; Hühnerfuss, Heinrich; and König, Wilfried A. 1991. Gas chromatographic separation of organic enantiomers of marine pollutants. 2. Enantioselective metabolism of (±)-alpha-1,2,3,4,5,6-hexachlorocyclohexane in different organs of eider duck. Angewandte Chemie. 103(3):328-329.
Rec #: 484
KEYWORDS: organochlorines; HCHs; common eider; *Somateria mollissima*; Baltic Sea; Germany; Europe; muscle; liver; kidney; biomagnification; elimination; metabolism
NOTES: SUPPLEMENTAL
ABSTRACT: HPLC on heptakis(3-O-butyryl-2,6-di-O-pentyl)- β-cyclodextrin was used for the detection of (+)-?-HCH and (-)-α-HCH in n-hexane extracts from the muscle, liver, and kidney of 6 healthy wild eider ducks (*S. mollissima*) from the Baltic Sea. The ratio of the (+)- to (-)-enantiomers in the muscle was higher than in the kidney (7.0 vs. 1.6, respectively). The differences in degradation rates of the α-HCH enantiomers may be due to the different physiological function of the organs.

Kannan, Kurunthachalam; Choi, Jae-Won; Iseki, Naomasa; Senthilkumar, Kurunthachalam; Kim, Dong Hoon; Masunaga, Shigeki; and Giesy, John P. 2002. Concentrations of perfluorinated acids in livers of birds from Japan and Korea. Chemosphere. 49(3):225-231.
Rec #: 370
KEYWORDS: PFOS; perfluorinated acids; Korea; Japan; Asia; liver; black-tailed gull; *Larus crassirostris;* spot-billed duck; *Anas poecilorhyncha;* black-headed gull; *Larus ridibundus*; black-eared kite; *Milvus lineatus*; grey heron; *Ardea cinerea*; cormorant; *Phalacrocorax carbo*; bar-tailed godwit*; Limosa lapponica*; common gull; *Larus canus*; black-necked grebe; *Podiceps nigricollis*; common tern; *Sterna hirundo*; great knot*; Calidris tenuirostris*; greenshank; *Tringa nebularia*; herring gull; *Larus argentatus*; sanderling; *Crocethia alba*; little egret; *Egretta garzetta*; spatial variations
ABSTRACT: Livers of birds collected from Japan and Korea (n=83) were analyzed to determine the concentrations of perfluorooctanesulfonate (PFOS), perfluorooctanesulfonamide (FOSA), perfluorooctanoic acid (PFOA) and perfluorohexanesulfonate (PFHS). PFOS was found in the livers of 95 % of the birds analyzed at concentrations greater than the limit of quantitation (LOQ) of 10 ng/g, wet weight. The greatest concentration of PFOS of 650 ng/g, wet weight, was found in the liver of a common cormorant from the Sagami River in Kanagawa Prefecture. Concentrations of PFOS in bird livers from Japan and Korea were within the ranges of values reported for those from the United States and certain European countries. PFOA and PFHS were found in 5-10 % of the samples analyzed. The greatest concentrations of PFOA and PFHS in bird livers were 21 and 34 ng/g, wet weight, respectfully. FOSA was found in all the samples (n=10) of cormorants collected from the Sagami River in Japan. The greatest concentration of FOSA in cormorant liver was 215 ng/g, wet weight. There was no significant correlation between the concentrations of PFOS and FOSA in cormorants collected from the Sagami River. These results suggested that the distribution of FOSA is localized. No age- or gender-specific differences in fluorochemical concentrations could be discerned in birds.

Kannan, Kurunthachalam and Falandysz, Jerzy. 1997. Butyltin residues in sediment, fish, fish-eating birds, harbor porpoise and human tissues from the Polish coast of the Baltic Sea. Mar.

Annotated Bibliography (Cont.).

Pollut. Bull. 34(3):203-207.
Rec #: 501
KEYWORDS: organometallics; organotins; liver; red-throated loon; *Gavia stellata*; razorbill; *Alca torda;* great crested grebe; *Podiceps cristatus*; cormorant; *Phalacrocorax carbo*; long-tailed duck; *Clangula hyemalis*; white tailed sea eagle*; Haliaeetus albicilla*; common murre; *Uria aalge*; Baltic Sea; Poland; Europe
ABSTRACT: Monobutyltin, dibutyltin, tributyltin concentrations in flounder, herring, eel, sea trout, turbot, cod, eelpout, pikeperch, mackerel, red throated diver, razorbill, great crested grebe, black cormorant, long tailed duck, white tailed eagle, guillemot, harbor porpoise and human tissues from the Polish coast of the Baltic Sea are given.

Kannan, Kurunthachalam; Franson, J. Christian; Bowerman, William W.; Hansen, Kris J.; Jones, Paul D.; and Giesy, John P. 2001. Perfluorooctane sulfonate in fish-eating water birds including bald eagles and albatrosses. Environ. Sci. Technol. 35(15):3065-3070.
Rec #: 363
KEYWORDS: PFOS; USA; Great Lakes; North America; Pacific Ocean; blood; egg; herring gull; *Larus argentatus;* double-crested cormorant; *Phalacrocorax auritus;* ring-billed gull; *Larus delawarensis*; bald eagle; *Haliaeetus leucocephalus*; common loon; *Gavia immer*; brown pelican; *Pelecanus occidentalis*; white pelican; *Pelecanus erythrorhynchos*; great egret; *Ardea alba*; snowy egret; *Egretta thula*; woodstork; *Mycteria americana*; white-faced ibis; *Plegadis chihi*; black-crowned night heron; *Nycticorax nycticorax*; Franklin's gull; *Larus pipixcan*; gannet; *Sula bassana*; great black-backed gull; *Larus marinus*; osprey; *Pandion haliaetus*; red-throated loon; *Gavia stellata*; Brandt's cormorant; *Phalacrocorax penicillatus*; great blue heron; *Ardea herodias*; Laysan albatross; *Diomedea immutabilis*; black-footed albatross; *Diomedea nigripes*; spatial variations
ABSTRACT: Perfluorooctane sulfonate (PFOS) was measured in 161 samples of liver, kidney, blood, or egg yolk from 21 species of fish-eating water birds collected in the United States including albatrosses from Sand Island, Midway Atoll, in the central North Pacific Ocean. concentrations of PFOS in the blood plasma of bald eagles collected from the midwestern United States ranged from 13-2220 ng/mL (mean: 330 ng/mL), except one sample that did not contain quantifiable concentrations of PFOS. Concentrations of PFOS were greater in blood plasma than in whole blood. Among 82 livers from various species of birds from inland or coastal U.S. locations, Brandt's cormorant from San Diego, CA, contained the greatest concentration of PFOS (1780 ng/g, wet weight). PFOS was also found in the sera of albatrosses from the central North Pacific Ocean at concentrations ranging from 3-34 ng/mL. Occurrence of PFOS in birds from remote marine locations suggests widespread distribution of PFOS and related fluorochemicals in the environment.

Kannan, Kurunthachalam; Hilscherova, Klara; Imagawa, Takashi; Yamashita, Nobuyoshi; Williams, Lisa L.; and Giesy, John P. 2001. Polychlorinated naphthalenes, -biphenyls, -dibenzo-p-dioxins, and -dibenzofurans in double-crested cormorants and herring gulls from Michigan waters of the Great Lakes. Environ. Sci. Technol. 35(3):441-447.
Rec #: 328
KEYWORDS: herring gull; *Larus argentatus*; double-crested cormorant; *Phalacrocorax auritus*; egg; organochlorines; PCDDs; PCDFs; PCBs; TEQs; toxicology; MI; USA; Great Lakes; North America

Annotated Bibliography (Cont.).

ABSTRACT: Concentrations of polychlorinated dibenzo-p-dioxins (PCDDs), dibenzofurans (PCDFs), naphthalenes (PCNs), and biphenyls (PCBs) were measured in eggs of double-crested cormorants and herring gulls collected from Michigan waters of the Great Lakes. Concentrations of PCNs in eggs of double-crested cormorants and herring gulls were in the ranges of 380-2400 and 83-1300 pg/g, wet weight, respectively. Concentrations of 2,3,7,8-substituted PCDDs and PCDFs were 10-200 times less than those of PCNs in eggs whereas those of total PCBs (380-7900 ng/g, wet weight) were 3-4 orders of magnitude greater. While the profile of PCB isomers and congeners between double-crested cormorants and herring gulls was similar, the PCN isomer profile differed markedly between these two species. PCN congeners 66/67 (1,2,3,4,6,7/1,2,3,5,6,7) accounted for greater than 90 % of the total PCN concentrations in herring gulls, whereas their contribution to total PCN concentrations in double-crested cormorants ranged from 18 to 40 % (mean, 31 %). The ratios of concentrations of PCDDs to PCDFs were greater in herring gulls than in double-crested cormorants collected from the same locations, suggesting the ability of the former to metabolize PCDF congeners relatively rapidly. 2,3,7,8-Tetrachlorodibenzo-p-dioxin (TCDD) equivalents (TEQs) contributed by PCNs in double-crested cormorant and herring gull eggs were 2-3 % of the sum TEQs of PCBs, PCDDs, PCDFs, and PCNs. PCB congener 126 (3,3',4,4',5-PeCB) accounted for 57-72 % of the total TEQs in double-crested cormorant and herring gull eggs.

Karlin, Antti; Rantamäki, Pirjo; and Lemmetyinen, Risto. 1985. Residues of DDT and PCBs in the eggs of the herring gull *Larus argentatus* in the archipelago of southwestern Finland. Ornis Fennica. 62:168-170.
Rec #: 10
KEYWORDS: egg; organochlorines; DDTs; PCBs; toxicology; herring gull; *Larus argentatus*; Finland; Europe
ABSTRACT: As DDT and PCBs have deleterious effects on the reproductive success of birds and also on chick survival, the authors measured DDT and PCB residues in eggs of herring gulls in the archipelago of southwestern Finland. The samples were taken in Kustavi and Sauvo in 1978. The distance between the colonies was about 100 km. The authors also examined whether the laying sequence had an effect on the levels of chlorinated hydrocarbons.

Karlog, O.; Elvestad, K.; and Clausen, B. 1983. Heavy metals (cadmium, copper, lead and mercury) in common eiders (*Somateria mollissima*) from Denmark. Nordisk Veterinaermedicin. 35(12):448-451.
Rec #: 468
KEYWORDS: metals; Cd; Cu; Pb; Hg; common eider; *Somateria mollissima*; Denmark; Europe; toxicology; bioaccumulation; gender differences
ABSTRACT: In Danish common eiders, the liver and kidney contents of Cd, Cu, Pb, and Hg were detected This species feeds almost exclusively on the common mussel (*Mytilus edulis*), which accumulates heavy metals. The concentrations recorded of Cd, Cu, and Hg were lower than those usually regarded as toxic for birds, except for 1 eider which carried >2000 mg Cu/kg liver tissue (dry weight). Of the 42 eiders analyzed for Pb, 4 had toxic levels, i.e., >7 mg/kg liver weight, and 2 others had increased levels, i.e., 3-7 mg Pb/kg liver wet weight. The mean Cd concentration was 3.3 mg/kg liver wet weight This means that by consumption of approximately 160 g liver from the common eider, the weekly tolerable intake of Cd suggested by the FAO/WHO would be exceeded.

Annotated Bibliography (Cont.).

Kawano, Masahide; Inoue, Tsuyoshi; Wada, Toyohito; Hidaka, Hideo; and Tatsukawa, Ryo. 1988. Bioconcentration and residue patterns of chlordane compounds in marine animals: Invertebrates, fish, mammals, and seabirds. Environ. Sci. Technol. 22:792-797.
Rec #: 64
KEYWORDS: biomagnification; organochlorines; chlordanes; DDTs; HCHs; fat; carcass; Pacific Ocean; Bering Sea; thick-billed murre; *Uria lomvia*
ABSTRACT: The bioconcentration and compositional patterns of chlordane compounds (CHLs: *cis*-chlordane, *trans*-chlordane, *cis*-nonachlor, *trans*-nonachlor, and oxychlordane) were investigated in organisms from two marine ecosystems. The bioconcentration factors (BCF: concentration in organism/concentration seawater) of CHLs in lower trophic organisms were in between the values obtained for HCHs (α-HCH, β-HCH, and γ-HCH) and DDTs (*p,p*'-DDE and *p,p*'-DDT). In the case of higher trophic organisms, the scatter in the biomagnification factors (BMF: concentration in organism/concentration in food) of CHLs was found to be wider than those observed for HCHs and DDTs. Also, there were remarkable differences in CHLs composition among higher trophic organisms. For example, the percent composition of oxychlordane, which is one of the persistent metabolites of CHLs in seabirds from both areas, was higher than those of marine mammals.

Kawano, Masahide; Matsushita, Sanae; Inoue, Tsuyoshi; Tanaka, Hiroyuki; and Tatsukawa, Ryo. 1986. Biological accumulation of chlordane compounds in marine organisms from the northern North Pacific and Bering Sea. Mar. Pollut. Bull. 17(11):512-516.
Rec #: 40
KEYWORDS: organochlorines; DDTs; HCHs; PCBs; chlordanes; fat; thick-billed murre; *Uria lomvia*; gland; fat; Pacific Ocean; Bering Sea; biomagnification; metabolism
ABSTRACT: Sum of chlordane compounds (Sigma CHL; cis-chlordane + trans-chlordane + cis-nonachlor + trans-nonachlor + oxychlordane) are concentrated gradually with trophic levels from zooplankton to Dall's porpoise (*Phocoenoides dalli*) through squid and fish. The order of bioconcentration factors (BCF: concentration in organism/concentration in seawater) in these organisms was Sigma DDT (p,p'-DDE + p,p'-DDD + p,p'-DDT)> Sigma CHL less than or equal to PCBs> Sigma HCH (alpha -HCH + beta -HCH + gamma -HCH). Calculation for the concentration factor against food, namely biomagnification factor (BMF: concentration in organism/concentration in its food), was made for Dall's porpoise and thick-billed murre (*Uria lomvia*). The BMFs of these chemicals in thick-billed murre were lower than those of Dall's porpoise, suggesting the degradation and/or excretion of organochlorines through the uropygial gland with lipids. Moreover, the lowest BMF of Sigma CHL in thick-billed murre among organochlorines may indicate that chlordane compounds (CHLs) are metabolized more rapidly by this seabird than Dall's porpoise.

Keith, J. A. and Gruchy, I. M. 1972. Residue levels of chemical pollutant in North American bird life. Proc. Inter. Ornithol. Congr. 15:437-454.
Rec #: 550
KEYWORDS: organochlorines; DDTs; PCBs; dieldrin; metals; Hg; egg; carcass; feather; fat; effects; eggshell thickness; AK; USA; Canada; Mexico; Atlantic Ocean; Pacific Ocean; North America; *Aechmophorus occidentalis;* western grebe; *Fulmarus glacialis;* northern fulmar; *Puffinus creatopus;* pink-footed shearwater; *Puffinus gravis;* great shearwater; *Puffinus griseus;* sooty shearwater; *Puffinus tenuirostris;* short-tailed shearwater; *Pterodroma cahow;* Bermuda

Annotated Bibliography (Cont.).

petrel; *Oceanodroma leucorhoa;* Leach's storm-petrel; *Oceanodroma homochroa;* ashy petrel; *Oceanodroma melania*; black petrel; *Oceanodroma microsoma;* least petrel; *Oceanites oceanicus;* Wilson's petrel; *Pelecanus erythrorhynchos;* white pelican; *Pelecanus occidentalis;* brown pelican; *Sula bassana;* gannet; *Sula leucogaster;* brown booby; *Phalacrocorax auritus;* double-crested cormorant; *Ardea herodias;* great blue heron; / *Anas platyrhynchos;* mallard; *Anas rubripes;* American black duck; *Accipiter striatus;* sharp-shinned hawk; *Accipiter cooperii;* Cooper's hawk; *Buteo jamaicensis;* red-tailed hawk; *Buteo swainsoni;* Swainson's hawk; *Buteo regalis;* Ferruginous hawk; *Aquila chrysaetos;* golden eagle; *Haliaeetus leucocephalus;* bald eagle; *Circus cyaneus;* marsh hawk; *Falco columbarius;* pigeon hawk; *Falco mexicanus;* prairie falcon; *Falco sparverius;* American kestrel; *Charadrius semipalmatus;* semipalmated plover; *Charadrius vociferus;* killdeer; *Pluvialis squatarola;* black-bellied plover; *Scolopax minor;* American woodcock; *Gallinago gallinago;* common snipe; *Numenius americanus;* long-billed curlew; *Actitis macularia;* spotted sandpiper; *Catoptrophorus semipalmatus;* willet; *Tringa flavipes;* lesser yellowlegs; *Calidris minutilla;* least sandpiper; *Calidris alpina;* dunlin; *Limnodromus griseus;* short-billed dowitcher; *Limosa fedoa;* marbled godwit; *Recurvirostra americana;* American avocet; *Phalaropus fulicarius;* red phalarope; *Phalaropus lobatus;* red-necked phalarope; *Larus occidentalis;* western gull; *Larus argentatus;* herring gull; *Larus californicus;* California gull; *Sterna hirundo;* common tern; *Sterna elegans;* elegant tern; *Uria aalge;* common murre; *Alle alle;* little auk; *Brachyramphus marmoratus;* marbled murrelet; *Synthliboramphus antiquus;* ancient murrelet; *Synthliboramphus craveri;* Craveri's murrelet; *Ptychoramphus aleuticus;* Cassin's auklet; *Cerorhinca monocerata;* rhinoceros auklet; *Fratercula arctica;* Atlantic puffin; *Bubo virginianus;* great horned owl; *Speotyto cunicularia;* burrowing owl; *Asio otus;* long-eared owl; *Asio flammeus;* short-eared owl; *Eremophila alpestris;* horned lark; *Sturnus vulgaris;* starling
ABSTRACT: A review with 35 references. This report concentrated on contaminant studies in North American birds published within the last 5 years. Data was chosen to compare contaminants between groups of birds as well as geographically across North America.

Kennedy, S. W.; Fox, G. A.; Trudeau, S.; Bastien, L. J.; and Jones, S. P. 1998. Highly carboxylated porphyrin concentration: a biochemical marker of PCB exposure in herring gulls. Mar. Environ. Res. 46(1-5):65-69.
Rec #: 427
KEYWORDS: herring gull; *Larus argentatus*; liver; organochlorines; PCBs; toxicology; effects; porphyrin; Great Lakes; Canada; North America
NOTES: FIGURES ONLY
ABSTRACT: Our previous studies showed that the concentration of highly carboxylated porphyrins (HCPs) was higher in livers of herring gulls (*Larus argentatus*) from the Great Lakes than in livers of herring gulls from a relatively uncontaminated area of the Atlantic coast of Canada. Since there are relatively few causes of elevated HCPs other than exposure to certain halogenated aromatic hydrocarbons (HAHs), we suggested that HCP concentration might offer promise as a biochemical marker of HAH exposure in herring gulls. We did not confirm if, or identify which, HAHs were the cause of elevated HCPs. Here we provide evidence that polychlorinated biphenyls (PCBs) are a likely cause of HCP accumulation in herring gulls. HCP concentration was measured in livers of adult gulls that were collected from several sites on the Great Lakes and from two reference sites. Non-polar extracts prepared from the livers were assayed for porphyrinogenic potency in a chicken embryo hepatocyte (CEH) bioassay, and

Annotated Bibliography (Cont.).

analyzed for concentrations of various HAHs, including PCBs. There was a good linear correlation between (a) liver HCP concentration and porphyrinogenic potency of liver extracts, and (b) liver HCP and PCB concentration Congener-specific analysis suggested that the mono-ortho substituted PCB congeners 2,3,3',4,4'-pentachlorobiphenyl (PCB 105) and 2,3',4,4',5-pentachlorobiphenyl (PCB 118) were the congeners which contributed most to elevated HCPs in gull livers. We conclude that HCP concentration is a good biochemical marker of PCB exposure in herring gulls.

Kennedy, S. W.; Lorenzen, A.; James, C. A.; and Norstrom, R. J. 1992. Ethoxyresorufin-O-deethylase (EROD) and porphyria induction in chicken embryo hepatocyte cultures - A new bioassay of PCB, PCDD, and related chemical contamination in wildlife. Chemosphere. 25(1-2):193-196.
Rec #: 96
KEYWORDS: organochlorines; PCBs; PCDDs; toxicology; EROD; porphyrin; herring gull; *Larus argentatus*; great blue heron; *Ardea herodias*; egg; effects; Great Lakes; British Columbia; Canada; North America
ABSTRACT: Ethoxyresorufin-O-deethylase (EROD) and porphyria induction potencies of extracts from polychlorinated biphenyl (PCB) and 2,3,7,8-tetrachloro dibenzo-p-dioxin (TCDD) contaminated eggs of Herring Gulls, *Larus argentatus* from the Great Lakes of North America, and Great Blue Herons, *Ardea herodias* from British Columbia were measured in primary cultures of chicken embryo hepatocytes. The results show that EROD and porphyria induction in these cells have potential for determining the toxic potencies of complex mixtures of halogenated aromatic hydrocarbons (HAHs) extracted from wildlife tissue.

Khan, R. A. and Ryan, P. 1991. Long term effects of crude oil on common murres (*Uria aalge*) following rehabilitation. Bull. Environ. Contam. Toxicol. 46(2):216-222.
Rec #: 244
KEYWORDS: oil; common murre; *Uria aalge*; effects; North America; Canada; Newfoundland; carcass; liver; kidney; intestine; toxicology
NOTES: SUPPLEMENTAL
ABSTRACT: The purpose of this communication is to report the long term effects of crude oil on common murres.

Kierkegaard, Amelie; Sellström, Ulla; Bignert, Anders; Olsson, Mats; Asplund, Lillemor; Jansson, Bo; and De Wit, Cynthia. 1999. Temporal trends of a polybrominated diphenyl ether (PBDE), a methoxylated PBDE, and hexabromocyclododecane (HBCD) in Swedish biota. Organohalogen Compounds. 40:367-370.
Rec #: 500
KEYWORDS: PBDEs; organochlorines; PCBs; HBCD; common murre; *Uria aalge*; egg; Baltic Sea; Sweden; Europe; temporal trends; biomagnification
NOTES: FIGURES ONLY
ABSTRACT: The temporal variations were studied of 2,2',4,4'-tetrabromodiphenyl ether (BDE 47), 2-methoxy-2',4,4',6-tetrabromodiphenyl ether (MeO-BDE47), and hexabromocyclodecane (HBCD) in pike and roach muscle from Swedish lakes and in guillemot (*Uria aalge*) eggs from the Baltic Sea. In pike, BDE47 concentrations increased from 1967 to the early 1980s. From 1984 to 1997, large variations occurred between years and between individuals within each year

Annotated Bibliography (Cont.).

with a stagnant trend. In roach, concentrations were lower with similar variations. Maximum concentrations appeared in 1988. BDE47 dominated over BDE99, which was, with few exceptions, below the quantification limit. For MeO-BDE47 in pike, the trend showed decreasing concentrations. The initial concentrations exceeded the highest BDE47 levels by factor 3 (490 ng/g lpw). In the 1990s, a reversed trend was observed with variations. HBCD concentrations in guillemot showed an increase over the entire time. Traces below the quantification limit were also found in pike and roach. The authors suggest that these higher HBCD levels may be due to biomagnification.

Kim, E. Y.; Goto, R.; Tanabe, S.; Tanaka, H.; and Tatsukawa, R. 1998. Distribution of 14 elements in tissues and organs of oceanic seabirds. Arch. Environ. Contam. Toxicol. 35(4):638-645.
Rec #: 190
KEYWORDS: metals; Li; V; Mn; Co; Cu; Zn; Se; Rb; Sr; Ag; Cd; Cs; Pb; Hg; brain; lung; heart; stomach; intestine; liver; pancreas; spleen; gallbladder; kidney; gonad; gland; fat; bone; muscle; skin; eyeball; feather; trachea; esophagus; northern fulmar; *Fulmarus glacialis*; royal albatross; *Diomedea epomophora*; black-footed albatross; *Diomedea nigripes;* black-browed albatross; *Diomedea melanophris*; white-capped albatross; *Diomedea cauta*; yellow-nosed albatross; *Diomedea chlororhynchos*; grey-headed albatross; *Diomedea chrysostoma*; light-mantled sooty albatross; *Phoebetria palpebrata*; northern giant petrel; *Macronectes halli*; grey petrel; *Procellaria cinerea*; white-chinned petrel; *Procellaria aequinoctialis*; Pacific Ocean; Indian Ocean
ABSTRACT: The concentrations of 14 trace elements (Li, V, Mn, Co, Cu, Zn, Se, Rb, Sr, Ag, Cd, Cs, Pb, and Hg) were determined in tissues and organs of three species and in the liver of 11 species of seabirds. Comparatively high concentrations of Li, Co, Sr, and V were found in the femur. Cd, Se, Cu, and Mn concentrations were relatively higher in the kidney than in other tissues and organs. Rb, Cs, and Pb concentrations were rather uniform among tissues. Concentrations of essential elements such as Mn, Cu, and Co were comparable among seabird species, except high Cu concentrations in northern giant petrel. Among nonessential elements, concentrations of Cd and Hg were variable according to seabird species. Pb levels were low in all the species. High Se levels (100 µg/g dry weight) were found in the liver of black-footed albatross and grey petrel. There were significant positive correlations between Se and Cd concentrations in three species and between Se and Hg in black-footed albatross, suggesting that Se has an antagonistic action on the toxic effects of Cd and Hg. Concentrations of Li, V, Ag, and Cs were usually low (less than 1 µg/g dry weight).

Kim, E. Y.; Murakami, T.; Saeki, K.; and Tatsukawa, R. 1996. Mercury levels and its chemical form in tissues and organs of seabirds. Arch. Environ. Contam. Toxicol. 30(2):259-266.
Rec #: 194
KEYWORDS: metals; Hg; organometallics; organic Hg; black-footed albatross; *Diomedea nigripes*; arctic tern; *Sterna paradisaea;* long-tailed duck; *Clangula hyemalis*; northern fulmar; *Fulmarus glacialis;* herring gull; *Larus argentatus*; royal albatross; *Diomedea epomophora*; laysan albatross; *Diomedea immutabilis;* white-chinned petrel; *Procellaria aequinoctialis*; brown booby; *Sula leucogaster;* body burden; liver; muscle; feather; kidney; Russia; Asia; Pacific Ocean; Indian Ocean
ABSTRACT: Liver, muscle, kidney, and feather samples from nine species of seabirds were

Annotated Bibliography (Cont.).

analyzed for total and organic (methyl) mercury (MM). Total mercury (TM) levels in liver showed great intra- and inter-species variations, with the concentrations varied from 306 µg/g (dry weight) in black-footed albatross (*Diomedea nigripes*) to 4.9 µg/g in arctic tern (*Sterna paradisaea*), while MM levels were less relatively variable. The order of MM concentrations in tissues of all the seabirds except oldsquaw (*Clangula hyemalis*) was as follows: liver > kidney > muscle. The mean percentage of MM in total was 35 %, 36 %, and 66 % in liver, kidney, and muscle, respectively, for all the species. Statistically significant negative correlations were found between the proportion of MM to TM and concentrations of TM in the liver and muscle of black-footed albatross and in the liver of laysan albatross. Furthermore, the percentage of MM decreased with an increase in TM concentrations in the liver, muscle, and kidney of all the species. Black-footed albatross had the highest concentration and burden of mercury in the liver, wherein more than 70 % of the TM occurred as inorganic mercury. On the other hand, the mercury burdens in feathers were less than 10 % of the body burdens, indicating that excretion of mercury by moulting is negligible. The results suggest that some seabirds are capable of demethylating MM in the tissues (mainly in liver), and store mercury as an immobilizable inorganic form in the liver. It is noteworthy that the species with a high degree of demethylation capacity and slow moulting pattern showed low mercury burdens in feathers.

Kim, E. Y.; Saeki, K.; Tanabe, S.; Tanaka, H.; and Tatsukawa, R. 1996. Specific accumulation of mercury and selenium in seabirds. Environ. Pollut. 94(3):261-265.
Rec #: 193
KEYWORDS: metals; Hg; Se; organometallics; organic Hg; elimination; liver; northern fulmar; *Fulmarus glacialis*; royal albatross; *Diomedea epomophora*; black-footed albatross; *Diomedea nigripes;* white-capped albatross; *Diomedea cauta*; yellow-nosed albatross; *Diomedea chlororhynchos*; light-mantled sooty albatross; *Phoebetria palpebrata*; northern giant petrel; *Macronectes halli*; grey petrel; *Procellaria cinerea*; white-chinned petrel; *Procellaria aequinoctialis*; brown booby; *Sula leucogaster*; Pacific Ocean; Indian Ocean
ABSTRACT: Total mercury (T-Hg), methyl mercury (MeHg) and selenium (Se) concentrations were determined to elucidate the relationship between Hg and Se levels in the liver of 10 seabird species. Highest concentrations of T-Hg (mean 267 µg/g dry weight), MeHg (mean 25.5 µg/g dry weight) and Se (mean 113 µg/g dry weight) were in the liver of black-footed albatross (*Diomedea nigripes*). An equivalent molar ratio of 1:1 between T-Hg and Se was found in the liver of individuals which contain over 100 µg Hg/g. However, such a relationship was unclear in other individuals which had relatively low Hg levels. This suggests that Se plays a role in Hg detoxification for those individuals with high Hg. In seabird tissues, Hg and Se levels should be a most important factor determining the relationship between big and Se, and fluctuation of Hg burden through molting and the species-specific demethylation capacity would also influence their relationships.

Kim, Eun-young; Ichihashi, Hideki; Atrashkevich, G. I.; Andreev, A. V.; Tanabe, Shinsuke; and Tatsukawa, Ryo. 1994. Accumulation and fluctuation of heavy metals in migrating birds in breeding area, Northeast Siberia. Kankyo Kagaku. 4(2):614-615.
Rec #: 385
KEYWORDS: metals; Fe; Mn; Zn; Cu; Cd; Hg; kidney; liver; feather; long-tailed duck; *Clangula hyemalis*; king eider; *Somateria spectabilis*; pintail; *Anas acuta*; herring gull; *Larus*

Annotated Bibliography (Cont.).

argentatus; glaucous gull; *Larus hyperboreus*; Sabine's gull; *Larus sabini*; Arctic tern; *Sterna paradisaea*; long-tailed jaeger; *Stercorarius longicaudus*; parasitic jaeger; *Stercorarius parasiticus*; red-throated loon; *Gavia stellata*; Arctic loon; *Gavia arctica*; Russia; Asia
NOTES: FIGURES ONLY
ABSTRACT: Studied were the accumulation and fluctuation of Fe, Mn, Zn, Cu, Cd, and Hg in kidney, liver, feather, etc. of northeast Siberia birds. The birds included Oldsquaw, King Eider, Pintail, Herring Gull, Glaucous Gull, Sabine's Gull, Arctic Tern, Long-tail Jaeger, Parasitic Jaeger, Red-throated Loon, and Arctic Loon.

Kim, Eun-Young; Ichihashi, Hideki; Saeki, Kazutoshi; Atrashkevich, Gennady; Tanabe, Shinsuke; and Tatsukawa, Ryo. 1996. Metal accumulation in tissues of seabirds from Chaun, northeast Siberia, Russia. Environ. Pollut. 92(3):247-252.
Rec #: 469
KEYWORDS: metals; Fe; Mn; Zn; Cu; Cd; Hg; Russia; Asia; long-tailed duck; *Clangula hyemalis*; king eider; *Somateria spectabilis*; pintail; *Anas acuta*; herring gull; *Larus argentatus*; glaucous gull; *Larus hyperboreus*; Sabine's gull; *Larus sabini*; Arctic tern; *Sterna paradisaea*; long-tailed jaeger; *Stercorarius longicaudus*; parasitic jaeger; *Stercorarius parasiticus*; red-throated loon; *Gavia stellata*; Arctic loon; *Gavia arctica*; muscle; liver; kidney; feather; bioaccumulation
ABSTRACT: Concentrations of four essential elements (Fe, Mn, Zn, and Cu) and two toxic metals (Cd and Hg) were detected in selected tissues of 11 seabird species collected in Chaun, northeast Siberia. In oldsquaw, arctic tern and herring gull, zinc concentrations were correlated with Cd concentrations. Cadmium concentrations in all the species were highest in kidney and Hg in liver. Cd levels in the liver and kidney of herring gulls were higher than those observed from other breeding areas. Similarly, Hg concentrations were also high in the liver of herring gull. High concentrations of Cd and Hg found in some birds from Chaun might have arisen from exposure on migration.

Koistinen J.; Koivusaari, J.; Nuuja, I.; and Paasivirta, J. 1995. PCDEs, PCBs, PCDDs and PCDFs in black guillemots and white-tailed sea eagles from the Baltic Sea. Chemosphere. 30(9):1671-1684.
Rec #: 320
KEYWORDS: black guillemot; *Cepphus grylle*; white-tailed sea-eagle; *Haliaeetus albicilla*; egg; Baltic Sea; Finland; Europe; organochlorines; HCB; HCHs; chlordanes; DDTs; PCBs; PCDEs; PCDDs; PCDFs; toxicology; TEQs
ABSTRACT: Concentrations and patterns of several chloro compounds including polychlorinated dibenzo-p-dioxins (PCDD), dibenzofurans (PCDF), biphenyls (PCB) and diphenyl ethers (PCDE) were determined in black guillemots (Cepphus grylle L.) and white-tailed sea-eagles (Haliaeetus albicilla L.) from the Baltic Sea environment. Three breast muscles of eagles were analyzed and had different concentrations and patterns of the studied compounds, whereas the three guillemot eggs were found to have more similar levels and patterns. The concentrations of individual PCDE congeners varied from <3 to 79 ng/g lipid weight (1w) in guillemots and from <5 to 13,000 ng/g 1w in eagles. Toxic PCDDs and PCDFs occurred at lower concentrations than toxic coplanar PCB congeners in both guillemots and eagles. The toxic load as TCDD equivalents (TEQ) for coplanar PCBs was on average 5 ng/g 1w in guillemots and ranged from 9 to 340 ng/g 1w in eagles.

Annotated Bibliography (Cont.).

Koslowski, Susan E.; Metcalfe, Christopher D.; Lazar, Rodica; and Haffner, G. Douglas. 1994. The distribution of 42 PCBs, including three coplanar congeners, in the food web of the western basin of Lake Erie. J. Great Lakes Res. 20(1):260-270.
Rec #: 333
KEYWORDS: herring gull; *Larus argentatus*; egg; organochlorines; PCBs; biomagnification; Great Lakes; North America
ABSTRACT: Non-ortho substituted PCBs, IUPAC numbers 77, 126, and 169, along with 39 other PCB congeners were quantified in samples of Lake Erie sediment and biota during the summer of 1991. Many PCB congeners were found at elevated levels, and biomagnification was apparent in all congeners but was more predominant in congeners with relatively high octanol-water partition coefficients. Congener 126 was found primarily in benthic fish and top predators, whereas there was no significant differences in the concentrations of congener 77 in fish species of the benthic and pelagic food web. Congener 169 was found at very low concentrations in the food web of Lake Erie. There were also significant differences in concentrations of coplanar PCBs among liver, egg, and muscle tissues, but trends were not consistent among all the species examined.

Koster, M. D.; Ryckman, D. P.; Weseloh, D. V. C.; and Struger, J. 1996. Mercury levels in Great Lakes herring gull (*Larus argentatus*) eggs, 1972-1992. Environ. Pollut. 93(3):261-270.
Rec #: 157
KEYWORDS: metals; Hg; egg; herring gull; *Larus argentatus*; North America; Great Lakes; temporal trends; reproductive success; effects
ABSTRACT: Since 1971, the herring gull (*Larus argentatus*) has been used as a sentinel species for monitoring the levels of persistent contaminants in the Great Lakes ecosystem. In this study, 21 herring gull colonies in the Great Lakes and connecting channels were sampled during 1972-1976, 1981-1983, 1985 and 1992. For each year, 10 eggs (usually) were collected from each colony site and analyzed for total mercury (μg/g, wet weight). Results indicated that eggs from Lake Ontario displayed the highest lake-wide mercury levels (0.28-0.73 μg/g), followed by Lake Superior (0.21-0.50 μg /g). Lake Erie typically displayed the lowest mercury levels (0.18-0.24 μg/g). Overall, mercury levels ranged from 0.12 μg/g in 1985 to 0.88 μg/g in 1982 for Channel Shelter Island (Lake Huron) and Pigeon Island (Lake Ontario), respectively. Generally, all colony sites showed peak egg mercury levels in 1982. A significant decline in egg mercury levels was observed in five colony sites for the period 1972-1992 and in three different colony sites for the period 1981-1992. Mercury levels in the eggs of herring gulls for the period of this study were below levels associated with acute toxic effects in this species but were within a range, for certain years, which potentially reduces hatchability in other avian species.

Krol, W. J.; Arsenault, T.; and Mattina, M. J. I. 2002. Persistent organochlorine pesticide contamination of birds collected in Connecticut during the year 2000. Bull. Environ. Contam. Toxicol. 69(3):452-458.
Rec #: 361
KEYWORDS: herring gull; *Larus argentatus*; organochlorines; DDTs; chlordanes; dieldrin; brain; gizzard; liver; CT; USA; North America
ABSTRACT: Residues of persistent organochlorine pesticides (POPs) were analyzed in avian tissues. Brain, gizzard, and liver samples were examined. Namely residues of chlordane-related

Annotated Bibliography (Cont.).

compounds were found in 43 % of the birds. The DDT metabolite 2,2-bis(p-chlorophenyl)-1,1-dichloroethylene, heptachlor epoxide, cis-nonachlor, trans-nonachlor, oxychlordane, and MC5, a component of tech. chlordane, were detected Dieldrin was found only in 1 bird. POPs were more prevalent in gizzard than in either the brain or liver tissue. Despite banning POPs in 1988, they continue to pervade environment.

Kubiak, T. J.; Harris, H. J.; Smith, L. M.; Schwartz, T. R.; Stalling, D. L.; Trick, J. A.; Sileo, L.; Docherty, D.; and Erdman, T. C. 1989. Microcontaminants and reporductive impairment of the Forster's tern on Green Bay, Lake Michigan-1983. Arch. Environ. Contam. Toxicol. 18:706-727. Rec #: 522
KEYWORDS: Forster's tern; *Sterna forsteri*; organochlorines; PCDDs; PCBs; Great Lakes; WI; USA; North America; egg; toxicology; AHH; effects; reproductive success
ABSTRACT: For the 1983 nesting season, Forster's tern (*Sterna forsteri*) reproductive success was significantly impaired on organochlorine contaminated Green Bay, Lake Michigan compared to a relatively uncontaminated inland location at Lake Poygan, Wisconsin. Compared with tern eggs from Lake Poygan, eggs from Green Bay had significantly higher median concentrations of 2,3,7,8-tetrachlorodibenzo-p-dioxin (TCDD), other polychlorinated dibenzo-p-dioxins (PCDDs), total polychlorinated biphenyls (PCBs), total (three congeners) non-ortho, ortho' PCBs, five individual PCB congeners known to induce aryl hydrocarbon hydroxylase (AHH) and several other organochlorine contaminants. Conversions of analytical concentrations of TCDD and PCB congeners based on relative AHH induction potencies allowed for estimation of total 2,3,7,8-TCDD equivalents. Two PCB congeners, 2,3,3',4,4'- and 3,3',4,4',5-pentachlorobiphenyl (PeCB) accounted for more than 90 % of the median estimated TCDD equivalents at both Green Bay and Lake Poygan. The median estimated TCDD equivalents were almost 11-fold higher in tern eggs from Green Bay than in eggs from Lake Poygan (2175 and 201 pg/g). The hatching success of Green Bay sibling eggs from nests where eggs were collected for contaminant analyses was 75 % lower at Green Bay than at Lake Poygan. Hatchability of eggs taken from other nests and artificially incubated was about 50 % lower for Green Bay than for Lake Poygan. Among hatchlings from laboratory incubation, those from Green Bay weighed approximately 20 % less and had a mean liver weight to body weight ratio 26 % greater than those from Lake Poygan. In both field and laboratory, mean minimum incubation periods were significantly longer for eggs from Green Bay compared to Lake Poygan (8.25 and 4.58 days, respectively). Mean minimum incubation time for Green Bay eggs in the field was 4.37 days longer than in the laboratory.
 Hatchability was greatly improved when Green Bay eggs were incubated by Lake Poygan adults in an egg-exchange experiment, but was sharply decreased in Lake Poygan eggs incubated in Green Bay nests. Nest abandonment and egg disappearance were substantial at Green Bay but nil at Lake Poygan. Thus, not only factors intrinsic to the egg, but also extrinsic factors (parental attentiveness), impaired reproductive outcome at Green Bay. The epidemiological evidence from this study strongly suggested that contaminants were a causal factor. AHH-active PCB congeners (intrinsic effects) and PCBs in general (extrinsic effects) appeared to be the only contaminants at the concentrations measured in eggs, capable of producing the effects that were observed at Green Bay.

Kucklick, John R.; Vander Pol, Stacy S.; Becker, Paul R.; Pugh, Rebecca S.; Simac, Kristin; York, Geoff W.; and Roseneau, David G. 2002. Persistent organic pollutants in murre eggs from

Annotated Bibliography (Cont.).

the Bering Sea and Gulf of Alaska. Organohalogen Compounds. 59:13-16.
Rec #: 496
KEYWORDS: organochlorines; PCBs; DDTs; HCB; chlordanes; dieldrin; mirex; egg; common murre; *Uria aalge*; thick-billed murre; *Uria lomvia*; AK; USA; North America; Gulf of Alaska; Bering Sea; spatial variations
ABSTRACT: The objective of this work was to generate persistent organic pollutant data on murre eggs from the five colonies so far sampled and to examine geographical differences in the persistent organic pollutant levels and patterns.

Kuhnlein, H. V. and Chan H. M. 2000. Environment and contaminants in traditional food systems of Northern indigenous peoples. Annu. Rev. Nutr. 20:595-626.
Rec #: 428
KEYWORDS: review; metals; As; Cd; Pb; Hg; organochlorines; aldrin; chlordanes; benzenes; DDTs; dieldrin; TCDDs; HCHs; PCBs; toxaphene; radionuclides; organometallics; organic Hg; Canada goose; *Branta canadensis*; thick-billed murre; *Uria lomvia*; white-winged scoter; *Melanitta fusca*; surf scoter; *Melanitta perspicillata*; effects; AK; AZ; NM; NY; WI; USA; British Columbia; Newfoundland; Northwest Territories; Ontario; Quebec; Saskatchewan; Yukon; Canada; North America; Atlantic Ocean; Pacific Ocean; Gulf of Alaska; Bering Sea; Great Lakes; Baltic Sea; North Sea; Greenland; Svalbard; Norway; Iceland; Finland; Sweeden; Europe
NOTES: SUPPLEMENTAL; NO TABLES
ABSTRACT: A review with 132 references Traditional food resources of indigenous peoples are now recognized as containing a variety of environmental contaminants which reach food species through local or long-range transport avenues. Data from published reports on contaminants contained in traditional foods in northern North America and Europe, as organochlorines, heavy metals, and radionuclides, are presented. Usually multiple contaminants are contained in the same food species. Measurement of dietary exposure to these environmental contaminants and major issues of risk assessment, evaluation and management are discussed. The dilemma faced by indigenous peoples in weighing the multiple nutritional and socioeconomic benefits of traditional food use against risk of contaminants in culturally important food resources is described.

Kuiken, Thijs; Fox, Glen A.; and Danesik, Karen L. 1999. Bill malformations in double-crested cormorants with low exposure to organochlorines . Environ. Toxicol. Chem. 18(12):2908-2913.
Rec #: 26
KEYWORDS: organochlorines; PCBs; DDTs; hydrocarbons; PAHs; metals; Pb; Cd; Hg; Se; ultraviolet radiation; toxicology; Canada; Saskatchewan; North America; double-crested cormorant; liver; kidney; *Phalacrocorax auritus*; abnormalities; effects
ABSTRACT: Eight of 20 newly hatched double-crested cormorants (*Phalacrocorax auritus*), captured at Dore Lake (Saskatchewan, Canada) and raised in captivity, developed malformed bills when they were 2 to 3 weeks old. Malformation was characterized by abnormal flexure and rotation of the maxilla and mandible, resulting in a crossed bill. By radiography, the premaxillary and dental bones were misshapen. Morphologically similar malformed bills in free-living cormorants have been attributed to exposure to polyhalogenated aromatic hydrocarbons. However, the concentrations of total PCBs in the livers of these captive cormorants with malformed bills and in their diet were lower than have been previously associated with such

Annotated Bibliography (Cont.).

malformations and were considered too low to have been the cause. The bill malformations may have been caused by deficiency of vitamin D_3, because the cormorants were kept indoors without exposure to ultraviolet light and were fed frozen fish that may have been deficient in this vitamin.

Kunisue, Tatsuya; Minh, Tu Binh; Fukuda, Kayo; Watanabe, Mafumi; Tanabe, Shinsuke; and Titenko, Alexei M. 2002. Seasonal variation of persistent organochlorine accumulation in birds from Lake Baikal, Russia, and the role of the South Asian region as a source of pollution for wintering migrants. Environ. Sci. Technol. 36(7):1396-1404.
Rec #: 362
KEYWORDS: organochlorines; PCBs; DDTs; HCHs; chlordanes; toxicology; TEQs; Russia; Asia; muscle; carcass; herring gull; *Larus argentatus*; common gull; *Larus canus*; black-headed gull; *Larus ridibundus*; seasonal variations
ABSTRACT: Concentrations of persistent organochlorines (OCs), such as polychlorinated biphenyls (PCBs), DDT and its metabolites (DDTs), hexachlorocyclohexane isomers (HCHs), and chlordane compounds (CHLs), were detected in whole body soft tissue homogenates and in muscles of resident and migratory birds collected from Lake Baikal, Russia. The residue pattern in both resident and migratory birds was in the following order: PCBs > DDTs > HCHs > CHLs. OC concentrations in migratory birds varied, depending on the feeding habits. The maximum levels of OCs were found in piscivores, followed by insectivores, omnivores, and herbivores. OC residue levels in Lake Baikal birds were lower than those in the Great Lakes region, as well as in other lakes in Europe and Japan. Concentrations of HCHs and DDTs in most of the migratory birds collected in the spring were higher than for those collected in the autumn, indicating a notable accumulation in wintering grounds. Compilation and analysis of the available data in fish and birds from Asia suggested that the tropical and subtropical regions in south Asian countries may be a source of pollution for the wintering accumulation of migratory birds from Lake Baikal. Relatively higher compositions of α- and γ-HCH in total HCHs and p,p'-DDT in total DDTs were observed in some migratory species, indicating recent exposure to HCHs and DDTs in Lake Baikal or wintering areas. PCB isomer patterns were different between residents and migrants, with the predominance of lower chlorinated congeners in migratory species, suggesting recent PCB accumulation in stopover sites during wintering. TEQ concentrations of toxic non- and mono-ortho coplanar PCBs in common terns from Lake Baikal were comparable to those reported in some species from Japan, the USA, and Europe. Relative contributions of non-ortho coplanar congeners to toxic equivalents (TEQs) were predominant, in which PCB-126 accounted for the highest toxicity contribution. Estimated TEQ concentrations in the common tern from Lake Baikal exceeded the levels associated with enzyme induction in bald eagles. To our knowledge, this is the first comprehensive study showing the seasonal variations of OC accumulation in the birds from Lake Baikal.

Kury, Channing R. 1969. Pesticide residues in a marine population of double-crested cormorants. J. Wildl. Manage. 33(1):91-95.
Rec #: 414
KEYWORDS: organochlorines; DDTs; double-crested cormorant; *Phalacrocorax auritus*; brain; gonad; heart; egg; ME; USA; North America; biomagnification; bioaccumulation; gender differences; reproductive success; temporal trends
ABSTRACT: The average levels of pesticide in tissues from 21 adult and 4 immature

Annotated Bibliography (Cont.).

cormorants (*Phalacrocorax auritus*) and 24 eggs which were collected from Muscongus Bay, Maine in 1966 were: brains, 1.5 ppm. DDE; gonads, 6.5 ppm. DDE; hearts, 3.0 ppm. DDE; eggs, 1.5 ppm. DDT, 0.7 ppm. DDD, and 6.2 ppm. DDE. Food samples (41) had no detectable amounts of pesticide (detectable limit 0.05 ppm.). In 1967, the 18 adult brains collected from Muscongus Bay had an average level of 0.34 ppm. DDE. Only 5 of 23 samples (representing 89 individuals) of brains of nestlings from both Muscongus Bay and Duck Island had measurable amounts of pesticide and these varied from a trace to 0.29 ppm. DDE. Eggs (11) from Duck Island had an average level of 4.5 ppm. DDT, 1.5 ppm. DDD, and 7.6 ppm. DDE.

Kuzyk, Zou Zou A.; Burgess, Neil M.; Stow, Jason P.; and Fox, Glen A. 2003. Biological effects of marine PCB contamination on black guillemot nestlings at Saglek, Labrador: Liver biomarkers. Ecotoxicology. 12(1-4):183-197.
Rec #: 351
KEYWORDS: black guillemot; *Cepphus grylle;* organochlorines; PCBs; PCDDs; PCDFs; chlordanes; HCB; DDTs; dieldrin; metals; As; Cu; Se; Zn; Hg; toxicology; TEQs; EROD; retinol; porphyrin; liver; Labrador; Canada; North America; effects
ABSTRACT: Black guillemots (*Cepphus grylle*) in Saglek Bay, Labrador have elevated polychlorinated biphenyl (PCB) concentrations due to marine sediment contamination around a former military site. We measured liver biomarkers and ΣPCB concentrations in 31 nestlings from three PCB-exposure groups: reference group (range: 15-46 ng/g liver, wet weight), moderately exposed Islands group (24-150 ng/g), and highly exposed Beach group (170-6200 ng/g). Biomarker responses were dose-dependent and in some cases sex-dependent. Livers of female Beach nestlings were enlarged 36 % relative to reference females. In both sexes, Beach nestlings had liver ethoxyresorufin-O-deethylase (EROD) activities elevated 79 % and liver retinol concentrations reduced 47 %. Retinyl palmitate concentrations were reduced 50 % but only among female nestlings. Island nestlings also exhibited EROD induction (57 %) and reductions in retinol and retinyl palmitate concentrations (28 and 58 %, respectively). Liver lipid content increased with ΣPCBs in both sexes, and correlated with liver mass in males. Malic enzyme activity and porphyrin concentrations showed little association with ΣPCBs. Although similar associations between liver biomarkers and organochlorine exposure in fish-eating birds are well documented, typically exposures involve multiple contaminants and there is uncertainty about specific PCB effects. Our findings indicate that liver biomarkers respond to relatively low PCB exposures (approximately 73 ng/g liver) in guillemots.

Lande, Eirik. 1977. Heavy metal pollution in Trondheimsfjorden, Norway, and the recorded effects on the fauna and flora. Environ. Pollut. 12(3):187-198.
Rec #: 512
KEYWORDS: metals; Cd; Cu; Fe; Ni; Cr; Ag; Zn; Hg; Norway; Europe; common eider; *Somateria mollissima*; lesser black-backed gull; *Larus fuscus*; muscle; liver; kidney; egg; spatial variations; biomagnification; effects
ABSTRACT: The extent of heavy metal pollution in Trondheimsfjorden was investigated throughout the period 1972-73. Samples of invertebrates and seaweeds from the intertidal zone were collected and parts of some pelagic and bottom-dwelling fish were also sampled for the analysis of Cd, Cu, Fe, Ni, Cr, Ag, Zn, and Hg. Heavy metal pollution was found in 2 distinct areas of the fjord. The qualitative and quantitative changes in the faunal assemblages of these 2

Annotated Bibliography (Cont.).

areas are described.

Langlois, Claude and Langis, René. 1995. Presence of airborne contaminants in the wildlife of northern Quebec. Sci. Total Environ. 160-161:391-402.
Rec #: 195
KEYWORDS: metals; organochlorines; North America; Canada; Quebec; biomagnification; common eider; *Somateria mollissima*; arctic tern; *Sterna paradisaea*; herring gull; *Larus argentatus*; Canada goose; *Branta canadensis;* green-winged teal; *Anas crecca*; pintail; *Anas acuta*; common merganser; *Mergus merganser*; red-breasted merganser; *Mergus serrator*; American black duck; *Anas rubripes*; common loon; *Gavia immer*; great black-backed gull; *Larus marinus*; willow ptarmigan; *Lagopus lagopus*; muscle; liver; feather; egg; Hg; As; Cd; Pb; Se; PCBs; DDTs; HCHs; chlordanes; mirex; dieldrin
ABSTRACT: As part of the environmental impact studies of the Great Whale and the Nottaway-Broadback-Rupert (NBR) hydroelectric projects, Hydro-Quebec collected data on the occurrence and levels of several contaminants present in wildlife from both regions between 1989 and 1991. The analyses performed included metals (mercury, arsenic, selenium, cadmium, lead, nickel and copper), polychlorinated biphenyls (PCBs as arochlors or the sum of 20-40 congeners) and organochlorine pesticides such as hexachlorobenzene (HCB), DDT, DDE, hexachlorocyclohexane (HCH), chlordane, mirex and dieldrin. Species sampled included fish (freshwater and marine), birds (waterfowl, gull and ptarmigan), terrestrial mammals (marten, mink and hare) and marine mammals (freshwater and marine seals, belugas). Most laboratory analyses were carried out on both muscle and liver tissues, but some were conducted on other tissues as well: feathers, eggs and blubber. The results indicate that numerous airborne contaminants were present in the wildlife of both the Great Whale and the NBR study areas and that their level of contamination was similar to that of other northern environments. Total mercury in muscle was high in piscivorous fish, birds and mammals (terrestrial and marine). We observed significant levels of cadmium and lead in the livers of some herbivorous terrestrial animals, such as ptarmigans and hares. Among the organochlorine contaminants analyzed, levels of PCBs and DDE in piscivorous birds (mergansers and loons) and in marine mammals (seals and belugas) were high. For some contaminants, such as mercury, cadmium, lead, PCBs and DDE, the levels observed in some species or tissues could be considered worrisome with regard to public health, if those species or tissues constitute an important part of traditional native diets.

Larson, Jill M.; Karasov, William H.; Sileo, Louis; Stromborg, Kenneth L.; Hanbidge, Barbara A.; Giesy, John P.; Jones, Paul D.; Tillitt, Donald E.; and Verbrugge, David A. 1996. Reproductive success, developmental anomalies, and environmental contaminants in double-crested cormorants (*Phalacrocorax auritus*). Environ. Toxicol. Chem. 15(4):553-559.
Rec #: 170
KEYWORDS: reproductive success; organochlorines; PCBs; PCDDs; toxicology; TEQs; EROD; WI; USA; Canada; Manitoba; Great Lakes; North America; effects; abnormalities; egg; double-crested cormorant; *Phalacrocorax auritus*; spatial variations
ABSTRACT: To test an association between environmental contaminants and the prevalence of congenital anomalies in colonial waterbirds, we collected representative eggs for chemical analysis from double-crested cormorant nests at colonies in Lake Michigan, Wisconsin, USA, and Lake Winnipegosis, Manitoba, Canada, and periodically revisited the nests to determine the

Annotated Bibliography (Cont.).

hatching success, survivorship of hatchlings, and number of deformed hatchlings in the remainder of each clutch. Total concentrations of polychlorinated biphenyls (PCBs) in eggs were determined by capillary gas chromatography. The combined activity of planar chlorinated hydrocarbons (PCHs) in the eggs was measured in an in vitro bioassay based on the induction of ethoxyresorufin-O-deethylase (EROD) activity in rat hepatoma cells. The combined EROD induction activity was expressed as 2,3,7,8-tetrachlorodibenzo-p-dioxin equivalents (TCDD-EQ). Total concentrations of PCBs and TCDD-EQ were seven to eight times greater in eggs from Lake Michigan (7.8 µg/g and 138 pg/g, respectively) than in those from Lake Winnipegosis (1.0 µg/g and 19 pg/g, respectively). The proportion of eggs hatching at the Lake Michigan colony (59 %) was less ($p < 0.05$) than at Lake Winnipegosis (70 %), and the prevalence of hatchlings with deformed bills was greater ($p < 0.001$) at Lake Michigan (0.79 vs. 0.06 %). However, within the Lake Michigan colony, concentrations of PCBs and TCDD-EQ were not correlated with either hatching success or the occurrence of deformities in nestlings.

Lebedev, A. T.; Poliakova, O. V.; Karakhanova, N. K.; Petrosyan, V. S.; and Renzoni, A. 1998. The contamination of birds with organic pollutants in the Lake Baikal region. Sci. Total Environ. 212(2-3):153-162.
Rec #: 204
KEYWORDS: biomagnification; metabolism; egg; organochlorines; PCDFs; hydrocarbons; PAHs; phenols; Russia; Asia; goose; *Anser anser*; grey heron; *Ardea cinerea*; domestic fowl; *Gallus gallus*; mallard; *Anas platyrhynchos*; pintail; *Anas acuta*; shoveler; *Anas clypeata*; tufted duck; *Aythya fuligula;* pochard; *Aythya ferina*; herring gull; *Larus argentatus*; common gull; *Larus canus*; black headed gull; *Larus ridibundus*; horned grebe; *Podiceps auritus*; common tern; *Sterna hirundo*; lapwing; *Vanellus vanellus*; marsh sandpiper; *Tringa stagnatilis*
ABSTRACT: Lake Baikal is considered to be the largest reservoir of fresh natural water in the world. Nevertheless industrial enterprises on its banks as well as river effluents contaminate this unique basin. In the present study birds' eggs (15 species) collected in the Baikal region (Selenga river estuary) have been analysed. Quantitative determination of more than 40 individual organic pollutants (polycyclic aromatic hydrocarbons, phenols, organochlorine compounds) has been carried out using GC-MS as an analytical tool. The results obtained demonstrated a wide range of toxicant concentrations (2-3 orders of magnitude) for various species. Very high levels of polycyclic aromatic hydrocarbons have been detected in the eggs of *Anas platyrhynchos* (mallard), *Tringa stagnatilis* (marsh sandpiper) and *Podiceps auritus* (slavonian grebe). These particular species also have the highest levels of other toxicants. Taking into account high rate of metabolism of certain of these compounds in birds, it has been proposed that the major route of transfer into higher trophic levels is via water and aquatic invertebrates.

Lee, Y-Z.; Leighton, F. A.; Peakall, D. B.; Norstrom, R. J.; O'Brien, P. J.; Payne, J. F.; and Rahimtula, A. D. 1985. Effects of ingestion of Hibernia and Prudhoe Bay crude oils on hepatic and renal mixed function oxidase in nestling herring gulls (*Larus argentatus*). Environ. Res. 36(1):248-255.
Rec #: 144
KEYWORDS: oil; effects; enzymes; herring gull; *Larus argentatus*; liver; kidney; toxicology; MFOs; cytochrome P_{450}; EROD; dosing study
NOTES: SUPPLEMENTAL

Annotated Bibliography (Cont.).

ABSTRACT: Oral administration of Prudhoe Bay crude or Hibernia crude to nestling herring gulls increased the hepatic cytochrome P_{450} content 4-fold. Concomitantly, there was an increase in various mixed-function oxidase and phase II enzyme activities. 7-Ethoxyresorufin O-deethylase was elevated 19-fold, benzo(a)pyrene 3-hydroxylase 6-fold, aniline hydroxylase 3-fold, and aminopyrine N-demethylase and uridine diphosphate glucuronyl transferase 2-fold. There was no change in reduced glutathione S-transferase activity. Renal mixed-function oxidase activities were also elevated. Herring gull livers contained very low levels of DT-diaphorase activity which was inducible 3- to 5-fold by oil administration.

Leighton, F. A.; Lee, Y. Z.; Rahimtula, A. D.; O'Brien, P. J.; and Peakall, D. B. 1985. Biochemical and functional disturbances in red blood cells of herring gulls ingesting Prudhoe Bay crude oil. Toxicol. Appl. Pharmacol. 81(1):25-31 .
Rec #: 151
KEYWORDS: blood; oil; toxicology; dosing study; effects; herring gull; *Larus argentatus*
NOTES: SUPPLEMENTAL
ABSTRACT: Heinz body hemolytic anemia developed in herring gull (*Larus argentatus*) nestlings given oral doses of 10 ml of Prudhoe Bay crude oil per kilogram of body weight per day for 5 days. Associated disturbances in red blood cells were increased amounts of reduced glutathione (GSH), peroxidation of membrane lipids, an increase in membrane permeability, and a decrease in the oxygen-carrying capacity of cyanomethemoglobin-convertible hemoglobin. Among groups of gulls given different cumulative doses of oil over a 6-day period, significant covariance with dose and dependence on dose was demonstrated for packed cell volume, hemoglobin, and red cell GSH. Rapid defecation of oil by gulls indicated that the effective dose was substantially less than the administered dose. Pronounced damage to red cells occurred in some birds administered oil for only 2 days. The data imply that the toxic effects of ingested oil may contribute significantly to the morbidity and mortality of oil-contaminated birds.

Leighton, Frederick A.; Peakall, David B.; and Butler, Ronald G. 1983. Heinz-Body hemolytic anemia from the ingestion of crude oil: A primary toxic effect in marine birds. Science. 220(4599):871-873.
Rec #: 140
KEYWORDS: blood; oil; effects; herring gull; *Larus argentatus*; Atlantic puffin; *Fratercula arctica*; toxicology; dosing study
NOTES: SUPPLEMENTAL
ABSTRACT: Hemolytic anemia developed in young herring gulls (*Larus argentatus*) and Atlantic puffins (*Fratercula arctica*) given daily oral doses of a Prudhoe Bay crude oil. Anemia developed 4 to 5 days after the initiation of oil ingestion and was accompanied by Heinz-body formation and a strong regenerative response. The data evince a toxic effect on circulating red blood cells involving an oxidative biochemical mechanism and the first clear evidence of a primary mechanism of toxicity from the ingestion of crude oil by birds.

Lemmetyinen, Risto and Rantamäki, Pirjo. 1980. DDT and PCB residues in the arctic tern (*Sterna paradisaea*) nesting in the archipelago of southwestern Finland. Annales Zoologici Fennici. 17(3):141-146.
Rec #: 471
KEYWORDS: organochlorines; DDTs; PCBs; liver; muscle; egg; Arctic tern; *Sterna*

Annotated Bibliography (Cont.).

paradisaea; Finland; Europe; gender differences; age variations; biomagnification; effects; abnormalities

ABSTRACT: Arctic terns of different ages and their eggs were collected in the archipelago of Southwestern Finland. DDT (I) [50-29-3] and PCB residues of the samples were analyzed by gas chromatography In the adults, there was a significant difference between the sexes, contamination being higher in the males, whose livers contained on average 31.4 mg/kg total I compounds and 80.1 mg/kg PCB compounds in extractable fat. In the females, the amounts of these residues were significantly smaller (21.4 and 38.7 mg/kg, respectively). The reason for the difference is probably that laying females are able to shed part of their pesticide loads, including total I and especially PCB, into the eggs. Total I contamination of tern eggs was 8.1 mg/kg of lipid weight and that of newly hatched chicks 10.5 mg/kg, the corresponding values for PCBs being 27.7 and 44.0 mg/kg, respectively In chicks 2-3 wk old, in contrast, contamination had decreased (mean total I 2.8 mg/kg and PCBs 10.8 mg/kg). Thus, when drawing up models of the accumulation of environmental pollutants along food chains, the stages of the life cycle at which samples were taken are important. No developmental defects in Arctic terns could be attributed to chlorinated hydrocarbons during the productivity study in 1965-1973.

Lemmetyinen, Risto; Rantamäki, Pirjo; and Karlin, Antti. 1982. Levels of DDT and PCB's in different stages of life cycle of the arctic tern *Sterna paradisaea* and the herring gull *Larus argentatus*. Chemosphere. 11(10):1059-1068.
Rec #: 177
KEYWORDS: Arctic tern; *Sterna paradisaea;* herring gull*; Larus argentatus*; organochlorines; DDTs; PCBs; Finland; Baltic Sea; Europe; egg; muscle; liver; age variations; bioaccumulation
ABSTRACT: Sigma DDT and PCB levels were analyzed in samples of arctic terns (*Sterna paradisaea*) and herring gulls (*Larus argentatus*) collected in the archipelago of southwestern Finland. The levels were nearly ten times higher in the herring gull. The highest loads were found in adult birds and in newly hatched chicks but the levels were much lower in chicks just before fledgling. The levels in young gulls remained low until the end of August at least. Therefore it is plausible that the high levels found in adult gulls are a consequence of their wintering in the southern Baltic. The level of Sigma DDT and PCB residues were significantly lower in female arctic terns than in male terns. Differences between the sexes were small in the herring gull. Thus it is possible that the female of the arctic tern in able to release pollutants, especially PCB residues, more effectively into eggs than the female of the herring gull.

Leonzio, C.; Focardi, S.; Fossi, C.; and Renzoni, A. 1986. Sea-birds as indicators of mercury pollution in the Mediterranean. FAO Fish. Rep. 325:116-121.
Rec #: 105
KEYWORDS: metals; Hg; Mediterranean Sea; Italy; Europe; egg; little tern; *Sterna albifrons*; black-necked grebe; *Podiceps nigricollis*; Cory's shearwater; *Calonectris diomedea*; Audouin's gull; *Larus audouinii*; mallard; *Anas platyrhynchos*; little egret; *Egretta garzetta*; black-headed gull; *Larus ridibundus;* slender-billed gull; *Larus genei;* herring gull; *Larus argentatus;* gull-billed tern; *Sterna nilotica;* black-winged stilt; *Himantopus himantopus;* avocet; *Recurvirostra avosetta;* coot; *Fulica atra;* cormorant; *Phalacrocorax carbo;* pygmy cormorant; *Phalacrocorax pygmeus;* great white pelican; *Pelecanus onocrotalus;* common tern; *Sterna hirundo;* little tern; *Sterna albifrons*; seasonal variations; biomagnification
ABSTRACT: An analysis was made of mercury in migratory water birds of the Mediterranean

Annotated Bibliography (Cont.).

comparing results of analyses made a few days after survival in winter quarters with those made a few days before the flock's departure for breeding areas. Eggs of species breeding the Mediterranean Basin were also analysed. The liver was found to be the most contaminated organ in the birds, which clearly reflect the amount of mercury available in their feeding grounds.

Leonzio, Claudio; Fossi, Cristina; and Focardi, Silvano. 1986. Lead, mercury, cadmium and selenium in two species of gull feeding on inland dumps, and in marine areas. Sci. Total Environ. 57:121-127.
Rec #: 106
KEYWORDS: metals; Pb; Hg; Se; Cd; black-headed gull; *Larus ridibundus*; herring gull; *Larus argentatus*; Italy; Europe; fat; gland; muscle; brain; kidney; liver; spatial variations
ABSTRACT: The influence of the foraging area on the intake of trace elements was evaluated in tissues of two species of gull: the black-headed gull (*Larus ridibundus*) and the herring gull (*Larus argentatus*). Mercury levels were higher in birds from the coastal environment than in those caught on inland dumps; birds found at one dump site showed high levels of lead. Interspecific differences in the levels of mercury and cadmium are discussed.

Lewis, S. A.; Becker, P. H.; and Furness, R. W. 1993. Mercury levels in eggs, tissues, and feathers of herring gulls, *Larus argentatus*, from the German Wadden Sea coast. Environ. Pollut. 80(3):293-299.
Rec #: 125
KEYWORDS: effects; bioaccumulation; metals; Hg; egg; Wadden Sea; Germany; Europe; herring gull; *Larus argentatus*; feather; muscle; liver; ovary; egg; elimination; gender differences
ABSTRACT: The relationships between mercury levels in eggs, tissues, and feathers of male and female herring gulls (*Larus argentatus*) caught at their nests at a colony on the German Wadden Sea Coast were investigated, and an assessment of mercury intake and excretion of these birds was made. Samples of the liver, ovary, pectoral muscle, and body feathers, as well as the primary feather (in some cases), and eggs were taken from 37 adult herring gulls. Analysis of total mercury in all samples showed that body-feather and tissue levels were independent of sex. There was, however, a significant difference between levels of mercury in the primary feathers of male and female birds. Egg levels were not correlated to pectoral muscle, ovary, or feather levels although they were positively correlated with mercury levels in the liver. Liver levels were in turn correlated with mercury levels in the ovary. The ratio of mean feather to mean egg concentration ranged from 3.7 to 5.5 according to which feather was used. It was estimated that herring gulls from the Wadden coast ingested between 825 and 1337 µg of mercury in the year prior to analysis. It was also estimated that female birds may excrete over 20 % more mercury via their eggs than could be excreted by male birds.

Lloyd, C.; Bogan, J. A.; Bourne, W. R. P.; Dawson, P.; Parslow, J. L. F.; and Stewart, A. G. 1974. Seabird mortality in the North Irish Sea and Firth of Clyde early in 1974. Mar. Pollut. Bull. 5(9):136-140.
Rec #: 508
KEYWORDS: organochlorines; PCBs; DDTs; dieldrin; common murre; *Uria aalge*; razorbill; *Alca torda;* herring gull; *Larus argentatus*; common eider; *Somateria mollissima*; black-headed gull; *Larus ridibundus*; oystercatcher; *Haematopus ostralegus*; black-legged kittiwake; *Rissa tridactyla*; common gull; *Larus canus;* cormorant; *Phalacrocorax carbo*; redshank; *Tringa*

Annotated Bibliography (Cont.).

totanus; red-breasted merganser; *Mergus serrator*; mute swan; *Cygnus olor*; liver; muscle; Irish Sea; Scotland; Europe; effects

ABSTRACT: Autopsy on seabirds from the Firth of Clyde showed the presence of high levels of dieldrin (I) [60-57-1], as well as polychlorinated biphenyls. Autopsy on seabirds from the North Irish Sea (especially *Uria aalge*) showed the presence of high levels of polychlorinated biphenyls. Probably, organochlorine contamination was a contributory factor to the heavy mortality of these birds.

Lock, J. W.; Thompson, D. R.; Furness, R. W.; and Bartle, J. A. 1992. Metal concentrations in seabirds of the New Zealand region. Environ. Pollut. 75(3):289-300.
Rec #: 57
KEYWORDS: New Zealand; age variations; bioaccumulation; toxicology; biomagnification; elimination; metals; Cd; Cu; Pb; Zn; Hg; organometallics; organic Hg; liver; kidney; feather; bone; wandering albatross; *Diomedea exulans*; royal albatross; *Diomedea epomophora*; black-browed albatross; *Diomedea melanophris*; white-capped albatross; *Diomedea cauta*; Salvin's albatross; *Diomedea cauta salvini;* grey-headed albatross*; Diomedea chrysostoma;* Buller's albatross; *Diomedea bulleri*; light-mantled sooty albatross; *Phoebetria palpebrata*; flesh-footed shearwater; *Puffinus carneipes*; short-tailed shearwater; *Puffinus tenuirostris*; Hutton's shearwater; *Puffinus huttoni*; little shearwater; *Puffinus assimilis*; diving petrel; *Pelecanoides urinatrix*; South Georgian diving petrel; *Pelecanoides georgicus*; grey petrel; *Procellaria cinerea;* westland petrel; *Procellaria westlandica*; Kerguelen petrel; *Pterodroma brevirostris*; white-chinned petrel; *Procellaria aequinoctialis*; cape pigeon; *Daption capense*; Antarctic petrel; *Thalassoica antarctica*; southern giant petrel; *Macronectes giganteus*; northern giant petrel; *Macronectes halli*; fairy prion; *Pachyptila turtur*; fulmar prion; *Pachyptila crassirostris*; Antarctic prion; *Pachyptila desolata*; blue petrel; *Halobaena caerulea*; Pycroft's petrel; *Pterodroma pycrofti*; Cook's petrel; *Pterodroma cookii*; black-winged petrel; *Pterodroma nigripennis*; Chatham Island petrel; *Pterodroma axillaris*; mottled petrel; *Pterodroma inexpectata*; grey-faced petrel; *Pterodroma macroptera*; white-headed petrel; *Pterodroma lessonii*; grey-backed storm-petrel; *Oceanites nereis*; black-bellied storm-petrel; *Fregetta tropica*; yellow-eyed penguin; *Megadyptes antipodes*; blue penguin; *Eudyptula minor*; rockhopper penguin*; Eudyptes chrysocome*; Fiordland crested penguin; *Eudyptes pachyrhynchus*; erect-crested penguin; *Eudyptes sclateri*; red-tailed tropicbird; *Phaethon rubricauda*; white-tailed tropicbird; *Phaethon lepturus*; brown booby; *Sula leucogaster*; cormorant; *Phalacrocorax carbo*; Australasian gannet*; Morus serrator*; little shag; *Phalacrocorax melanoleucos*; Bounty Island shag; *Leucocarbo ranfurlyi*; king shag; *Leucocarbo atriceps*; blue shag; *Stictocarbo punctatus steadi*; spotted shag; *Stictocarbo punctatus punctatus*; long-tailed jaeger; *Stercorarius longicaudus*; great frigatebird; *Fregata minor*; brown skua; *Catharacta lonnbergi*; parasitic jaeger; *Stercorarius parasiticus*; southern black-backed gull; *Larus dominicanus*; red-billed gull; *Larus novaehollandiae*; white-fronted tern; *Sterna striata*; sooty tern; *Sterna fuscata*; white-capped noddy; *Anous tenuirostris*; Arctic tern; *Sterna paradisaea*

ABSTRACT: Concentrations of the heavy metals cadmium, copper, lead, zinc, mercury and, in some individuals, methyl mercury were determined in a range of tissues of 64 tropical, subtropical, subantarctic and antarctic seabird taxa mostly from the New Zealand region. Although apparently natural, levels of cadmium and mercury in some species greatly exceed those known to have toxic effects in some terrestrial birds. Copper and zinc levels exhibited less

Annotated Bibliography (Cont.).

inter-species variation than the non-essential metals cadmium and mercury. Cadmium concentrations were highest in kidney tissues but uniformly low in feathers. Total mercury concentrations showed most inter-species variation. Mean methyl mercury levels in liver tissues of several large procellariiforms represented less than 5 % of the corresponding mean total mercury level. Lead concentrations were generally low or below the limits of detection, but elevated levels were measured in some coastal or scavenging species. In a significant number of species, mean concentrations of liver cadmium and mercury and kidney cadmium were greater in adults than in young birds. The reverse was true for copper. Mean zinc levels in liver did not differ between adults and young. High levels of cadmium in some species seem likely to be due to diet, whereas high levels of mercury probably reflect more closely the moult intervals which constrain the ability of birds to eliminate methyl mercury.

Lorentsen, S-H. and Anker-Nilssen, T. 1993. Behaviour and oil vulnerability of fulmars *Fulmarus glacialis* during an oil spill experiment in the Norwegian Sea. Mar. Pollut. Bull. 26(3):144-146.
Rec #: 243
KEYWORDS: oil; northern fulmar; *Fulmarus glacialis*; Norway; Europe; effects
NOTES: SUPPLEMENTAL
ABSTRACT: The behaviour and oil vulnerability of Fulmars (*Fulmarus glacialis*) was studied when 30 t of crude oil was experimentally released at Haltenbanken, 150 km off the coast of Central Norway in July 1989. Evidence strongly suggests that the Fulmars deliberately avoided to settle on sea surface that was polluted with heavy oil. However, about 4 % of the Fulmars in the area were slightly oiled, probably because they had been attracted to surrounding blueshine areas by food remains thrown overboard from the research vessel.

Lorenzen Angela; Moon, Thomas W.; Kennedy, Sean W.; and Fox, Glen A. 1999. Relationships between environmental organochlorine contaminant residues, plasma corticosterone concentrations, and intermediary metabolic enzyme activities in Great Lakes herring gull embryos. Environ. Health Perspect. 107(3):179-186.
Rec #: 327
KEYWORDS: herring gull; *Larus argentatus*; egg; blood; liver; kidney; Great Lakes; North America; metabolism; organochlorines; PCBs; PCDDs; PCDFs; TEQs; toxicology
NOTES: FIGURES ONLY
ABSTRACT: Experiments were conducted to survey and detect differences in plasma corticosterone concentrations and intermediary metabolic enzyme activities in herring gull (*Larus argentatus*) embryos environmentally exposed to organochlorine contaminants in ovo. Unincubated fertile herring gull eggs were collected from an Atlantic coast control site and various Great Lakes sites in 1997 and artificially incubated in the laboratory. Liver and/or kidney tissues from approximately half of the late-stage embryos were analyzed for the activities of various intermediary metabolic enzymes known to be regulated, at least in part, by corticosteroids. Basal plasma corticosterone concentrations were determined for the remaining embryos. Yolk sacs were collected from each embryo and a subset was analyzed for organochlorine contaminants. Regression analysis of individual yolk sac organochlorine residue concentrations, or 2,3,7,8-tetrachlorodibenzo-p-dioxin equivalents (TEQs), with individual basal plasma corticosterone concentrations indicated statistically significant inverse relationships for polychlorinated dibenzo-p-dioxins/polychlorinated dibenzofurans (PCDDs/PCDFs), total

Annotated Bibliography (Cont.).

polychlorinated biphenyls (PCBs), non-ortho PCBs, and TEQs. Similarly, inverse relationships were observed for the activities of two intermediary metabolic enzymes (phosphoenolpyruvate carboxykinase and malic enzyme) when regressed against PCDDs/PCDFs. Overall, these data suggest that current levels of organochlorine contamination may be affecting the hypothalamo-pituitary-adrenal axis and associated intermediary metabolic pathways in environmentally exposed herring gull embryos in the Great Lakes.

Ludwig, James P.; Giesy, John P.; Summer, Cheryl L.; Bowerman, William; Aulerich, Richard; Bursian, Steven; Auman, Heidi J.; Jones, Paul D.; Williams, Lisa L.; Tillitt, Donald E.; and Gilbertson, Michael. 1993. A comparison of water quality criteria for the Great Lakes based on human and wildlife health. J. Great Lakes Res. 19(4):789-807.
Rec #: 447
KEYWORDS: herring gull; *Larus argentatus*; double-crested cormorant; *Phalacrocorax auritus*; Caspian tern; *Sterna caspia*; egg; Great Lakes; North America; organochlorines; PCBs; PCDDs; TEQs; toxicology; review; effects
NOTES: SUPPLEMENTAL
ABSTRACT: Water quality criteria (WQC) can be derived in several ways. The usual techniques involve hazard and risk assessment procedures. For non-persistent, non-biomagnified compounds and elements, WQC are experimentally derived from their acute and chronic toxicity to aquatic organisms. For those persistent chlorinated hydrocarbons (PCHs) that are bioaccumulated and biomagnified, these traditional techniques have not been effective, partly because effects higher in the food web were not considered. Polychlorinated biphenyls (PCBs) are the bioaccumulative synthetic chemicals of primary toxicological significance to the Great Lakes biota which have caused widespread injury to wildlife. In the Laurentian Great Lakes, the primary emphasis of hazard assessments has been on the potential for adverse effects in humans who eat fish. The primary regulatory endpoint of traditional hazard and risk assessments underlying current WQC are the probabilities of additional cancers occurring in the human population. The analysis presented here indicates that this is not adequate to restore sensitive wildlife species that are highly exposed to PCBs, especially those that have suffered serious population declines. Because WQC are legal instruments, the methods of deriving WQC have large implications for remediation, litigation, and damage assessments. Here WQC are derived for six species based on the responses of wildlife in the field or produced by feeding fish to surrogate species, rather than projecting a potential of increased cancer rates in humans. If the most sensitive wildlife species are restored and protected for very sensitive reproductive endpoints, then all components of the ecosystem, including human health, should be more adequately protected. The management of Great Lakes wildlife requires an understanding of the injury and causal relationships to persistent toxic substances.

Ludwig, James P.; Kurita-Matsuba, Hiroko; Auman, Heidi J.; Ludwig, Matthew E.; Summer, Cheryl L.; Giesy, John P.; Tillitt, Donald E.; and Jones, Paul D. 1996. Deformities, PCBs, and TCDD-equivalents in double-crested cormorants (*Phalacrocorax auritus*) and Caspian terns (*Hydroprogne caspia*) of the upper Great Lakes 1986-1991: Testing a cause-effect hypothesis. J. Great Lakes Res. 22(2):172-197.
Rec #: 169
KEYWORDS: organochlorines; PCBs; PCDDs; TEQs; toxicology; EROD; reproductive success; effects; abnormalities; North America; Great Lakes; egg; double-crested cormorant;

Annotated Bibliography (Cont.).

Phalacrocorax auritus; Caspian tern; *Sterna caspia*
ABSTRACT: Deformities have been reported in many species of colonial waterbirds from several localities on the Laurentian Great Lakes. The hypothesis that deformities were caused by either polychlorinated biphenyls (PCBs) or contaminants measured as 2,3,7,8-tetrachlorodibenzo-p-dioxin equivalents (TCDD-EQs) is tested in this review of available data on concentrations of contaminants in eggs and observed deformities in embryos and chicks of double-crested cormorants (*Phalacrocorax auritus*) and Caspian terns (*Hydroprogne caspia*) between 1986 and 1991. Hatched chicks, live and dead eggs retrieved from 37 colonies in the upper Great Lakes were assessed for gross anatomical deformities. Rates of embryo death from seven regions of the upper Great Lakes were measured annually between 1986-1991. Half the embryos found dead in eggs were deformed. Nineteen types of abnormalities or deformities were observed. Subcutaneous edema in cormorants and gastroschisis in terns were the most common abnormalities in live or dead eggs. One of ten crossed-billed cormorant embryos survived to hatch. No bill-deformed terns hatched, although tern embryos had a greater rate of crossed-bills than cormorants. The suite of deformities and abnormalities found was similar to that produced in chickens by exposure to planar polychlorinated biphenyl (pPCB) and dioxin congeners. Hatching and deformity rates were correlated with concentrations of pPCBs and TCDD-EQs. Planar PCB congeners that contributed most of the TCDD-EQs were present at concentrations sufficient to cause the observed effects. TCDD-EQs measured by H4IIE rat hepatoma cell 7-ethoxyresorufin O-deethylase (EROD) bioassay were highly correlated with deformity rates observed in cormorant chicks, live and dead eggs, and egg death rates. Similar correlations of TCDD-EQs with deformity rates were found in hatched tern chicks, dead eggs, and egg death rates, but not in live eggs. TCDD-EQs were more highly correlated to deformity and embryo death rates than total PCBs. The weight of evidence and these data are sufficient to reject the null hypothesis that there is no causal relationship between the incidence of deformities in cormorants and terns and exposure to planar halogenated compounds measured as TCDD-EQs or total PCBs in the Great Lakes.

MacDonald, C. R.; Norstrom, R. J.; and Turle, R. 1992. Application of pattern recognition techniques to assessment of biomagnification and sources of polychlorinated multicomponent pollutants, such as PCBs, PCDDs and PCDFs. Chemosphere. 25(1-2):129-134.
Rec #: 232
KEYWORDS: organochlorines; PCBs; PCDDs; PCDFs; mirex; dieldrin; chlordanes; egg; herring gull; *Larus argentatus*; North America; Great Lakes; temporal trends; biomagnification
NOTES: FIGURES ONLY
ABSTRACT: Principal component analysis of PCB congeners and organochlorines in herring gull eggs from four of the Great Lakes reveals that, with the exception of Lake Erie, the two colonies within each lake contain distinctive patterns which did not change significantly between 1983 and 1990. Lake Ontario colonies were modelled by high percentages of mirex and photomirex, while high proportions of dieldrin, heptachlor epoxide and oxychlordane grouped the colonies in lakes Huron and Superior. The Middle Island colony in the western end of Lake Erie showed high relative amounts of the higher chlorinated PCB congeners, due to continuing contamination from the Detroit River, or resuspension of contaminated sediments. PCDD/F patterns in the same colonies were dominated by 1,2,3,6,7,8 HxCDD and 1,2,3,7,8 PeCDD in lakes Erie, Huron and Superior while Lake Ontario patterns were dominated by 2,3,7,8 TCDD.

Annotated Bibliography (Cont.).

Lake trout and walleye in the same lakes showed high levels of 2,3,7,8 TCDF which does not accumulate in herring gulls.

Malcolm, H. M.; Osborn, D.; Wright, J.; Wienburg, C. L.; and Sparks, T. H. 2003. Polychlorinated biphenyl (PCB) congener concentrations in seabirds found dead in mortality incidents around the British coast. Arch. Environ. Contam. Toxicol. 45(1):136-147.
Rec #: 441
KEYWORDS: common murre; *Uria aalge*; shag; *Phalacrocorax aristotelis*; black-legged kittiwake; *Rissa tridactyla*; Atlantic puffin; *Fratercula arctica*; razorbill; *Alca torda*; northern fulmar; *Fulmarus glacialis*; gannet; *Sula bassana*; herring gull; *Larus argentatus*; liver; organochlorines; PCBs; Great Britain; Europe; toxicology; TEQs; biomagnification; temporal trends; spatial variations; effects
ABSTRACT: Livers from 121 birds killed in mortality incidents in U.K. coastal waters between 1991 and 1996 were analyzed for 16 PCB congeners (IUPAC numbers 8, 18, 28, 31, 52, 77, 101, 118, 126, 128, 138, 149, 153, 169, 170, and 180). Species analyzed were guillemot (*Uria aalge*), shag (*Phalacrocorax aristotelis*), kittiwake (*Rissa tridactyla*), puffin (*Fratercula arctica*), razorbill (*Alca torda*), fulmar (*Fulmarus glacialis*), gannet (*Sula bassana*), and herring gull (*Larus argentatus*). This is the first report of PCBs in U.K. seabird tissues to be presented since 1983. Mean concentrations of total PCBs ranged from 0.47-15.8 mg/kg WW, similar to concentrations reported for North Sea birds during the 1970s and 1980s and lower than those reported for the same species in the Irish Sea during the late 1960s and early 1970s. The congeners generally present in the highest concentrations were (in decreasing order) 153, 138, 180, 118, and 170. Despite the birds having similar diets, both inter- and intraspecies differences in PCB congener profile were found. In guillemots from several sites on the east coast, the dominant congener was 138, compared to guillemots from the south coast, which had the more typical congener pattern found in seabirds. Some of the differences in total PCB values could be due to different causes of death, with a subsequent effect on body lipid levels.

Martin, Jonathan W.; Smithwick, Marla M.; Braune, Birgit M.; Hoekstra, Paul F.; Muir, Derek C. G.; and Mabury, Scott A. 2004. Identification of long-chain perfluorinated acids in biota from the Canadian Arctic. Environ. Sci. Technol. 38(2):373-380.
Rec #: 475
KEYWORDS: PFOS; perfluorinated acids; Nunavut; Yukon; Northwest Territories; Quebec; Canada; USA; North America; common loon; *Gavia immer*; northern fulmar; *Fulmarus glacialis*; black guillemot; *Cepphus grylle*; liver; biomagnification; spatial variations
ABSTRACT: Recently it was discovered that humans and animals from urban and remote global locations contained a novel class of persistent fluorinated pollutants, the most pervasive of which was perfluorooctane sulfonate (PFOS). Lower concentrations of perfluorooctanoate, perfluorohexane sulfonate, and heptadecafluorooctane sulfonamide were also detected in various samples. Although longer perfluoroalkyl carboxylates (PFCA) are used in industry and have been detected in fish following a spill of aqueous film forming foam, no studies have been conducted to examine the widespread occurrence of long-chain PFCA, e.g., $CF_3(CF_2)xCOO-$, where x >6. To preliminarily assess fluorinated pollutants, including PFCA, in the Canadian Arctic, polar bears, ringed seals, arctic fox, mink, common loons, northern fulmars, black guillemots, and fish were collected at various locations in the circumpolar region. PFOS was the major pollutant detected in most samples; in polar bear liver, it was the most prominent

Annotated Bibliography (Cont.).

organohalogen (mean PFOS = 3.1 µg/g wet weight) vs. individual polychlorinated biphenyl congeners, chlordane, or hexachlorocyclohexane-related chemicals in fat. Using 2 independent mass spectral methods, it was confirmed that all samples also contained ng/g concentrations of a homologous series of PFCA, ranging in length from 9 to 15 carbons. Sum concentrations of PFCA (ΣPFCA) were lower than total PFOS equivalents (ΣPFOS) in all samples except mink. In mink, perfluorononanoate (PFNA) concentrations exceeded PFOS concentrations, indicating PFNA and other PFCA should be considered in future risk assessments. Mammals feeding at higher trophic levels had greater PFOS and PFCA concentrations than mammals feeding at lower trophic positions. Generally, odd-length PFCA exceeded the concentration of even-length PFCA; concentrations decreased with increasing chain length in mammals. PFOS and PFCA concentrations were much lower for animals living in the Canadian Arctic than for the same species living in mid-latitude regions of the USA. Future studies should continue to monitor all fluorinated pollutants and examine absolute and relative toxicities for this novel suite of PFCA.

Martin, Pamela A.; Weseloh, D. Vaughn; Bishop, Christine A.; Legierse, Karin; Braune, Birgit; and Norstrom, Ross J. 1995. Organochlorine contaminants in avian wildlife of Severn Sound. Water Qual. Res. J. Canada. 30(4):693-711.
Rec #: 229
KEYWORDS: biomagnification; egg; muscle; Canada; Great Lakes; North America; organochlorines; DDTs; dieldrin; chlordanes; HCHs; benzenes; HCB; OCS; PCBs; PCDFs; PCDDs; herring gull; *Larus argentatus;* Caspian tern; *Sterna caspia;* common tern; *Sterna hirundo*; red-winged blackbird; *Agelaius phoeniceus*; tree swallow; *Tachycineta bicolor*; lesser scaup; *Aythya affinis*; mallard; *Anas platyrhynchos*
ABSTRACT: Organochlorine contaminants were measured in pooled egg samples of colonial waterbirds, red-winged blackbirds and tree swallows breeding in or near Severn Sound, Lake Huron, an area designated by the International Joint Commission as an Area of Concern, during 1991. Breast muscle samples of staging waterfowl were also collected the preceding autumn. PCBs, DDE and mirex were the most prevalent organochlorine contaminants present in eggs. Generally, the following interspecific gradient in contaminant levels in bird eggs was found: herring gulls = Caspian terns > common terns > tree swallows > blackbirds.

Matheson, R. A. F.; Hamilton, E. A.; Trites, A.; and Whitehead, D. 1980. Chlorinated benzenes in herring gull and double-crested cormorant eggs from three locations in the Maritime Provinces. Halifax, Nova Scotia, Canada: Environmental Protection Service, Atlantic Region. EPS-5-AR-80-1. 37 pp.
Rec #: 407
KEYWORDS: organochlorines; HCB; benzenes; herring gull; *Larus argentatus*; double-crested cormorant; *Phalacrocorax auritus*; egg; Nova Scotia; New Brunswick; Canada; North America; spatial variations
ABSTRACT: Concentrations of hexachlorobenzene [118-74-1] in extracts from eggs of herring gulls (*Larus argentatus*) and double-crested cormorants (*Phalacrocorax auritus*) were higher (8.3-20 ng/g) than those of other chlorinated benzene isomers. 1,2,3,5-Tetrachlorobenzene [634-90-2], 1,2,3,4-tetrachlorobenzene [634-66-2], and pentachlorobenzene [608-93-5] were found in most of the samples, whereas 1,2,4-trichlorobenzene [120-82-1] and 1,2,3-trichlorobenzene [87-61-6] were detected less frequently. The egg samples from the maritime provinces had less chlorinated benzenes than similar samples from Lake Ontario.

Annotated Bibliography (Cont.).

Meadows, J. C.; Tillitt, D. E.; Schwartz, T. R.; Schroeder, D. J.; Echols, K. R.; Gale, R. W.; Powell, D. C.; and Bursian, S. J. 1996. Organochlorine contaminants in double-crested Cormorants from Green Bay, WI: I. Large-scale extraction and isolation from eggs using semi-permeable membrane dialysis. Arch. Environ. Contam. Toxicol. 31(2):218-224.
Rec #: 24
KEYWORDS: egg; organochlorines; PCBs; WI; USA; Great Lakes; North America; double-crested cormorant; *Phalacrocorax auritus*
ABSTRACT: A 41.3-kg sample of double-crested cormorant (*Phalacrocorax auritus*) egg contents was extracted, yielding over 2 L of egg lipid. The double-crested cormorant (DCC) egg extract, after clean-up and concentration, was intended or use in egg injection studies to determine the embryotoxicity of the organic contaminants found within the eggs. Large-scale dialysis was used as a preliminary treatment to separate the extracted contaminants from the co-extracted sample lipids. The lipid was dialyzed in 80×5 cm semi-permeable membrane devices (SPMDs) in 50-ml aliquants. After the removal of 87 g of cholesterol by freeze-fractionation, the remaining lipid carryover (56 g) was removed by 100 routine gel permeation chromatography (GPC) operations. A 41,293-g sample was thus extracted and purified to the extent that it could easily be placed at a volume of 5 ml, the volume calculated to be necessary for the egg injection study. Analyses were performed comparing contaminant concentrations in the final purified extract to those present in the original egg material, in the extract after dialysis and cholesterol removal, and in the excluded materials. Recoveries of organochlorine pesticides through dialysis and cholesterol ranged from 96 % to 135 %. Total polychlorinated biphenyls in the final extract were 96 % of those measured in the original egg material. Analysis of excluded lipid and cholesterol indicated that 92 % of the polychlorinated dibenzo-dioxins and-furans were separated into the final extract.

Medvedev, Nikolai and Markova, Lubov. 1995. Residues of chlorinated pesticides in the eggs of Karelian birds, 1989-90. Environ. Pollut. 87(1):65-70.
Rec #: 373
KEYWORDS: egg; Russia; Asia; common gull; *Larus canus;* herring gull; *Larus argentatus;* black-headed gull; *Larus ridibundus;* common tern; *Sterna hirundo*; crow; *Corvus cornix*; organochlorines; DDTs; HCHs; reproductive success; effects; eggshell thickness
ABSTRACT: Eggs (n = 52) of 4 aquatic bird species (*Larus canus, L. argentatus, L. ridibundus, Sterna hirundo*) and crow (*Corvus cornix*), collected in Southern Karelia in 1989 and 1990, contained DDE and lindane. The highest mean levels of these pollutants were in herring gull and common tern eggs; lowest mean levels of pesticides were in crow eggs. In all eggs, DDE concentrations were lower than the critical threshold and it seems, cannot influence reproductive success.

Mehlum, Fridtjof and Daelemans, F. F. 1995. PCBs in Arctic seabirds from the Svalbard Region. Sci. Total Environ. 160/161:441-446.
Rec #: 77
KEYWORDS: organochlorines; PCBs; Svalbard; Norway; Europe; liver; glaucous gull; *Larus hyperboreus*; black guillemot; *Cepphus grylle*; common eider; *Somateria mollissima*; thick-billed murre; *Uria lomvia*; spatial variations; biomagnification
ABSTRACT: Previous studies have indicated the presence of high levels of organochlorines, especially PCBs, in some species of arctic seabirds. The glaucous gull (*Larus hyperboreus*) in

Annotated Bibliography (Cont.).

particular has shown high levels of organochlorine contamination. We present data on total PCB and isomer specific PCBs in liver samples from four species of seabirds from the Svalbard region. Two of the species were sampled both in the vicinity of the mining town Longyearbyen, where PCBs have been used in the past, and in the remote region of Nordaustlandetect We compared the levels obtained from these two localities in order to test for any indication of local contamination in the Longyearbyen area. No significant difference was found in total hepatic PCB between glaucous gulls collected at Longyearbyen and those collected at Nordaustlandet (grand mean ± S.D.: 15.59 ± 21.53 µg/g wet weight; n = 22). Similarly, no difference was found in black guillemots (*Cepphus grylle*) from these two regions (grand mean ± S.D.: 0.14 ± 0.05 µg/g wet weight; n = 20). However, in both species, we found significantly higher levels of higher chlorinated biphenyls in the Longyearbyen samples, and higher levels of lower chlorinated biphenyls in the samples from Nordaustlandetect. This finding indicates that the birds sampled at the two localities might have been contaminated by different sources. Local contamination in the Longyearbyen area is one of several possible explanations for this difference. Total PCB levels in common eiders (*Somateria mollissima*) and Brünnich's guillemots (*Uria lomvia*) were in the same order of magnitude as those for the black guillemots (mean ± S.D.: 0.04 ± 0.04 µg/g; n = 11 and 0.08 ± 0.04 µg/g; n = 8, respectively).

Miller, D. S.; Butler, R. G.; Trivelpiece, W. Z.; Janes-Butler, S.; Green, S.; Peakall, B.; Lambert, G.; and Peakall, D. B. 1980. Crude oil ingestion by seabirds: Possible metabolic and reproductive effects. Bull. Mt. Desert Isl. Biol. Lab. 20:137-138.
Rec #: 126
KEYWORDS: oil; effects; toxicology; metabolism; reproductive success; herring gull; *Larus argentatus*; dosing study
NOTES: SUPPLEMENTAL; NO TABLES
ABSTRACT: The authors present here the results of initial experiments concerned with two problems related to oil toxicity in birds (*Larus argentatus*). First, does ingested oil affect metabolism? Second, does oral dosing of adult seabirds affect the growth and survival of their young?

Miller, D. S.; Hallett, D. J.; and Peakall, D. B. 1982. Which components of crude oil are toxic to young seabirds? Environ. Toxicol. Chem. 1(1):39-44.
Rec #: 150
KEYWORDS: oil; effects; toxicology; dosing study; herring gull; *Larus argentatus*
NOTES: SUPPLEMENTAL
ABSTRACT: Studies from this laboratory have focused on sublethal physiological effects of small amounts of ingested crude oil in young herring gulls (*Larus argentatus*). Clearly, the most striking effect of certain oils in gulls is the marked reduction in rates of weight gain found after administration of a single 0.2-1 ml oral dose. SLC-76 was split on an alumina column into aliphatic and aromatic fractions; only the aromatic fraction reduced gull rates of weight gain. Using a Prudhoe Bay crude (as toxic to gulls as SLC-76), the aromatic fraction was split into two subfractions on Sephadex LH-20. Analyses showed that one subfraction contained those aromatics with three or less rings and the other contained those with four or more rings. Only the second fraction (high molecular weight aromatics) reduced gull weight gain. The findings clearly show that oil composition is a major determinant of oral toxicity to young seabirds, and that the

Annotated Bibliography (Cont.).

higher molecular weight aromatics are most effective in reducing gull weight gain.

Miller, David S.; Kinter, William B.; and Peakall, David B. 1979. Effects of crude oil ingestion on immature Pekin ducks (*Anas platyrhynchos*) and herring gulls (*Larus argentatus*). Animals as Monitors of Environmental Pollutants. Symposium on Pathobiology of Environmental Pollutants: Animal Models and Wildlife as Monitors, University of Connecticut, 1977. 27-40.
Rec #: 134
KEYWORDS: effects; oil; dosing study; mallard; *Anas platyrhynchos*; herring gull; *Larus argentatus*
NOTES: SUPPLEMENTAL
ABSTRACT: A single small oral dose of either Kuwait or South Louisiana crude oil was administered to Pekin ducks and herring gull chicks. Crude oil dosing caused significant osmoregulatory impairment in seawater-stressed ducks. In seawater-adapted gulls osmoregulation was only slightly impaired by crude oil; however, growth was substantially inhibited. These findings suggest that crude oil causes sublethal effects which might impair a bird's ability to survive in a coastal or marine environment.

Miller, David S.; Kinter, William B.; Peakall, David B.; and Risebrough, Robert W. 1976. DDE feeding and plasma osmoregulation in ducks, guillemots, and puffins. Am. J. Physiol. 231(2):370-376.
Rec #: 256
KEYWORDS: kidney; liver; blood; gland; North America; ME; USA; Newfoundland; Canada; organochlorines; DDTs; mallard; white Pekin ducks; *Anas platyrhynchos*; black guillemot; *Cepphus grylle;* Atlantic puffin; *Fratercula arctica*; dosing study; osmoregulation; effects
NOTES: SUPPLEMENTAL
ABSTRACT: To assess the possibility that organochlorine pesticide disruption of osmoregulation is responsible for recent large kills of young seabirds, we have studied the effects of DDE feeding (10-250 ppm) on plasma osmoregulation and nasal gland function in the following species: mallard and white Pekin ducks (both *Anas platyrhynchos*), black guillemot (*Cepphus grylle*), and Atlantic puffin (*Fratercula arctica*). Other investigators have recently reported that dietary DDE (10-1,000 ppm) inhibits nasal gland secretion in freshwater-maintained mallards; our initial experiments with white Pekins showed no such inhibition during either freshwater or seawater maintenance. Moreover, DDE had minimal effects on plasma electrolyte levels and total nasal gland Na-K-ATPase activities in all species studied. Liver DDE levels in experimental ducks and guillemots were comparable to those reported for seabirds found dead after kills; levels in starved experimental puffins were much higher. Thus DDE at environmental levels does not affect osmoregulation or nasal gland Na-K-ATPase either in ducks or in two species of oceanic birds.

Miller, David S.; Peakall, David B.; and Kinter, William B. 1978. Ingestion of crude oil: Sublethal effects in herring gull chicks. Science. 199(4326):315-317.
Rec #: 128
KEYWORDS: effects; oil; herring gull; *Larus argentatus*; dosing study; toxicology
NOTES: SUPPLEMENTAL
ABSTRACT: A single small oral dose of Kuwait or South Louisiana crude oil caused cessation of growth, osmoregulatory impairment, and hypertrophy of hepatic, adrenal, and nasal gland

Annotated Bibliography (Cont.).

tissue in herring gull (*Larus argentatus*) chicks living in a simulated marine environment. These findings suggest that ingesting crude oil causes multiple sublethal effects that might impair a bird's ability to survive at sea.

Moisey, John; Fisk, Aaron T.; Hobson, Keith A.; and Norstrom, Ross J. 2001. Hexachlorocyclohexane (HCH) isomers and chiral signatures of α-HCH in the Arctic marine food web of the Northwater Polynya. Environ. Sci. Technol. 35:1920-1927.
Rec #: 353
KEYWORDS: organochlorines; HCHs; Northwater Polynya; Greenland; Canada; North America; Atlantic Ocean; biomagnification; fat; little auk; *Alle alle*; thick-billed murre; *Uria lomvia*; black guillemot; *Cepphus grylle*; black-legged kittiwake; *Rissa tridactyla*; ivory gull; *Pagophila eburnea*; glaucous gull; *Larus hyperboreus*; northern fulmar; *Fulmarus glacialis*
ABSTRACT: Concentrations of hexachlorocyclohexane (HCH) isomers (α, β, andγ) and enantiomer fractions (EFs) of α-HCH were determined in the Northwater Polynya Arctic marine food web. Relative food web structure was established using trophic level models based on organic $\delta^{15}N$ values. Concentrations of HCH in the samples collected, including water, sediment, benthic invertebrates (four species), pelagic zooplankton (six species), Arctic cod, seabirds (seven species), and ringed seal, were in the range previously reported for the Canadian Arctic. The relative proportion of the HCH isomers varied across the food web and appeared to be related to the biotransformation capacity of each species. For invertebrates and fish the biomagnification factors (BMFs) of the three isomers were >1 and the proportion of each isomer and the EFs of α-HCH were similar to water, suggesting minimal biotransformation. Seabirds appear to readily metabolize γ- and α-HCH based on low BMFs for these isomers, high proportions of β-HCH (62- 96 %), and high EFs (0.65-0.97) for α-HCH. The α- and β-HCH isomers appear to be recalcitrant in ringed seals based on BMFs >1 and near racemic EFs for α-HCH. The β isomer appears to be recalcitrant in all species examined and had an overall food web magnification factor of 3.9. EFs of α-HCH and the proportion of β-HCH in Σ-HCH in the food web were highly correlated ($r^2 = 0.92$) suggesting that EFs were a good indicator of a species capability to biotransform α-HCH.

Moksnes, Milica T. and Norheim, Gunnar. 1986. Levels of chlorinated hydrocarbons and compositions of PCB in herring gull *Larus argentatus* eggs collected in Norway in 1969 compared to 1979 - 81. Environ. Pollut. (B. Chem. Phys.). 11(2):109-116.
Rec #: 98
KEYWORDS: organochlorines; PCBs; DDTs; OCS; HCB; egg; Norway; Europe; herring gull; *Larus argentatus*; temporal trends
ABSTRACT: Eggs from herring gull *Larus argentatus* were collected at seven different locations along the Norwegian coast in 1969 and again in 1979-81. Hexachlorobenzene (HCB), octachlorostyrene (OCS), DDE, PCB and decachlorobiphenyl (DCB) were determined in each of the 200 eggs. Mean levels of HCB in 1969 and 1979-81 were 0 multiplied by 19 and 0 multiplied by 12 $\mu g\ g^{-1}$ wet weight, respectively. When excluding the results from Telemark, an area with local contamination with HCB, OCS, and DCB, there were no statistical differences in HCB levels between the two sampling periods. OCS and DCB were only detected in eggs from Telemark. Significant decreases in DDE levels and DDE/PCB ratios but not in PCB levels were observed between 1969 and 1979-81. Only for the most persistent of the PCB isomers recorded

Annotated Bibliography (Cont.).

was there a small increase in the relative amount from 1969 to 1979-81.

Monteiro, L. R. and Furness, R. W. 1995. Seabirds as monitors of mercury in the marine environment. Water Air Soil Pollut. 80(1-4):851-870.
Rec #: 165
KEYWORDS: review; metals; Hg; organometallics; organic Hg; feather; bioaccumulation; effects; toxicology; great white egret; *Egretta alba*; black-headed gull; *Larus ridibundus*; common tern; *Sterna hirundo*; Cory's shearwater; *Calonectris diomedea*; herring gull; *Larus argentatus*; great skua; *Catharacta skua;* mallard; *Anas platyrhynchos*; roseate tern; *Sterna dougallii*; common murre; *Uria aalge*; red-billed gull; *Larus novaehollandiae*; body burden; feather; liver; carcass; feces; egg; muscle; kidney; blood; spatial variations; temporal trends; Great Britain; Ireland; Germany; North Sea; Mediterranean Sea; Baltic Sea; Europe; Azores; Atlantic Ocean
NOTES: FIGURES ONLY
ABSTRACT: The oceans play a major role in global cycling of mercury and widespread contamination of marine ecosystems has been demonstrated in recent years. Monitoring mercury in the marine environment is a priority and biomonitoring has featured prominently in this respect. Seabirds, as top predators, present high mercury levels due to food chain amplification and thus will reflect slight variations in environmental mercury and its hazards to humans better than do most invertebrates and cold blood vertebrates. There is experimental evidence that levels of mercury in seabirds show a dose-response relationship, so that increased contamination of the environment causes a corresponding increase in the level in birds. This coupled with current knowledge on the dynamics of mercury in birds gives a good basis for the use of seabird as monitors of mercury. Internal tissues, blood, eggs, feathers and chicks have been used as monitoring units. Feathers are the most attractive amongst them. They are both chemically and physically stable, accumulate higher mercury levels than other tissues and their sampling is non-destructive. However, it is essential to sample a consistent feather area from all birds to minimise the effects of moult and body feathers are the most adequate. Feathers from birds in museum collections offer a great potential for the study of synoptic geographical and historical of changes in mercury levels on a global scale with large sample sizes. For example, studies with time series of feather samples from seabirds provide evidence of a 3-fold increase of mercury contamination in the marine ecosystem of North-eastern Atlantic over the last 100 years and little increase in mercury contamination in the Southern hemisphere during the same period.

Moore, N. W. and Tatton, J. O. G. 1965. Organochlorine insecticide residues in eggs of the sea birds. Nature. 207(4992):42-43.
Rec #: 528
KEYWORDS: organochlorines; dieldrin; DDTs; HCHs; chlordanes; egg; North Sea; Ireland; Europe; biomagnification; effects; reproductive success; little tern; *Sterna albifrons*; shelduck; *Tadorna tadorna*; oystercatcher; *Haematopus ostralegus*; common tern; *Sterna hirundo*; sandwich tern; *Sterna sandvicensis*; black-headed gull; *Larus ridibundus*; black-legged kittiwake; *Rissa tridactyla*; herring gull; *Larus argentatus*; Atlantic puffin; *Fratercula arctica*; common murre; *Uria aalge*; razorbill; *Alca torda*; shag; *Phalacrocorax aristotelis*; black-legged kittiwake; *Rissa tridactyla*; cormorant*; Phalacrocorax carbo*
ABSTRACT: Eggs of 13 species of sea birds were collected at random on the North Sea Coast and from Grand Saltee Island. Dieldrin and p,p'-DDE were detected in all the eggs. Residues of

Annotated Bibliography (Cont.).

p,p'-DDT and (or) its metabolite, p,p'-TDE, were found in 77 % of the eggs and BHC, usually the β-isomer, and heptachlor epoxide were found in 52 % of the eggs. Ten fish and 8 mollusks were collected at Scolt Head. All except one cockle contained insecticide residues. Results show small amounts of organochlorine insecticides to be widely distributed in British waters. There is some evidence that the eggs of birds which feed on large fish contain higher amounts of residues than those which feed on small fish or invertebrates. The quantities of residues found should not affect the hatching of eggs. There was a slight decrease in contamination from 1963 to 1964. The eggs of sea birds which do not range widely may be a useful indicator for detecting insecticidal contamination.

Mora, M. A. and Anderson, D. W. 1995. Selenium, boron, and heavy metals in birds from the Mexicali Valley, Baja California, Mexico. Bull. Environ. Contam. Toxicol. 54(2):198-206.
Rec #: 403
KEYWORDS: Baja; Mexico; North America; metals; Se; B; Cd; Zn; Co; Cr; double-crested cormorant; *Phalacrocorax auritus*; mourning dove; *Zenaida macroura;* cattle egret; *Bubulcus ibis*; great-tailed grackle; *Quiscalus mexicanus*; red-winged blackbird; *Agelaius phoeniceus*; seasonal variations; liver; biomagnification
ABSTRACT: Analysis of samples from bird on the Mexicali Valley, northeastern Baja California, Mexico, showed that there were no significant differences in the concentrations of trace elements between summer and winter for most species of birds, except for mourning doves, which had significantly higher concentrations of Cr, Cu, and Zn in summer than in winter. Selenium concentrations were significantly higher in double-crested cormorants and lower in mourning doves than in other species. Zn levels did not differ significantly; Cd was greater in red-winged blackbirds than in other species. Copper levels were lower in cattle egrets than in other species, which had similar concentrations. Cr was significantly higher in mourning dives and red-winged blackbirds, and boron was significantly higher in mourning doves than in any other species. High concentrations of boron may result from brine discharges from the Cerro Prieto geothermal energy plant. The results for fish suggest increased exposure to Se, B, and Zn. The concentrations of Zn, Cd, Cu, and Cr in livers seemed well below threshold for biological effects in birds.

Mora, Miguel A. and Anderson, Daniel W. 1991. Seasonal and geographical variation of organochlorine residues in birds from Northwest Mexico. Arch. Environ. Contam. Toxicol. 21(4):541-548.
Rec #: 415
KEYWORDS: Mexico; North America; organochlorines; PCBs; DDTs; HCHs; HCB; chlordanes; dieldrin; endrin; double-crested cormorant; *Phalacrocorax auritus*; olivaceous cormorant; *Phalacrocorax olivaceus*; mourning dove; *Zenaida macroura;* cattle egret; *Bubulcus ibis;* great-tailed grackle; *Quiscalus mexicanus*; red-winged blackbird; *Agelaius phoeniceus;* white-winged dove; *Zenaida asiatica;* carcass; seasonal variations; spatial variations
ABSTRACT: Eight species of birds (129 individuals) were collected from three agricultural areas with long histories of pesticide use in northwestern Mexico. Plucked carcasses were analyzed for organochlorine (OC) pesticides and polychlorobiphenyls (PCBs). DDE was found in all of the samples and at higher levels than other OCs. Mean (geometric) DDE concentrations varied from 0.04 (microgram/g) ppm in mourning doves (*Zenaida macroura*) to 5.05 ppm in double-crested cormorants (*Phalacrocorax auritus*). Hexachlorocyclohexane (HCH) was

Annotated Bibliography (Cont.).

detected in 95 % of the samples, but at lower levels than DDE. Hexachlorobenzene (HCB) residues were detected more frequently in birds from Mexicali (62 %, p less than 0.05) than in those from Yaqui and Culiacan. HCH and HCB concentrations were significantly higher in birds from Mexicali during the winter than in the summer (p less than 0.05), indicating accumulation of these compounds during that period. Other OCs such as DDT, DDD, dieldrin, oxychlordane, heptachlor epoxide, endosulfan, and endrin were found at lower levels and less frequently. PCBs (quantitated as Aroclor 1260) were found mostly in cattle egrets (*Bubulcus ibis*) and cormorants at the three locations. Overall, concentrations of OCs were higher for Mexicali than for Yaqui and Culiacan (p less than 0.01). In a few cases, DDE levels were above those that might adversely affect birds.

Moriarty, F.; Bell, A. A.; and Hanson, H. 1986. Does p,p'-DDE thin eggshells? Environ. Pollut. (A Ecol. Biol.). 40(3):257-286.
Rec #: 394
KEYWORDS: organochlorines; DDTs; eggshell thickness; egg; gannet; *Sula bassana;* shag; *Phalacrocorax aristotelis*; grey heron; *Ardea cinerea*; Great Britain; Europe; effects
ABSTRACT: Data on egg size (measured by maximum length), shape (measured by the ratio of maximum breadth to maximum length), shell thickness (measured indirectly by Ratcliffe's index, I), and p,p'-DDE (I) [72-55-9] content (log µg/g fresh weight) were examined in samples of eggs taken from 3 species: 2 samples from the gannet *Sula bassana*, 3 samples from the shag *Phalacrocorax aristotelis*, and 6 samples from the heron *Ardea cinerea*. The value of I is virtually unaffected by changes of egg size but is affected by egg shape, and this variable bias in Ratcliffe's index did in some instances materially affect the conclusions to be drawn from the data. A revised index (J), derived from this prolate spheroid, was therefore developed. The value of J is virtually unaffected by changes of egg size and shape, and yields a much more accurate estimate of the product of the mean shell diameter and thickness. The conventional negative linear regression of Ratcliffe's index on p,p'-DDE content occurred in many, but not all, of the egg samples that had at least a 10-fold range of p,p'-DDE concentrations. Two other samples both contained 2 eggs with p,p'-DDE concentrations of not >0.1 µg/g fresh weight The shell index I increased, or at least did not decrease, until the p,p'-DDE content exceeded approximately 0.1-0.2 µg/g. Thus a curvilinear relations with a maximum turning point is probably a common physiological response to pollutants.

Muir, D.; Braune, B.; DeMarch, B.; Norstrom, R.; Wagemann, R.; Lockhart, L.; Hargrave, B.; Bright, D.; Addison, R.; Payne, J.; and Reimer, K. 1999. Spatial and temporal trends and effects of contaminants in the Canadian Arctic marine ecosystem: A review. Sci. Total Environ. 230:83-144.
Rec #: 307
KEYWORDS: review; Canada; North America; organochlorines; benzenes; HCHs; chlordanes; DDTs; dieldrin; PCBs; toxaphene; metals; Hg; Pb; Cd; Se; temporal trends; thick-billed murre; *Uria lomvia*; northern fulmar; *Fulmarus glacialis*; black-legged kittiwake; *Rissa tridactyla*; black guillemot; *Cepphus grylle*; glaucous-winged gull; *Larus glaucescens*; common eider; *Somateria mollissima*; king eider; *Somateria spectabilis*; egg; carcass; spatial variations; effects; biomagnification; gender differences; age variations; bioaccumulation; toxicology; cytochrome P_{450}; EROD; CYP1A; metallothionein

Annotated Bibliography (Cont.).

ABSTRACT: Recent studies have added substantially to our knowledge of spatial and temporal trends of persistent organic pollutants and heavy metals in the Canadian Arctic marine ecosystem. This paper reviews the current state of knowledge of contaminants in marine biota in the Canadian Arctic and where possible, discusses biological effects. The geographic coverage of information on contaminants such as persistent organochlorines (OCs) (PCBs, DDT- and chlordane-related compounds, hexachlorocyclohexanes, toxaphene) and heavy metals (mercury, selenium, cadmium, lead) in tissues of marine mammal and sea birds is relatively complete. All major beluga, ringed seal and polar bear stocks along with several major sea bird colonies have been sampled and analysed for OC and heavy metal contaminants. Studies on contaminants in walrus are limited to Foxe Basin and northern Québec stocks, while migratory harp seals have only been studied recently at one location. Contaminant measurements in bearded seal, harbour seal, bowhead whale and killer whale tissues from the Canadian Arctic are very limited or non-existent. Many of the temporal trend data for contaminants in Canadian Arctic biota are confounded by changes in analytical methodology, as well as by variability due to age/size, or to dietary and population shifts. Despite this, studies of OCs in ringed seal blubber at Holman Island and in sea birds at Prince Leopold Island in Lancaster Sound show declining concentrations of PCBs and DDT-related compounds from the 1970s to 1980s then a leveling off during the 1980s and early 1990s. For other OCs, such as chlordane, HCH and toxaphene, limited data for the 1980s to early 1990s suggests few significant declines in concentrations in marine mammals or sea birds. Temporal trend studies of heavy metals in ringed seals and beluga found higher mean concentrations of mercury in more recent (1993/1994) samples than in earlier collections (1981-1984 in eastern Arctic, 1972-1973 in western Arctic) for both species. Rates of accumulation of mercury are also higher in present day animals than 10Ž20 years ago. Cadmium concentrations in the same animals (eastern Arctic only) showed no change over a 10-year period. No temporal trend data are available for metals in sea birds or polar bears. There have been major advances in knowledge of specific biomarkers in Canadian Arctic biota over the past few years. The species with the most significant risk of exposure to PCBs and OC pesticides may be the polar bear which, based on comparison with EROD activity in other marine mammals (beluga, ringed seal), appears to have elevated CYP1A-mediated activity. The MFO enzyme data for polar bear, beluga and seals suggest that even the relatively low levels of contaminants present in Arctic animals may not be without biological effects, especially during years of poor feeding.

Murphy, Stephen M.; Day, Robert H.; Wiens, John A.; and Parker, Keith R. 1997. Effects of the Exxon Valdez oil spill on birds: Comparisons of pre- and post-spill surveys in Prince William Sound, Alaska. Condor. 99(2):299-313.
Rec #: 246
KEYWORDS: oil; AK; USA; North America; effects; red-necked grebe; *Podiceps grisegena*; pelagic cormorant; *Phalacrocorax pelagicus*; harlequin duck; *Histrionicus histrionicus*; common merganser; *Mergus merganser*; bald eagle; *Haliaeetus leucocephalus*; black oystercatcher; *Haematopus bachmani*; Bonaparte's gull; *Larus philadelphia*; common gull; *Larus canus;* glaucous-winged gull; *Larus glaucescens;* black-legged kittiwake; *Rissa tridactyla*; pigeon guillemot; *Cepphus columba*; reproductive success
NOTES: SUPPLEMENTAL
ABSTRACT: We used data from pre- and post-spill surveys to assess the effects of the Exxon Valdez oil spill on the abundance and distribution of birds in Prince William Sound, Alaska. We

Annotated Bibliography (Cont.).

conducted post-spill surveys during mid-summer (1989-1991) in 10 bays that had been surveyed prior to the spill (1984-1985) and that had experienced different levels of initial oiling from the spill (unoiled to heavily oiled). We evaluated whether there were changes in overall abundance across all bays between the pre-spill and post-spill sampling periods, and changes in abundance in unoiled/lightly oiled bays versus moderately/heavily oiled bays that would suggest oiling impacts. Of 12 taxa examined for changes in overall abundance, 7 showed no significant change, 2 (Bald Eagle and Glaucous-winged Gull) increased in abundance, and 3 (Red-necked Grebe, Pelagic Cormorant, and Pigeon Guillemot) decreased in abundance during all three post-spill years. Of the 11 taxa examined for differences in use of oiled versus unoiled habitats, 7 showed no significant response, 1 (Black-legged Kittiwake) exhibited a positive response to oiling, and 3 (Pelagic Cormorant, Black Oystercatcher, and Pigeon Guillemot) exhibited negative responses to initial oiling. We conclude that the impacts of this oil spill on abundance and distribution of birds were most evident in 1989, the year of the spill, and were most pronounced for Pigeon Guillemots. By 1991, signs of recovery were evident for all taxa that showed initial oiling impacts.

Murvoll, K. M.; Skaare, J. U.; Nilssen, V. H.; Bech, C.; Ostnes, J. E.; and Jenssen, B. M. 1999. Yolk PCB and plasma retinol concentrations in shag (*Phalacrocorax aristotelis*) hatchlings. Arch. Environ. Contam. Toxicol. 36(3):308-315.
Rec #: 266
KEYWORDS: Norway; Europe; organochlorines; PCBs; toxicology; TEQs; vitamin A; retinol; blood; egg; shag; *Phalacrocorax aristotelis*
ABSTRACT: To evaluate the possibilities of applying plasma retinol as a biomarker of response in seabirds exposed to chronic low levels of organochlorines, the relationship between yolk content of polychlorinated biphenyls (PCBs) and plasma retinol levels were studied in newly hatched shag chicks (Phalacrocorax aristotelis) from the coast of central Norway. The mean concentration of 29 PCB-congeners (ΣPCB) in the yolk sac was 1.22 microgram/g ww (wet weight basis) (SD = 0.57, n = 10), or 17.99 ng/g lw (lipid weight basis) (SD = 6.26, n = 10). Expressed as TCDD-equivalents (ΣTEQ), the exposure in the yolk sac was 43.9 pg/g ww (SD = 19.5, n = 10), or 637.1 pg/g lw (SD = 240.8, n = 10), considerably lower than the levels that have been associated with clear-cut lethal and sublethal effects such as egg mortality, hatchability, or live deformity in Phalacrocoracidae species. There were significant negative correlations between ΣPCB ww and the variables egg volume, yolk mass, and hatchling mass. We suggest that these relationships are passive causes of a higher lipid concentration in small eggs, rather than the PCB affecting the variables. Analyses showed that there was a borderline significant positive correlation between ΣPCB lw in yolk and plasma retinol concentration. Although the results indicate that plasma retinol level alone is a poor indicator of PCB exposure in shag hatchlings, the result may be related to the low level of contaminant exposure and the low sample size of the study.

Nettleship, D. N. and Peakall, D. B. 1987. Organochlorine residue levels in three high Arctic species of colonially-breeding seabirds from Prince Leopold Island. Mar. Pollut. Bull. 18(8):434-438.
Rec #: 44
KEYWORDS: organochlorines; PCBs; DDTs; Northwest Territories; Canada; North America;

Annotated Bibliography (Cont.).

bioaccumulation; liver; egg; thick-billed murre; *Uria lomvia*; northern fulmar; *Fulmarus glacialis*; black-legged kittiwake; *Rissa tridactyla*; temporal trends
ABSTRACT: Levels of DDE and PCB residues in Northern Fulmar *Fulmarus glacialis*, black-legged kittiwake *Rissa tridactyla* and thick-billed Murre *Uria lomvia* collected at Prince Leopold Island, Northwest Territories, between 1975 and 1977 are relatively low. Organochlorine levels were highest in all species in 1976, a year when lipid levels in all tissue examined (adult and chick liver, egg) were highest. Annual variation of DDE and PCB may thus be largely due to differences in nutritional and associated environmental conditions. The unusually high PCB/DDE ratio for the kittiwake cannot be explained by low DDE concentrations rather than by high concentrations of PCBs as suggested elsewhere. Data from Prince Leopold Island birds indicate both that DDE concentrations are low and that PCB concentrations are high.

Newton, I.; Haas, Margaret B.; and Bell, A. A. 1981. Pollutants in guillemot eggs. Institute of Terrestrial Ecology Annual Report 1981. 57-59.
Rec #: 309
KEYWORDS: egg; common murre; *Uria aalge*; Great Britain; Europe; organochlorines; DDTs; dieldrin; PCBs; metals; Hg; spatial variations; temporal trends
ABSTRACT: In 1980, 10 guillemot (*Uria aalge*) egg were collected from each of three colonies in the UK. The eggs were analyzed for DDE, HEOD, PCBs, and Hg and compared to results obtained in 1969-1972.

Nielsen, Christian Overgaard and Dietz, Rune. 1989. Heavy metals in Greenland seabirds. Meddelelser Om Grønland, Bioscience 29. 1-26.
Rec #: 300
KEYWORDS: Greenland; North America; metals; Zn; Cd; Hg; Se; muscle; liver; kidney; black guillemot; *Cepphus grylle;* thick-billed murre; *Uria lomvia;* little auk; *Alle alle*; common eider; *Somateria mollissima*; king eider; *Somateria spectabilis*; long-tailed duck; *Clangula hyemalis;* red-breasted merganser; *Mergus serrator;* Iceland gull; *Larus glaucoides;* glaucous gull; *Larus hyperboreus;* black-legged kittiwake; *Rissa tridactyla;* ivory gull; *Pagophila eburnea;* northern fulmar; *Fulmarus glacialis;* cormorant; *Phalacrocorax carbo;* pomarine jaeger; *Stercorarius pomarinus*; gender differences; age variations; bioaccumulation; spatial variations;
ABSTRACT: From six Greenland districts we report the concentration of Zn, Cd, Hg, and Se in muscle (pectoral), liver and kidneys for 320 seabirds of the following species: *Cepphus grylle, Uria lomvia, Alle alle, Somateria mollissima, Somateria spectabilis, Clangula hyemalis, Mergus serrator, Larus glaucoides, L. hyperboreus, Rissa tridactyla, Pagophila eburnea, Fulmarus glacialis, Phalacrocorax carbo,* and *Stercorarius pomarinus*.
Concentrations vary widely within species. Yearlings are low in Cd and Hg. Concentrations tend to increase with age. No significant differences between sexes were found.
On a wet weight basis, the Zn concentration in liver and kidney is c. three times that of muscle. Gulls and the fulmar possess significantly more Zn in muscle than do other seabirds. The Cd concentration in liver and kidney is c. 20 and 80 times higher than in muscle, whereas the Hg concentration liver and kidney is three to five times higher than that of muscle. The Se concentration in liver and kidney is c. five times the muscle concentration. Muscle, liver, and kidney concentrations tend to correlate positively for Cd, Hg, and Se. For Zn only liver and kidney concentrations correlate mutually.
On a molar basis, the three organs of all species have a large excess of Se or Hg. The intra-organ

Annotated Bibliography (Cont.).

association of elements is strongest for Zn and Cd in liver and kidney, and for Hg and Se generally.
All four elements show consistently higher concentrations in birds from NW and NE Greenland than in those from S Greenland. For *C. grylle* from Avanersuaq, NW Greenland, the Cd concentration is twice that of birds from S Greenland, the difference being highly significant. Hg concentrations are not significantly different.

Norheim, Gunnar. 1987. Levels and interactions of heavy metals in sea birds from Svalbard and the Antarctic. Environ. Pollut. 47(2):83-94.
Rec #: 514
KEYWORDS: metals; Cu; Zn; Cd; Pb; Hg; Se; liver; kidney; Svalbard; Europe; glaucous gull; *Larus hyperboreus;* northern fulmar; *Fulmarus glacialis;* thick-billed murre; *Uria lomvia*; little auk; *Alle alle;* common eider; *Somateria mollissima*; macaroni penguin; *Eudyptes chrysolophus;* brown skua; *Catharacta lonnbergi*; spatial variations; toxicology
ABSTRACT: Samples of liver and kidney from 92 sea birds of 10 species collected on Spitsbergen and in the Antarctic, were analyzed for their content of Cu, Zn, Cd, Pb, Hg, and Se. Significantly higher levels of Cu and Zn were observed in birds from Spitsbergen than in those from the Antarctic, while the opposite was true for Se. The highest Cd levels were found in fulmar *Fulmarus glacialis* and macaroni penguin *Eudyptes chrysolophus*. A possibility of kidney damage due to Cd exists. The highest Hg levels were recorded in brown skua *Catharacta lonnbergi* collected at Bouvetoya. Pb was not detected in any of the birds. Significant correlations were observed between levels of several of the metals studied, especially between Cd and Zn and between Hg and Se. However, for all birds, the highest correlation coefficients were observed when the molar concentrations of Cd plus Hg, and Se plus Zn were used in the calculations. Thus, several protective mechanisms may operate to diminish effects of heavy metal contaminants.

Norheim, Gunnar and Borch-Iohnsen, Berit. 1990. Svalbard: Trace elements in liver from eider. In: Låg, J. (Ed.) Excess and Deficiency of Trace Elements in Relation to Human and Animal Health in Arctic and Subarctic Regions. Norwegian Academy of Science and Letters. 217-219.
Rec #: 527
KEYWORDS: metals; liver; Svalbard; Norway; Europe; histopathology; gender differences; common eider; *Somateria mollissima*
ABSTRACT: During egg laying and brooding female common eiders (*Somateria mollissima*) are known to lose liver weight which, along with translocation of body iron, may result in an increase of liver iron. Brooding females and males were collected from Svalbard, Norway and their livers were examined for copper, iron, selenium, and zinc. The total body weight of the females was reduced 25 % compared to males. Iron levels were 5.1 times higher and zinc levels were twice as high in females. Copper and selenium levels were not significantly different. Massive siderosis was seen in females that cannot be explained by liver weight loss alone.

Norheim, Gunnar and Kjos-Hanssen, Bjørn. 1984. Persistent chlorinated hydrocarbons and mercury in birds caught off the west coast of Spitsbergen. Environ. Pollut. (A Ecol. Biol.). 33(2):143-152.
Rec #: 438
KEYWORDS: organochlorines; DDTs; PCBs; HCB; metals; Hg; glaucous gull; *Larus*

Annotated Bibliography (Cont.).

hyperboreus; northern fulmar; *Fulmarus glacialis*; thick-billed murre; *Uria lomvia*; little auk; *Alle alle;* common eider; *Somateria mollissima;* Norway; Europe; liver; fat; biomagnification
ABSTRACT: The tissue concentrations of DDE (I) [72-55-9], PCB, HCB [118-74-1] and Hg were detected in 5 species of migrating seabirds: glaucous gull *Larus hyperboreus*; fulmar *Fulmarus glacialis*; Brünnich's guillemot *Uria lomvia*; little auk *Alle alle* and eider *Somateria mollissima*. These birds nest on Svalbard and were shot in May 1980 off the west coast of Spitsbergen. The highest levels of DDE, PCB and HCB were found in glaucous gull, whereas low levels were found in Brünnich's guillemot, little auk and, especially, eider. Fulmars were intermediate. Highly significant correlations were found between the concentrations of HCB, DDE and PCB, indicating that the Gulf Stream may be a common source of these substances. The highest Hg levels were found in the fulmar; glaucous gull and eider were intermediate, whereas the lowest Hg levels were found in Brünnich's guillemot and little auk. There was no connection between the nutritional condition and concentrations of the pollutants detected However, there seems to be a close relationship between the levels of chlorinated hydrocarbons and the trophic level of the birds in the food chain. A comparison between the present results and analyses of Antarctic seabirds indicates that the aquatic food chain in the Arctic is more loaded with persistent chlorinated hydrocarbons than in the Antarctic, whereas more Hg seems to be found in Antarctic birds.

Norstrom, R. J.; Fox, G. A.; Jeffrey, D. A.; and Gilman, A. P. 1983. Dynamics of organochlorine compounds in herring gulls. Proceedings of the 26th Conference on Great Lakes Research. May 23-27, 1983, State University of New York at Oswego. 39 pp.
Rec #: 221
KEYWORDS: herring gull; *Larus argentatus*; egg; blood; carcass; organochlorines; PCBs; DDTs; mirex; HCB; dieldrin; chlordanes; Great Lakes; North America; elimination
NOTES: ABSTRACT ONLY - NOT AVAILABLE
ABSTRACT: Herring gull (*Larus argentatus*) eggs have been collected annually since 1973 in all of the Great Lakes for monitoring organochlorine trends in the ecosystems. In order to make meaningful quantitative interpretation of these data, a knowledge of the pharmacodynamics of the compounds in female gulls is required. Whole-body clearance of p,p-DDE, one of the major residues in the Great Lakes, has been measured in wild gulls by dosing with ^{14}C-labeled DDE and recapturing one year later. Assuming first-order kinetics, a half-life of 220 plus or minus 30 d was determined. Preliminary data from analysis of plasma from dosed, caged gulls show that half-lives of photomirex, mirex and PCBs were similar to that of DDE. Half-lives of other compounds were in the range of 50-100 d for hexachlorobenzene, dieldrin, oxychlordane and octachlorostyrene, 25 d for p,p-DDD and < 10 d for gamma -hexachlorocyclohexane and t-chlordane. The consequences of these results to the interpretation of the egg monitoring data is discussed.

Norstrom, R. J.; Gilman, A. P.; and Hallett, D. J. 1981. Total organically-bound chlorine and bromine in Lake Ontario herring gull eggs, 1977, by instrumental neutron activation and chromatographic methods. Sci. Total Environ. 20(3):217-230.
Rec #: 536
KEYWORDS: organochlorines; PCBs; DDTs; mirex; HCB; chlordanes; dieldrin; HCHs; herring gull; *Larus argentatus*; egg; Great Lakes; North America; spatial variations
ABSTRACT: In order to determine the extent to which organically-bound chlorine in Herring

Annotated Bibliography (Cont.).

Gull eggs from Lake Ontario can be accounted for by gas chromatographic analysis, comparison was made with values obtained for total chlorine using instrumental neutron activation analysis (INAA). Total chlorine and bromine (mg/kg fresh weight of egg) was determined by INAA on crude extract (Cl, 65 ± 35; Br, $1/03 \pm 1.00$), Florisil-chromatography treated extracts (Cl, 46 ± 10; Br, 0.93 ± 0.82) and H2SO4-treated extracts (Cl, 43 ± 11; Br, 0.44 ± 0.22) of eggs collected from seven colonies around Lake Ontario in 1977. Levels of chlorine were also determined by gas chromatography using the Hall electrolytic conductivity detector (51 ± 11 mg/kg) and estimated by conversion of levels of individual residues determined by electron-capture gas chromatography (61 ± 12 mg/kg). The agreement between the various determinations indicated that PCBs, DDE, mirex and photomirex accounted for most of the organically-bound chlorine. Two colonies had total chlorine levels in crude extracts 2–4 times higher than could be accounted for by known compounds. The "excess" chlorine was removed by H_2SO_4-treatment or Florisil clean-up. The same two samples had abnormally high bromine levels, possibly indicating the presence of compounds formed during aqueous chlorination processes.

Norstrom, R. J.; Hebert, C. E.; Fox, G. A.; Kennedy, S.; and Weseloh, D. V. 1995. The herring gull as a biomonitor of trends in levels and effects of halogenated contaminants in Lake Ontario: A 25-year case history. In: Proceedings of the 38th Conference of the International Association for Great Lakes Research, East Lansing, MI (USA), 28 May-1 Jun 1995. 32 pp.
Rec #: 236
KEYWORDS: Great Lakes; North America; herring gull; *Larus argentatus*; bioaccumulation; egg; organochlorines; TEQs; PCBs; PCDDs; HCB; DDTs; reproductive success; temporal trends; effects
NOTES: ABSTRACT ONLY - NOT AVAILABLE
ABSTRACT: Poor reproductive success of fish-eating birds in the Great Lakes in the late 1960s-early 1970s, ascribed to the effects of Halogenated Aromatic Hydrocarbons (HAHs) such as DDE, PCBs, and 2,3,7,8-TCDD, precipitated serious concern about HAHs in the Great Lakes ecosystem. Studies on herring gulls from Lake Ontario over the last 20-25 years have produced one of the best documented case studies on a seabird which integrates biology (feeding ecology, energetics) and effects of HAHs (behavioural, reproductive, and biochemical), temporal trends in HAH levels (including whole-lake chemical budgets) and TEQs, sources of contamination (PCDDs), identification of new contaminants (photomirex), experimental clearance studies and modelling of HAH food-chain bioaccumulation. The herring gull is a hardy, resident, largely piscivorous species which maintained breeding colonies on Lake Ontario throughout the period of highest contamination, despite low reproductive success (ca. 0.2 fledglings/nest). Reproductive success improved dramatically to >1 fledglings/nest between 1975-1977, during a period of rapid decline in HAH concentrations in eggs. Evidence points to PCB/TCDD and/or HCB embryotoxicity as the main cause of reproductive failure, but other contaminants may have contributed. We have identified over 150 HAH and other halogenated contaminants in eggs.

Norstrom, R. J.; Simon, M.; Whitehead, P. E.; Kussat, R.; and Garrett, C. 1988. Levels of polychlorinated dibenzo-p-dioxins (PCDDs) and polychlorinated dibenzofurans (PCDFs) in biota and sediments near potential sources of contamination in British Columbia, 1987. Canadian Wildlife Service Analytical Report CRD-88-5.
Rec #: 551
KEYWORDS: organochlorines; PCDDs; PCDFs; British Columbia; Canada; North America;

Annotated Bibliography (Cont.).

pulp mill; great blue heron; *Ardea herodias;* double-crested cormorant; *Phalacrocorax auritus*; egg; spatial variations; biomagnification

ABSTRACT: This report is intended to bring together in one place all of the chemical data obtained as part of the joint CWS/HQ and C&P, Pacific Region Pestfund projects on levels in biota and sediments and sources of PCDDs and PCDFs in the Strait of Georgia area in British Columbia in 1987. PCDD and PCDF contamination was discovered in pooled eggs of the Great Blue Heron (*Ardea herodius*) from a colony on the University of British Columbia in 1982. The monitoring program was expanded to 3 more colonies in 1983 including a "control" site and two sites near bleached kraft mills, sawmills, and lumber mills. By 1987, the program had expanded to 17 sites and included Dungeness Crabs (*Cancer magister*), Staghorn Sculpin (*Leptocottus armatus*), bivalves (oysters, mussels or clams), Prickly Sculpin (*Cottus asper*), Double-Crested Cormorant (*Phalacrocorax auritus*), and sediment samples. The results of the chemical analysis are given and briefly discussed.

Norstrom, Ross J.; Clark, Thomas P.; Jeffrey, Deborah A.; Won, Henry T.; and Gilman, Andrew P. 1986. Dynamics of organochlorine compounds in herring gulls (*Larus argentatus*): I. Distribution and clearance of [^{14}C] DDE in free-living herring gulls (*Larus argentatus*). Environ. Toxicol. Chem. 5(1):41-48.
Rec #: 166
KEYWORDS: organochlorines; DDTs; bioaccumulation; elimination; herring gull; *Larus argentatus*; carcass; muscle; liver; brain; blood; egg; Great Lakes; North America; gender differences

ABSTRACT: Radiolabelled [^{14}C] DDE was used as a model compound to determine important factors in the clearance of persistent lipophilic compounds in free-living herring gulls (*L. argentatus*). Adult, breeding male and female gulls were dosed orally during incubation and then captured one week and one year later. After one week, [^{14}C] DDE had equilibrated with native DDE in all tissues. The ratios of DDE levels on a lipid weight basis relative to whole body were as follows: muscle, 0.8 to 0.9; liver, 0.5 to 0.7; egg, 0.4; and brain. 0.1. The plasma/whole body lipid partition coefficient was 0.0041 plus or minus 0.0014. The whole body annual average clearance rate was 0.95 plus or minus 0.51 year^{-1}(half-life = 264 d). Native DDE levels in males were twice those in females, but no differences were found for [^{14}C] DDE after one year.

Norstrom, Ross J.; Simon, Mary; Moisey, John; Wakeford, Bryan; and Weseloh, D. V. Chip. 2002. Geographical distribution (2000) and temporal trends (1981-2000) of brominated diphenyl ethers in Great Lakes herring gull eggs. Environ. Sci. Technol. 36(22):4783-4789.
Rec #: 358
KEYWORDS: herring gull; *Larus argentatus;* egg; PBDEs; Great Lakes; North America; spatial variations; temporal trends

ABSTRACT: The geographical distribution of brominated di-Ph ether (BDE) flame retardants in the North American Great Lakes ecosystem in 2000 was detected by analysis of herring gull eggs (13 egg pools) from a network of 15 monitoring colonies scattered throughout the lakes and connecting channels. ΣBDEs were found at concentrations ranging from 192 to 1400 µg/kg, mean of 662 ± 368 µg/kg (wet weight of egg contents). The highest concentrations were found in northern Lake Michigan and Toronto Harbor (Ontario) (1000-1400 µg/kg) and the lowest in

Annotated Bibliography (Cont.).

Lake Huron and Lake Erie (192-340 µg/kg). The distribution suggested that input from large urban/industrial areas through air or water emissions contributes local contamination of the herring gull food web in addition to background levels from regional/global transport. The congener composition was similar among sampling sites. Major congeners were BDE-47 (43 %), BDE-99 (26 %), BDE-100 (13 %) BDE-153 (11 %), BDE-154 (4 %), BDE-183 (2 %), and BDE-28 (1 %). Temporal trends of BDE contamination, 1981-2000, were established by analysis of archived herring gull eggs (10 egg pools) from colonies in northern Lake Michigan, Saginaw Bay, Lake Huron and eastern Lake Ontario. BDE-47, -99 and -100, and BDE-153, -154 and -183 concentrations were grouped separately for analysis because these two groups had different trends and are primarily associated with the penta-BDE and octa-BDE flame retardant formulations, respectively. ΣBDE47,99,100 concentrations were 5-12 µg/kg (wet weight) in 1981-1983 and then increased exponentially ($p < 0.00001$) at all three sites to 400-1100 µg/kg over the next 17 years. Doubling times were 2.6 years in Lake Michigan, 3.1 years in Lake Huron, and 2.8 years in Lake Ontario. ΣBDE154, 153, 183 concentrations generally increased but varied in an erratic fashion among sites and decreased as a fraction of ΣBDE over time. Concentrations of ΣBDE154, 153, 183 were 100-200 µg/kg in eggs from all three colonies in 2000. Therefore, most of the dramatic increases in ΣBDE concentrations observed over the past 20 years in the Great Lakes aquatic ecosystem seem to be connected with the penta-BDE formulation, which is mainly used as a flame retardant in polyurethane foam in North America. If present rates of change continue, concentrations of ΣBDEs will equal or surpass those of ΣPCBs (polychlorinated biphenyls) in Great Lakes herring gull eggs in 10-15 years.

Nounou, Pierre. 1979. Oil pollution of oceans (La pollution petroliere des oceans). Recherche. 10 (97): 147-156.
Rec #: 268
KEYWORDS: review; oil; effects
NOTES: SUPPLEMENTAL
ABSTRACT: A review of the processes by which marine environments that have been subjected to oil pollution regain their equilibrium is presented. There are 3 stages of evolution of hydrocarbons in the marine environment, (1) spreading over the ocean surface which is accompanied by evaporation, and dispersion by solution or the formation of an emulsion; (2) dispersion of the emulsion, progressive elimination of light hydrocarbons and eventual sedimentation. The majority of the sediment is degraded by chemical and biological processes within a few months; and (3) the remaining heavy hydrocarbons form semi stable tar agglomerates. The effects of using detergents to disperse oil spillages are discussed. The geomorphology of coastlines is important in determining the susceptibility to oil pollution and the various factors involved are considered. Immediate effects of oil spillage on marine life and seabirds is discussed.

Odsjö, Tjelvar; Bignert, Anders; Olsson, Mats; Asplund, Lillemor; Eriksson, Ulla; Häggberg, Lisbeth; Litzén, Kerstin; de Wit, Cynthia; Rappe, Christoffer; and Lslund, Kerttu. 1997. The Swedish Environmental Specimen Bank--Application in trend monitoring of mercury and some organohalogenated compounds. Chemosphere. 34(9-10):2059-2066.
Rec #: 263
KEYWORDS: Sweden; Europe; common murre; *Uria aalge;* egg; blood; organochlorines;

Annotated Bibliography (Cont.).

PCBs; DDTs; PCDDs; PCDFs; metals; Hg; radionuclides; toxicology; TEQs; vitamin A; retinol; effects; temporal trends
NOTES: FIGURES ONLY
ABSTRACT: The Environmental Specimen Bank (ESB) at the Swedish Museum of Natural History, Stockholm, constitutes a base for ecotoxicological research as well as for spatial and trend monitoring of contaminants in Swedish fauna. Since the 1960s, tissue samples from more than 150000 organisms have been collected from different groups of animals, habitats and types of landscape. Samples from the ESB have been utilized for retrospective studies of trace elements, organohalogenated compounds and radionuclides. Among many matrices utilized, eggs of guillemot (*Uria aalge*) have proven to be an appropriate matrix for assessment of the contamination of the Baltic Sea. Results from time trend studies based on this material showing trends in concentrations of DDT, PCB, PCDD/F, and mercury are presented in this paper.

Ohlendorf, Harry M.; Bartonek, James C.; Divoky, George J.; Klass, Erwin E.; and Krynitsky, Alexander J. 1982. Organochlorine residues in eggs of Alaskan seabirds. United States Department of the Interior, Fish and Wildlife Service, Special Scientific Report - Wildlife No. 245, Washington, DC. 42 pp.
Rec #: 294
KEYWORDS: AK; USA; North America; Pacific Ocean; organochlorines; DDTs; PCBs; HCB; chlordanes; toxaphene; mirex; dieldrin; egg; common loon; *Gavia immer*; glaucous-winged gull; *Larus glaucescens;* fork-tailed storm petrel; *Oceanodroma furcata*; Leach's storm-petrel; *Oceanodroma leucorhoa*; tufted puffin; *Fratercula cirrhata*; horned puffin; *Fratercula corniculata*; pelagic cormorant; *Phalacrocorax pelagicus*; red-faced cormorant; *Phalacrocorax urile*; Canada goose; *Branta canadensis*; common gull; *Larus canus*; Arctic tern; *Sterna paradisaea*; Aleutian tern; *Sterna aleutica*; pigeon guillemot; *Cepphus columba*; ancient murrelet; *Synthliboramphus antiquuus*; northern fulmar; *Fulmarus glacialis;* common murre; *Uria aalge*; black-legged kittiwake; *Rissa tridactyla*; double-crested cormorant; *Phalacrocorax auritus*; thick-billed murre; *Uria lomvia*; spatial variations; biomagnification
ABSTRACT: One egg from each of 440 clutches of eggs of 19 species of Alaskan seabirds collected in 1973-76 was analyzed for organochlorine residues. All eggs contained DDE; 98.9 % contained PCB's; 84.3 %, oxychlordane; and 82.7 %, HCB. Endrin was found in only one egg, but DDD, DDT, dieldrin, heptachlor, epoxide, mirex, *cis*-chlordane (or *trans*-nonachlor), *cis*-nonachlor, and toxaphene each occurred in at least 22 % of the samples.
Concentrations of organochlorines in the samples were generally low. Mean concentrations of eight compounds were highest in eggs of glaucous-winged gulls (*Larus glaucescens*) from three sites: DDE (5.16 ppm, wet weight), dieldrin (0.214 ppm), *cis*-chlordane (0.075 ppm), and HCB (0.188 ppm) in eggs form Buldir island: and *cis*-nonachlor (0.026 ppm) in eggs from the Semidi Islands. Highest concentrations of DDD (0.157 ppm), DDT (0.140 ppm), and toxaphene (0.101 ppm) were in eggs of fork-tailed storm petrel (*Oceanodroma furcata*) from Buldir island, and the highest concentration of mirex (0.044 ppm) was in fork-tailed storm petrel eggs from Barren Islands.
Both frequency of occurrence and concentration of residues in the eggs differed geographically and by species, apparently reflecting non-uniform distribution of organochlorines in the environment, dissimilar feeding habits and migration patterns of the species, or metabolic differences among the species.
The overall frequency of residue occurrence was highest in eggs from the Pribilof Islands, but

Annotated Bibliography (Cont.).

only three species were represented in the samples collected there. Detectable residues also were more frequent in eggs from the Gulf of Alaska colonies than elsewhere, and the lowest frequency was in eggs from nesting colonies on or near the Seward Peninsula. Regionally, concentrations of DDE and PCB's were usually higher than average in eggs from the Gulf of Alaska and lower than average in eggs fro the Aleutian Islands and Bristol Bay. However, within some species there were exceptions to ths general pattern, and mean concentrations of most chemicals differed from one site to another within the same region.

Among eggs of species collected in two or more regions, residue frequencies were highest in those of fork-tailed storm-petrel, tufted puffin (*Lunda cirrhata*), horned puffin (*Fratercula corniculata*), pelagic cormorant (*Phalacrocorax pelagicus*), and northern fulmar (*Fulmarus glacialis*), and lowest in those of common murre (*Uria aalge*), black-legged kittiwake (*Rissa tridactyla*), double-crested cormorant (*Phalacrocorax auritus*), and thick-billed murre (*Uria lomvia*).

On a regional basis, mean concentrations of DDE and PCB's varied significantly among species, but there were few consistent patterns of species differences, except that levels of DDE were always lowest in black-legged kittiwakes and concentrations of PCB's were usually lowest in murres. Also, concentrations of both chemicals (except PCB's in the Gulf of Alaska) were usually higher in northern fulmars than in other species, and the highest concentrations of both DDE and PCB's found in this study were in glaucous-winged gulls in the Aleutian Islands.

Olsson, M.; Jensen, S.; and Renber, L. 1973. PCB in coastal areas of the Baltic. PCB Conf. II, 1972, Suppl. 4E. Solina, Sweeden: Swed. Environ. Prot. Board. 321-326.
Rec #: 545
KEYWORDS: organochlorines; PCBs; biomagnification; Baltic Sea; Europe; muscle; common goldeneye; *Bucephala clangula;* long-tailed duck; *Clangula hyemalis;* red-breasted merganser*; Mergus serrator;* common merganser*; Mergus merganser;* cormorant*; Phalacrocorax carbo;* black guillemot*; Cepphus grylle;* common murre*; Uria aalge;* herring gull*; Larus argentatus;* white-tailed sea eagle*; Haliaeetus albicilla;* eagle owl*; Bubo bubo*
ABSTRACT: Invertebrates living in the littoral zone of the Baltic Sea showed mainly low levels of PCB and fish feeding mainly on these invertebrates also had rather low levels, <10 ppm in extractable fat. Plankton samples from the pelagic zone had rather high levels, approximately 5 ppm, considering their low position in the food chain. Herring and jellyfish which feed on plankton had PCB levels of approximately 30 ppm. Birds feeding mainly on invertebrates had 10-20 ppm PCB, while fish-eating birds showed 87-650 ppm PCB. Fish-eating mammals had approximately 70 ppm PCB, while birds of prey showed 10,000 ppm PCB. The main entrance of PCB into the food chain is probably through plankton.

Olsson, Mats and Reutergårdh, Lars. 1986. DDT and PCB pollution trends in the Swedish aquatic environment. Ambio. 15:103-109.
Rec #: 308
KEYWORDS: organochlorines; DDTs; PCBs; Baltic Sea; Sweden; Europe; egg; common murre; *Uria aalge*; temporal trends
ABSTRACT: Since the end of the 1960's samples of biota have been collected annually in various parts of Sweden and analyzed for presence of DDT and PCB substances. During this period various steps have been taken by governments to reduce the environmental burden of pollutants. How fast have the Baltic environment and the Swedish fresh water reacted? Trend

Annotated Bibliography (Cont.).

monitoring is one way to get a time perspective on environmental pollution.

Osborn, D.; Harris, M. P.; and Nicholson, J. K. 1979. Comparative tissue distribution of mercury, cadmium and zinc in three species of pelagic seabirds. Comp. Biochem. Physiol. 64C(1):61-67.
Rec #: 162
KEYWORDS: Atlantic puffin; *Fratercula arctica;* northern fulmar; *Fulmarus glacialis;* Manx shearwater; *Puffinus puffinus*; Scotland; Europe; metals; Cd; Hg; Zn; liver; kidney; brain; gut; pancreas; gonad; muscle; skin; feather; blood
ABSTRACT: The levels of Hg, Cd and Zn were measured in tissues of puffin (*Fratercula arctica*), fulmar (*Fulmarus glacialis*) and Manx shearwater (*Puffinus puffinus*). The highest levels of Zn and Cd (up to 480 mg/kg Cd, dry weight) were found in kidney, liver, pancreas, gonad, and intestine. Substantial quantities of Hg _ mostly methyl mercury _ were found in the liver of the fulmar and Manx shearwater (up to 450 mg/kg Hg, dry weight). Puffin feathers contained more Hg (7.94 mg/kg Hg, dry weight) than the liver and kidney. The possibility that since these birds were breeding they were not suffering any adverse effects of the metals is discussed.

Osborn, D.; Harris, M. P.; and Young, W. J. 1987. Relationships between tissue contaminant concentrations in a small sample of seabirds. Comp. Biochem. Physiol. 87C(2): 415-420.
Rec #: 226
KEYWORDS: metals; Cd; Hg; organochlorines; dieldrin; DDTs; PCBs; Atlantic puffin; *Fratercula arctica;* Manx shearwater; *Puffinus puffinus*; northern fulmar; *Fulmarus glacialis*; Scotland; Europe; liver; kidney; muscle; fat
ABSTRACT: 1. The metal and organochlorine contaminant concentrations were measured in a small sample of three seabird species collected from St Kilda, Scotland, UK, and the relationship between concentrations of different contaminants within and between tissues was compared. 2. There was little evidence that the concentration of a contaminant in one tissue was indicative of the concentration in any other tissue. 3. Generally, different contaminants had not co-accumulated in tissues; this was so even for the lipophilic compounds (DDE and PCBs), with the exception of puffin fat. 4. There was a tendency for the species containing the most fat to contain the highest concentrations of lipophilic contaminants and there was some suggestion that the fat in muscle could play a role in determining PCB concentrations.

Oxynos, K.; Schmitzer, J.; and Kettrup, A. 1993. Herring gull eggs as bioindicators for chlorinated hydrocarbons (contribution to the German Federal Environmental Specimen Bank). Sci. Total Environ. 139-140:387-398.
Rec #: 334
KEYWORDS: herring gull; *Larus argentatus*; egg; organochlorines; HCB; HCHs; chlordanes; dieldrin; DDTs; PCBs; Germany; Wadden Sea; Europe; temporal trends; spatial variations
ABSTRACT: From its position in the marine food chain, the herring gull (*Larus argentatus*) is a suitable indicator for the level of contamination of its habitats with lipophilic chemicals, especially the chlorinated hydrocarbons. The gull's utility as an indicator is demonstrated by investigations performed at Trischen island in the Elbe estuary and Alte Mellum island situated in the Weser estuary, when judged by their ability to reflect variations in pollution levels. The use of the analytical technique developed and standardized for the German Environmental

Annotated Bibliography (Cont.).

Specimen Bank gives comparable results for both temporal and spatial variations for contamination levels on the different islands in the Wadden Sea are (North Sea coast) of Germany.

Paasivirta, Jaakko; Särkkä, Jukka; Pellinen, Jukka; and Humppi, Tarmo. 1981. Biocides in eggs of aquatic birds. Completion of a food chain enrichment study for DDT, PCB and mercury. Chemosphere. 10(7):787-794.
Rec #: 375
KEYWORDS: egg; Finland; Europe; metals; Hg; organochlorines; HCB; DDTs; PCBs; temporal trends; spatial variations; herring gull; *Larus argentatus;* common gull; *Larus canus*
ABSTRACT: Eggs of 7 aquatic bird species of Lake Paijanne, Finland were analyzed for their contents of HCB [118-74-1], DDE [72-55-9], PCB, Hg, and methylmercury. Methylmercury was 85.5 % which is the same as in fish. Also the levels of total Hg were of the same order of magnitude in eggs and in fish of Paijanne, but the levels of chlorinated hydrocarbons were much higher in eggs. Differences between species were highly significant but the annual and regional differences were mostly insignificant. The chlorinated hydrocarbons correlated positive with each other and with fat contents but insignificantly with Hg contents. Calculation of total residue amounts in eggs and juvenile and adult bird specimens exposed an enrichment of DDT [50-29-3] to 3, PCB to 5, and Hg to all 7 species.

Page, G. W.; Stenzel, L. E.; and Ainley, D. G. 1982. Beached bird carcasses as a means of evaluating natural and human-caused seabird mortality. Final Report. DOE/EV/10254-T1. NTIS Order No.: DE82016736; Contract AC03-79EV10254. 150 pp.
Rec #: 270
KEYWORDS: oil; effects; CA; USA; North America; carcass
NOTES: SUPPLEMENTAL
ABSTRACT: A model was developed to determine the extent to which a large spill, by directly killing birds, could reduce the size of bird populations in the Gulf of the Faralones and its adjacent California estuaries. An attempt was then made to apply the model to a large spill that occurred in this area during 1971. To add perspective, the problem of the low but constant level of oil pollution that exists in developed coastal areas is discussed, and the way in which mortality from oil pollution compares with that from other factors is considered. (ERA citation 07:055223)

Parslow, J. L. F. and Jefferies, D. J. 1973. Relation between organochlorine residues in livers and whole bodies of guillemots. Environ. Pollut. 5(2):87-101.
Rec #: 509
KEYWORDS: organochlorines; PCBs; DDTs; dieldrin; common murre; *Uria aalge*; liver; carcass; Irish Sea; Europe; effects
ABSTRACT: Analyses of the liver and whole body levels of organochlorine insecticides and polychlorinated biphenyl residues in guillemots from the Irish Sea region indicated a significant relation between liver and whole body values. Birds found dead on the beach following a period of starvation had lost about 40 % of their body weight and, due to the mobilization of depot fat, organochlorine residues in their livers were much higher than those in the livers of shot birds. Polychlorinated biphenyl residues, together with natural stress factors, may have been involved in the death of large numbers of guillemots in 1969.

Annotated Bibliography (Cont.).

Peakall, D. B. 1988. Known effects of pollutants on fish-eating birds in the Great Lakes of North America. In: Schmidtke, Norbert W. (Ed), Toxic Contamination in Large Lakes. Volume 1. Chronic Effects of Toxic Contaminants in Large Lakes. 39-54.
Rec #: 127
KEYWORDS: review; effects; organochlorines; DDTs; Great Lakes; North America; egg; eggshell thickness; double-crested cormorant; *Phalacrocorax auritus;* black-crowned night heron; *Nycticorax nycticorax*; bald eagle; *Haliaeetus leucocephalus*; osprey; *Pandion haliaetus*; herring gull; *Larus argentatus*; common tern; *Sterna hirundo*; Forster's tern; *Sterna forsteri*; abnormalities; reproductive success; temporal trends; toxicology; MFOs; thyroid hormones; porphyrin
ABSTRACT: By the mid 1970s we had become aware that fish-eating birds had serious problems in the Great Lakes of North America. Marked declines of the population of the two fish-eating raptors - Bald Eagle and Osprey - and Double-crested Cormorant had occurred. Herring Gulls were experiencing poor reproductive success on Lake Michigan and Lake Ontario. There were reports of abnormalities in young common terns. In the case of the raptors and cormorant, eggshell thinning leading to reproductive failure appears to have been the major cause of the population declines. There is good evidence that the causative agent was DDE. Marked eggshell thinning was not observed in gulls and terns. Detailed studies on herring gulls showed the presence of both behavioral and embryotoxic effects. Reproductive success improved rapidly in the late 1970s and is now normal for all species except Foster's tern in Lake Michigan. Nevertheless, detailed investigations into the biochemistry and physiology of herring gulls still show significant differences along the pollutant gradient within the Great Lakes when compared to marine colonies. These parameters include induction of mixed function oxidases, thyroid function, and levels of hepatic porphyrins.

Peakall, D. B.; Hallett, D.; Miller, D. S.; Butler, R. G.; and Kinter, W. B. 1980. Effects of ingested crude oil on black guillemots: A combined field and laboratory study. Ambio. 9(1):28-30.
Rec #: 130
KEYWORDS: toxicology; oil; effects; dosing study; black guillemot; *Cepphus grylle*
NOTES: SUPPLEMENTAL
ABSTRACT: A new procedure for testing the physiological effects of pollutants in seabirds in their natural environment is described. With this procedure it was found that a single 0.1-0.5ml oral dose of weathered crude oil caused a transient rise in plasma sodium, nasal and adrenal gland hypertrophy and an overall decrease in body weight gain in nesting black guillemots (*Cepphus grylle*).

Peakall, D. B.; Hallett, D. J.; Bend, J. R.; Foureman, G. L.; and Miller, D. S. 1982. Toxicity of Prudhoe Bay crude oil and its aromatic fractions to nestling herring gulls. Environ. Res. 27(1):206-215.
Rec #: 132
KEYWORDS: toxicology; effects; oil; herring gull; *Larus argentatus*; dosing study
NOTES: SUPPLEMENTAL
ABSTRACT: The physiological effects of a single ingested dose of Prudhoe Bay crude oil (PBC), its aromatic fractions, and PBC/Clorexit emulsion were studied in nestling herring gulls (*Larus argentatus*). The data showed that the high-molecular-weight aromatic compounds were

Annotated Bibliography (Cont.).

responsible for retardation of growth and increases in adrenal and nasal gland weight. Little difference was found between PBC and the PBC/Clorexit emulsion although the latter did have a somewhat more marked effect on plasma sodium levels.

Peakall, D. B.; Jeffrey, D. A.; and Boersma, D. 1987. Mixed-function oxidase activity in seabirds and its relationship to oil pollution. Comp. Biochem. Physiol., C. 88C(1):151-154.
Rec #: 143
KEYWORDS: enzymes; oil; herring gull; *Larus argentatus*; Leach's storm-petrel; *Oceanodroma leucorhoa*; common murre; *Uria aalge*; Atlantic puffin; *Fratercula arctica*; liver; dosing study; effects; toxicology; EROD; MFOs
NOTES: SUPPLEMENTAL
ABSTRACT: The hepatic activity of epoxide hydrolase, aldrin epoxidase, aminopyrine N-demethylase, 7-ethoxyresorufin O-deethylase, benzo(a)pyrene 3-hydroxylase and UDP glucuronyl transferase was determined in adult herring gulls (*Larus argentatus*) at various stages of the breeding season. MFO activity was measured for adult Leach's storm-petrels (*Oceanodroma leucorhoa*), guillemot (*Uria aalge*) and Atlantic puffins (*Fratercula arctica*). For most assays the values were highest for the puffin. MFO activity in both nestling and adult Atlantic puffins was determined. The degree of induction caused by a single internal dose of Prudhoe Bay crude oil in adult puffins and that caused by multiple internal doses in nestling puffins was measured.

Peakall, D. B.; Miller, D. S.; and Kinter, W. B. 1983. Toxicity of crude oils and their fractions to nestling herring gulls. 1. Physiological and biochemical effects. Mar. Environ. Res. 8(2):63-71.
Rec #: 135
KEYWORDS: toxicology; oil; PAHs; hydrocarbons; dosing study; effects; herring gull; *Larus argentatus*
NOTES: SUPPLEMENTAL
ABSTRACT: The physiological and biochemical effects of ingested crude oil and their aromatic and aliphatic fractions were studied in nestling herring gulls (*Larus argentatus*). Single doses of 0 multiplied by 2 - 1 multiplied by 0 ml of two batches of South Louisiana Crude oil (SLC-76 and SLC-78) were used. SLC-76 caused marked retardation of growth, some disruption of plasma osmoregulation and hypertrophy of adrenal and nasal salt gland tissue; SLC-78 had no detectable effects on gull chicks. Weathering of SLC-76 for 36 h did not reduce its toxicity and when SLC-76 was fractionated, the aromatic fraction was substantially more toxic then the aliphatic fraction.

Peakall, D. B.; Norstrom, R. J.; Jeffrey, D. A.; and Leighton, F. A. 1989. Induction of hepatic mixed function oxidases in the herring gull (*Larus argentatus*) by Prudhoe Bay crude oil and its fractions. Comp. Biochem. Physiol., C. 94C(2):461-463.
Rec #: 142
KEYWORDS: oil; liver; enzymes; toxicology; MFOs; effects; herring gull; *Larus argentatus*; dosing study; EROD
NOTES: SUPPLEMENTAL
ABSTRACT: The hepatic activity in herring gull (*Larus argentatus*) chicks of aminopyrine N-demethylase, benzo(a)pyrene 3-hydroxylase, 7-ethoxyresorufin O-deethylase and aldrin epoxidase were all induced 24 hours after a single oral dose of Prudhoe Bay Crude Oil (PBCO).

Annotated Bibliography (Cont.).

Aminopyrine N-demethylase was maximally induced by a dose of 0.1 ml; the other enzymes responded in a dose-dependent manner over the range 0.1-1.0 ml. The activity of all enzymes had decreased significantly after 72 hours indicating that hepatic enzymes are a sensitive but short lived, biological indicator of exposure to oil. Fractionation of PBCO showed that the induction was largely caused by the aromatic fraction.

Peakall, D. B; Norstrom, R. J.; Rahimtula, A. D.; and Butler, R. G. 1986. Characterization of mixed-function oxidase systems of the nestling herring gull and its implications for bioeffects monitoring. Environ. Toxicol. Chem. 5(4):379-385.
Rec #: 145
KEYWORDS: liver; muscle; organochlorines; toxicology; MFOs; cytochrome P_{450}; EROD; organochlorines; HCB; DDTs; PCBs; chlordanes; dieldrin; effects; enzymes; herring gull; *Larus argentatus*; Newfoundland; Canada; North America
ABSTRACT: The hepatic mixed-function oxidase system was characterized in embryonic and nestling herring gulls (*Larus argentatus*). The activity of aminopyrine N-demethylase decreased significantly with the age of the nestlings. No consistent changes in 7-ethyoxyresorufin O-deethylase, benzo(a)pyrene 3-hydroxylase or cytochrome P_{450} levels with age were found. Levels of organochlorines were determined in individual livers for the 21-d-old nestlings. Correlations among organochlorine residue levels were good, and the activities of aminopyrine N-demethylase and benzo(a)pyrene 3-hydroxylase correlated with each other and with cytochrome P_{450} levels. No correlations were found with 7-ethoxyresorufin O-deethylase activity, and none of the enzyme levels correlated with the organochlorine residue levels. However, cytochrome P_{450} levels correlated with both hexachlorobenzene and oxychlordane levels.

Peakall, D. B.; Tremblay, J.; and Kinter, W. B. 1981. Endocrine dysfunction in seabirds caused by ingested oil. Environ. Res. 24(1):6-14.
Rec #: 133
KEYWORDS: oil; gland; herring gull; *Larus argentatus*; Leach's storm petrel; *Oceanodroma leucorhoa*; black guillemot; *Cepphus grylle*; dosing study; effects; toxicology
NOTES: SUPPLEMENTAL
ABSTRACT: In laboratory and field experiments, a single oral dose (0.1-1.0 ml) of certain crude oils or aromatic fractions caused elevated plasma corticosterone and thyroxine levels in nestling herring gulls and black guillemots. In gulls, plasma corticosterone levels were elevated within 1 day after dosing; the maximal effect was observed after 4 days and levels returned to control values after 2 wk. Thyroxine levels did not increase until 6 days after dosing and remained elevated after 2 wk. Since only those oils which reduced seabird growth rates affected hormone levels, the data suggest that disruption of endocrine balance is one underlying cause of depressed growth in oil-dosed birds.

Peakall, D. B.; Wells, P. G.; and Mackay, D. 1987. A hazard assessment of chemically dispersed oil spills and seabirds. Mar. Environ. Res. 22(2):91-106.
Rec #: 137
KEYWORDS: review; oil; toxicology; reproductive success; effects; dosing study; mallard; *Anas platyrhynchos*; Leach's storm petrel; *Oceanodroma leucorhoa;* herring gull; *Larus argentatus*
NOTES: SUPPLEMENTAL

Annotated Bibliography (Cont.).

ABSTRACT: The effects of dispersants on both the exposure to and toxicity of oil to seabirds are considered in order to assess the hazard. Ideally the dispersant mixes with oil and disperses it into the water column. This process is rapid but generally incomplete. The toxicology of one dispersant (Corexit 9527), for which data are available, shows that the toxicity of oil-Corexit mixtures is similar to that of oil alone. The effect of 2 feeding regimes, pursuit diving and surface diving, is considered. Calculations indicate that the amount of oil that is likely to be taken up by the bird while moving through the water column is small. It is concluded that there is little evidence of synergistic effects between oil and dispersant. The major oiling of birds occurs at the surface and thus dispersants must be highly effective to reduce the exposure of birds to oil.

Peakall, David. 1992. Animal Biomarkers as Pollution Indicators. Ecotoxicol. Ser., Vol. 1, New York, NY: Chapman and Hall. 291 pp.
Rec #: 148
KEYWORDS: review; effects; toxicology; eggshell thickness; reproductive success; enzymes; MFOs; thyroid hormones; porphyrin; vitamin A; retinol; age variations; seasonal variations; gender differences; spatial variations; biomagnification; organochlorines; DDTs; dieldrin; HCB; PCBs; PCDDs; oil; organometallics; organic Hg; organophosphates; malathion; methylparathion; phosmethylan; liver; egg; herring gull; *Larus argentatus;* starling; *Sturnus vulgaris;* red-winged blackbird; *Agelaius phoeniceus;* Caspian tern; *Sterna caspia;* wedge-tailed shearwater; *Puffinus pacificus;* Leach's storm-petrel; *Oceanodroma leucorhoa;* brown pelican; *Pelecanus occidentalis;* great egret; *Ardea alba;* black-crowned night-heron; *Nycticorax nycticorax;* mute swan; *Cygnus olor;* Canada goose; *Branta canadensis;* white-fronted goose; *Anser albifrons;* mallard; *Anas platyrhynchos;* canvasback duck; *Aythya valisineria;* sparrowhawk; *Accipiter nisus;* bald eagle; *Haliaeetus leucocephalus;* osprey; *Pandion haliaetus;* peregrine falcon; *Falco peregrinus;* pigeon hawk; *Falco columbarius;* American kestrel; *Falco sparverius;* Japanese quail; *Coturnix coturnix;* burrowing owl; *Athene cunicularia;* chukar; *Alectoris graeca;* bobwhite quail; *Colinus virginianus;* California quail; *Callipepla californica;* sharp-tailed grouse; *Tympanuchus phasianellus;* grey partridge; *Perdix perdix;* chicken; *Gallus domesticus;* pheasant; *Phasianus colchicus;* turkey; *Meleagris gallopavo;* glaucous-winged gull; *Larus glaucescens;* lesser black-backed gull; *Larus fuscus;* western gull; *Larus occidentalis;* California gull; *Larus californicus;* ring-billed gull; *Larus delawarensis;* red-billed gull; *Larus novaehollandiae;* laughing gull; *Larus atricilla;* common tern; *Sterna hirundo;* Forster's tern; *Sterna forsteri;* common murre; *Uria aalge;* black guillemot; *Cepphus grylle;* Atlantic puffin; *Fratercula arctica;* common pigeon; *Columba livia;* ring dove; *Streptopelia risoria;* barn swallow; *Hirundo rustica;* redstart; *Phoenicurus phoenicurus;* red-billed quelea; *Quelea quelea;* common grackle; *Quiscalus quiscalus;* white-throated sparrow; *Zonotrichia albicollis;* Bengalese finch; *Lonchura domestica;* zebra finch; *Taeniopygia guttata;* Great Lakes; North America
NOTES: SUPPLEMENTAL
ABSTRACT: The core of this book is a consideration of the major biomarkers that are available. At the beginning of each section there is a brief overview of the biochemistry and physiology of the system under consideration. Natural factors that influence the measurements such as age, sex, seasonal variations, etc. are then discussed briefly. The summary of experimental studies that follows is necessarily high selective. Chapter 3 looks at biomarkers of the nervous system; reproduction is considered in Chapter 4, which looks first at field investigations, secondly at using embryos, thirdly on effects on the hormone systems, and finally at receptors. The techniques available for examining adduct formation of pollutants with DNA

Annotated Bibliography (Cont.).

and the effects on DNA structure are considered in Chapter 5. The mixed function oxidase system, one of the key defenses against xenobiotics, is discussed in Chapter 6. A series of other well-established biomarkers is examined in Chapter 7: the thyroid hormones and function, retinol, porphyrins and other effects on heme and finally serum enzymes. Effects of toxic chemicals on behavior are looked at in Chapter 8 from the viewpoint of their relationship to parallel biochemical changes. The final chapter on specific biomarkers discusses the immune system.

Peakall, David B. and Fox, Glen A. 1987. Toxicological investigations of pollutant-related effects in Great Lakes gulls. Environ. Health Perspect. 71:187-193.
Rec #: 146
KEYWORDS: toxicology; reproductive success; effects; Great Lakes; North America; herring gull; *Larus argentatus;* egg; organochlorines; DDTs; mirex; PCBs; HCB; PCDDs; temporal trends; spatial variations
ABSTRACT: Reproductive failure of a number of fish-eating birds was observed on the Great Lakes in the mid-1960s to mid-1970s. The herring gull (*Larus argentatus*) has been used as the primary monitoring species. The low hatching success observed in this species on Lake Ontario in the mid-1970s was due to loss of eggs and failure of eggs to hatch. Egg exchange experiments demonstrated that this was due both to the incubation behavior of adults and to direct embryotoxic effects. Decrease of nest attentiveness was demonstrated using telemetered eggs, but attempts to reproduce the embryonic effects by injection of pollutant mixtures into eggs were not successful. Reproductive success improved rapidly during the late 1970s and was normal by the end of the decade. Recent studies have focused on cytogenetic and biochemical changes and detailed analytical chemistry of residues. The geographic pattern of these changes indicates that they are caused by xenobiotics, but is has not been possible to relate the changes to a specific chemical.

Pearce, P. A.; Elliott, J. E.; Peakall, D. B.; and Norstrom, R. J. 1989. Organochlorine contaminants in eggs of seabirds in the Northwest Atlantic, 1968-1984. Environ. Pollut. 56(3):217-235.
Rec #: 387
KEYWORDS: organochlorines; PCBs; DDTs; dieldrin; HCB; chlordanes; HCHs; egg; double-crested cormorant; *Phalacrocorax auritus*; Leach's storm-petrel; *Oceanodroma leucorhoa*; Atlantic puffin; *Fratercula arctica*; Atlantic Ocean; Newfoundland; Canada; North America; temporal trends; spatial variations
ABSTRACT: Eggs of three seabird species, double-crested cormorant (*Phalacrocorax auritus*), Leach's storm-petrel (*Oceanodroma leucorhoa*), and Atlantic puffin (*Fratercula arctica*) were collected at four-year intervals from 1968 to 1984, from colonies in eastern Canada and analyzed for organochlorines. This monitoring study was established to provide data on contamination of the marine environment and possible implications for seabird health. Long-term trend data are presented for PCBs, DDE, dieldrin, HCB, oxychlordane, heptachlor epoxide, HCH, and mirex. DDE and PCBs declined significantly in all species from the Bay of Fundy. DDE declined significantly in puffins and petrels whereas PCBs declined only in petrels from the Atlantic coast of Newfoundland. Generally DDE declined more than PCBs. Dieldrin, oxychlordane, HCH, and mirex levels decreased at some locations but were stable at others. Hexachlorobenzene and heptachlor epoxide levels remained steady or increased significantly, depending on the species

Annotated Bibliography (Cont.).

and location. Organochlorine levels in cormorants from the St. Lawrence River estuary showed no significant trends.

Pearce, Peter A.; Peakall, David B.; and Reynolds, Lincoln M. 1979. Shell thinning and residues of organochlorines and mercury in seabird eggs, Eastern Canada, 1970-1976. Pestic. Monit. J. 13(2):61-68.
Rec #: 21
KEYWORDS: egg; organochlorines; DDTs; PCBs; dieldrin; metals; Hg; Quebec; Newfoundland; New Brunswick; Nova Scotia; Bay of Fundy; Canada; North America; Leach's storm-petrel; *Oceanodroma leucorhoa*; double-crested cormorant; *Phalacrocorax auritus*; common eider; *Somateria mollissima*; common tern; *Sterna hirundo*; razorbill; *Alca torda*; common murre; *Uria aalge*; black guillemot; *Cepphus grylle*; Atlantic puffin; *Fratercula arctica*; temporal trends; spatial variations; eggshell thickness; reproductive success
ABSTRACT: Organochlorine and mercury concentrations are reported for 252 eggs of Leach's storm-petrel (*Oceanodroma leucorhoa*), double-crested cormorant (*Phalacrocorax auritus*), common eider (*Somateria mollissima*), common tern (*Sterna hirundo*), razorbill (*Alca torda*), common murre (*Uria aalge*), black guillemot (*Cepphus grylle*), and Atlantic puffin (*Fratercula arctica*) from the Bay of Fundy, the Gulf of St. Lawrence, and the open Atlantic shore of Canada during 1970-76. Concentrations of all organochlorines except DDE and polychlorinated biphenyls (PCBs) were low. DDE, PCBs, and mercury residues were highest in cormorant and petrel, intermediate in alcids, and lowest in eider and tern. Temporal and spatial aspects of contamination patterns are discussed. Authors conclude that only in cormorants were DDE residues high enough to cause, through eggshell thinning, population declines.

Pekarik, C. and Weseloh, D. V. 1995. Contaminant trends in herring gull eggs from the Great Lakes: Regression analysis of long-term and recent data, 1974-1993. Proceedings of the 38th Conference of the International Association of Great Lakes Research., International Association for Great Lakes Research, 2200 Bonisteel Boulevard, Ann Arbor, Mi 48109-2099 (USA). 84-85.
Rec #: 202
KEYWORDS: North America; Great Lakes; egg; organochlorines; PCBs; DDTs; temporal trends; spatial variations; herring gull; *Larus argentatus*
NOTES: ABSTRACT ONLY - NOT AVAILABLE
ABSTRACT: The levels of 14 organochlorine contaminants in herring gull eggs (*Larus argentatus*) were measured at 13 Great Lakes colonies. Linear regression was applied to every transformed contaminant value at each site in order to establish statistical patterns. Long-term analysis (1974-1993) consisted of finding an appropriate change point model, comparing its efficacy with a single line model and choosing the most appropriate. Ninety percent of the analyses required a change point in the model, the most common change points were 1984 and 1988. Of 169 total analyses 94 % showed significant decrease, 5 % showed an increase, and 1 % showed no significant change in the level of the compound. Rising levels were found in Lake Huron, Lake Michigan, and Lake Superior. Recent trends for total PCB congeners and DDE were established by single line regression over three time periods: 1988-1993, 1989-1993, and 1990-1993. In 82 % of the analyses (n=78) levels showed no statistical change, 9 % showed a decrease, and 9 % showed significant increase. Rising levels were found in Lake Huron, Lake Michigan, and Lake Superior. Decreasing levels were found in the Niagara River, Lake Erie, and Detroit River.

Annotated Bibliography (Cont.).

Pekarik, C. and Weseloh, D. V. 1998. Organochlorine contaminants in herring gull eggs from the Great Lakes, 1974-1995: Change point regression analysis and short-term regression. Environ. Monit. Assess. 53(1):77-115.
Rec #: 93
KEYWORDS: Great Lakes; North America; egg; organochlorines; chlordanes; benzenes; HCB; DDTs; HCHs; PCBs; dieldrin; herring gull; *Larus argentatus*; temporal trends; spatial variations
NOTES: FIGURES ONLY
ABSTRACT: The temporal trends (1974-1995) of 11 organochlorine contaminants in herring gull (*Larus argentatus*) eggs from 13 colonies throughout the Great Lakes were statistically analyzed using two regression methods on logarithmically transformed data. Change point analysis was used to determine if there had been significant year to year fluctuations in contaminant levels and/or changes in long-term trends. Short-term regressions were conducted on 6 major compounds for two time periods (early 1990s and early 1980s) to compare the rates of decline. Overall, change point analyses indicated that for most of the comparisons (75 %) there had been significant year to year fluctuations in contaminant levels. They also indicated that for most of the comparisons (67 %) the rate of decline after the change point was as fast as or faster than before the change point, this pattern was most common for dieldrin and heptachlor epoxide at certain locations. In 19 % of the comparisons the rates of decline had slowed or stabilized, this was most common for PCB and pentachlorobenzene. In 14 % of the comparisons there were no significant temporal trends, this was most common for photomirex and mirex. Results for short-term regression showed that out of 78 comparisons for each time period, 5 (6 %) were declining significantly in the early 1990s and 11 (14 %) were declining significantly in the early 1980s. Both types of regression indicated that, for most of the herring gull egg contaminant database, recent logarithmic rates of decline were similar to those seen previously in the sampling period. For PCB 1254:1260, a group of compounds of particular toxicological importance, change point analyses indicated that the logarithmic rates of decline in herring gull eggs from western Lake Ontario were slower from 1987-1995 than they were from 1974-1986. At both Lake Superior colonies and the Niagara River colony PCB 1254:1260 concentrations ceased to decline in the mid-1980s. At the colony in Green Bay, Lake Michigan, PCB 1254:1260 levels have shown no significant temporal trend since 1976. At the remaining 8 colonies, PCB 1254:1260 levels continue a logarithmic decline in recent years at the same rate as or faster than previously.

Pekarik, C; Weseloh, D. V; Barrett, G. C.; Simon, M.; Bishop, C. A; and Pettit, K. E. 1998. An atlas of contaminants in eggs of fish-eating colonial birds of the Great Lakes (1993-1997). Technical Report Series No. 321. Burlington, Ontario, Canada: Canadian Wildlife Service Ontario Region. 245 pp.
Rec #: 520
KEYWORDS: double-crested cormorant; *Phalacrocorax auritus*; great black-backed gull; *Larus marinus*; herring gull; *Larus argentatus*; ring-billed gull; *Larus delawarensis*; organochlorines; dieldrin; HCHs; mirex; chlordanes; OCS; DDTs; benzenes; HCB; PCBs; PCDDs; PCDFs; egg; Great Lakes; North America; temporal trends; spatial variations
ABSTRACT: During 1993-1997, Canadian Wildlife Service (Ontario Region) collected 1252 eggs from 32 sites. Four species of fish-eating colonial waterbirds were sampled: Double-crested Cormorant (*Phalacrocorax auritus*), Great Black-backed Gull (*Larus marinus*), Herring Gull (*Larus* argentatus), Ring-billed Gull (*Larus delawarensis*). The purpose was to measure the

Annotated Bibliography (Cont.).

levels of the following compounds: organochlorine pesticides, chlorinated benzenes, polychlorinated biphenyls, dioxins and furans, lipid and moisture. These data were generated as part of a monitoring program started in 1970 to understand the temporal and spatial trends of environmental contaminant levels in biota of the Great Lakes. Since the 1970s the levels of most chlorinated hydrocarbons have decreased significantly at most colonies on the Great Lakes. A regression model that applied two temporal trends to the log-transformed values of most organochlorine pesticides and PCB:1254-1260 found that the rates of decline in recent years were similar to those seen in the years shortly after sampling began (Di Maio et al., In Press; Pekarik and Weseloh, 1998).

The data from 1993-97 are summarized in two volumes. Volume I contains contaminant data for all (4) species summarized by location, and non-coplanar PCB data for Herring Gull eggs from 14 annual monitoring colonies. Volume II contains contaminant data for all (4) species summarized by compound. Both volumes contain sample locations and the means and standard deviations or pooled analysis values for organochlorine pesticides, chlorinated benzenes, polychlorinated biphenyls, dioxins and furans, lipid and moisture. Additionally, data for mercury from Herring Gull eggs collected in 1992 have been listed, since they were inadvertently omitted from Pettit et al., l994a; b. The publication of previous years' data has resulted in independent statistical analyses and publications of Herring Gull egg contaminant data (Smith, 1995; Stow, 1995). Within the Canadian Wildlife Service additional analyses have been published on the current dataset of contaminant levels in fish-eating birds of the Great Lakes (Hebert et al., 1997.; Koster et al., 1997.; Pekarik and Weseloh, 1998.; Ryckman et al., 1998).

Pettit, K. E.; Bishop, C. A; Weseloh, D. V; and Norstrom R. J. 1994. An atlas of contaminants in eggs of fish-eating colonial birds of the Great Lakes (1989-1992). Technical Report Series No. 193. Burlington, Ontario, Canada: Canadian Wildlife Service Ontario Region. 400 pp.
Rec #: 367
KEYWORDS: herring gull; *Larus argentatus*; double-crested cormorant; *Phalacrocorax auritus*; Caspian tern; *Sterna caspia*; common tern; *Sterna hirundo*; black-crowned night-heron; *Nycticorax nycticorax*; Forster's tern; *Sterna forsteri*; ring-billed gull; *Larus delawarensis*; organochlorines; dieldrin; HCHs; mirex; chlordanes; OCS; DDTs; benzenes; HCB; PCBs; PCDDs; PCDFs; egg; Great Lakes; North America; temporal trends; spatial variations
ABSTRACT: During 1989-1992, Canadian Wildlife Service (Ontario) collected a total of 1495 eggs from fish-eating colonial birds from 50 colonies throughout the Great Lakes to measure the levels of 86 chlorinated hydrocarbon compounds, and the lipid concentrations present. These data were generated as part of a monitoring program started in 1970 to understand the temporal and spatial trends in environmental contaminant levels in biota of the Great Lakes. During 1989-1992, the levels of chlorinated hydrocarbons in colonial waterbird eggs have remained relatively stable within colonies across the Great Lakes. This is consistent with trends occurring in the mid-1980s in fish-eating colonial bird eggs from the Great Lakes as reported in An atlas of contaminants in eggs of mistreating colonial birds of the Great Lakes (1970-1988) Volume I, Accounts by Species and An atlas of contaminants in eggs of fish-eating colonial birds of the Great Lakes (1970-1988) Volume II, Accounts by Chemical (Bishop et al., 1992a; 1992b).
The data from 1989-92 are summarized in two volumes. Volume I contains contaminant data summarized by location. Non-coplanar PCB congener levels, patterns and interpretation of these patterns are also included in Volume I for Herring Gull eggs from 14 annual monitoring colonies. Volume II contains contaminant data summarized by compound analyzed. Both

Annotated Bibliography (Cont.).

volumes contain sample locations and number of samples collected for each species each year, and pooled values or means and standard deviations for organochlorine pesticide, polychlorinated biphenyls, dioxin and furan concentrations.

Pfaffenberger, Bernd; Hühnerfuss, Heinrich; Kallenborn, Roland; Köhler-Guenther, Angela; König, Wilfried A.; and Krüner, Günter. 1992. Chromatographic separation of the enantiomers of marine pollutants. Part 6: comparison of the enantioselective degradation of α-hexachlorocyclohexane in marine biota and water. Chemosphere. 25(5):719-725.
Rec #: 482
KEYWORDS: common eider; *Somateria mollissima*; organochlorines; HCHs; Germany; North Sea; Europe; liver; metabolism
ABSTRACT: The enantiomeric ratios of the chiral marine pollutant α-hexachlorocyclohexane [α-HCH] were detected in blue mussels (*Mytilus edulis*), flounders (*Platychthys flesus*), common eider ducks (*Somateria mollissima*), and in North Sea water at experimentally sites in the German Bight by means of capillary gas chromatography using β-cyclodextrin derivatives as chiral stationary phases. Different enzymic degradation pathways are revealed by opposite enantioselective enrichments of the α-HCH enantiomers: while in the liver of the common eider ducks the preferable degradation of (-)-α-HCH leads to an enrichment of (+)-α-HCH [i.e., (+)-α-HCH/(-)-α-HCH ≈ 1.4 - ∞], the enantiomeric ratios measured in the liver samples of flounders [(+)-α-HCH/(-)-α-HCH ≈ 0.80 - 0.94], in blue mussels [(+)-α-HCH/(-)-α-HCH ≈ 0.84-0.93] and in the North Sea water samples [(+)-α-HCH/(-)-α-HCH ≈ 0.84] imply enzymic processes in the course of which common structural elements are preferred represented by (+)-α-HCH.

Piatt, John F. and Anderson, Paul. 1996. Response of common murres to the *Exxon Valdez* oil spill and long-term changes in the Gulf of Alaska marine ecosystem. Am. Fish Soc. Symp. 18:720-737.
Rec #: 17
KEYWORDS: oil; effects; common murre; *Uria aalge*; thick-billed murre; *Uria lomvia*; North America; USA; AK
NOTES: SUPPLEMENTAL
ABSTRACT: Short-term effects of the 1989 TV Exxon Valdez oil spill on seabirds were dramatic and well documented. Seabird populations at sea in the spill zone were immediately depressed, and more than 30,000 dead, oiled seabirds were recovered from beaches within months of the spill. It is estimated that 250,000 seabirds were killed by oil, of which 74 % were murres *Uria spp*. Based on comparisons of prespill (1970s) and postspill (1989-1994) data, long-term effects on murres attributed to oil pollution included population declines, reduced breeding success, and delayed breeding phenology. Populations remained depressed, but breeding success and phenology gradually returned to normal levels by 1993. An alternative hypothesis to explain these long-term effects is that murres were responding to natural events in their marine environment. Reduced flow of the Alaska Coastal Current (ACC) in 1989 may have reduced and delayed biological productivity in the ACC. On a broader time scale, marked changes in marine fish communities have occurred during the past 20 years. Coincident with cyclical fluctuations in seawater temperatures, the abundance of small forage species (e.g., northern pink shrimp *Pandalus borealis*, capelin *Mallotus villosus*, and Pacific sandfish *Trichodon trichodon*) declined precipitously in the late 1970s while populations of large predatory fish (e.g., walleye pollock

Annotated Bibliography (Cont.).

Theragra chalcogramma, Pacific cod *Gadus macrocephalus*, and flatfish) increased dramatically. Correspondingly, seabird diets shifted from mostly capelin in the 1970s to mostly Pacific sand lance *Ammodytes hexapterus* and juvenile pollock in the late 1980s. Furthermore, a variety of seabirds and marine mammals both inside and outside of the oil spill zone exhibited signs of food stress (population declines, reduced productivity, die-offs) throughout the 1980s and early 1990s. We conclude that available data are inadequate to distinguish between long-term effects of the Exxon Valdez oil spill on murres and a natural response of murres to long-term changes in their marine environment.

Potts, G. R. 1968. Success of eggs of the shag on the Farne Islands, Northumberland, in relation to their content of dieldrin and p,p'-DDE. Nature. 217(5135):1282-1284.
Rec #: 392
KEYWORDS: organochlorines; dieldrin; DDTs; egg; effects; reproductive success; shag; *Phalacrocorax aristotelis*; Great Britain; Europe
ABSTRACT: No significant correlation was found between clutch size or brood size and the concentration of either p,p'-DDE or dieldrin in shag, *Phalacrocorax aristotelis*, eggs from the Farne Islands. Total clutch or brood failure was not correlated with the concentration of p,p'-DDE, but a significant correlation, apparently of the threshold type, was found between the total clutch-brood failure and the dieldrin level. No evidence was found for an unusual proportion of deaths of chicks at hatching in unsuccessful shag clutches containing 2-3 ppm dieldrin. Concentrations of approximately 2.5 ppm dieldrin in the eggs reflected concentrations of approximately 9 ppm in the livers of shags of breeding age; no evidence has been found of dieldrin poisoning of adult shags from the Farne Islands. Although recent increases in the rate of egg breaking in the sparrow hawk (*Accipiter nisus*), peregrine (*Falco peregrinus*), and the golden eagle (*Aquila chrysaetos*) have been considered to be due to the sublethal effects of organochlorine insecticide poisoning, approximately 33 % of egg breakages in the shag, and most cases of repeated breakage, were due to interference by extra females resulting from simultaneous polygyny at one nest.

Quick, M. P. 1993. Suspected hydrocarbon poisoning of swans and guillemots. J. Forensic Sci. Soc. 33(3):143-148.
Rec #: 504
KEYWORDS: organochlorines; PCBs; hydrocarbons; liver; common murre; *Uria aalge*; Great Britain; Europe
ABSTRACT: A case of sub-acute hydrocarbon poisoning of swans and several separate incidents of suspected chronic hydrocarbon poisoning of guillemots were investigated. The birds' tissues were saponified and extracted with pentane and analyzed by gas chromatography-mass spectrometry. The residue levels of pristane in the swan tissues and aliphatic hydrocarbons in the guillemots' livers are reported and discussed.

Rattner, Barnett A.; Eroschenko, Victor P.; Fox, Glen A.; Fry, D. Michael; and Gorsline, Jane. 1984. Avian endocrine responses to environmental pollutants. J. Exp. Zool. 232(3):683-689.
Rec #: 131
KEYWORDS: review; effects; adrenal cortex; organochlorines; DDTs; PCBs; mirex; toxicology; gland; oil; reproductive success; Great Lakes; North America; California gull; *Larus californicus*; Japanese quail; *Coturnix coturnix*; ring dove; *Streptopelia risoria*; herring gull;

Annotated Bibliography (Cont.).

Larus argentatus; mallard; *Anas platyrhynchos*; Leach's storm petrel; *Oceanodroma leucorhoa*; wedge-tailed shearwater; *Puffinus pacificus*; American kestrel; *Falco sparverius*; bobwhite quail; *Colinus virginianus*; dosing study
NOTES: NO TABLES
ABSTRACT: Many environmental contaminants are hazardous to populations of wild birds. Chlorinated hydrocarbon pesticides and industrial pollutants are thought to be responsible for population declines of several species of predatory birds through eggshell thinning. Studies have demonstrated that these contaminants have estrogenic potency and may affect the functioning of the gonadal and thyroidal endocrine subsystems. Petroleum crude oil exerts toxicity externally, by oiling of plumage, and internally, by way of ingestion of oil while feeding or preening. Extensive ultrastructural damage to the inner zone of the adrenal, diminished adrenal responsiveness to adrenocorticotrophic hormone, and reduced corticosterone secretion rate suggest that low levels of plasma corticosterone reflect a direct effect of petroleum on the adrenal gland. Suppressive effects of oil on the ovary and decreases in circulating prolactin have been associated with impaired reproductive function. Large-scale field studies of free-living seabirds have confirmed some of the inhibitory effects of oil on reproduction that have been observed in laboratory studies. Organophosphorus insecticides, representing the most widely used class of pesticides in North America, have been shown to impair reproductive function, possibly by altering secretion of luteinizing hormone and progesterone. Relevant areas of future research on the effects of contaminants on avian endocrine function are discussed.

Riget, F. and Dietz, R. 2000. Temporal trends of cadmium and mercury in Greenland marine biota. Sci. Total Environ. 245(1-3):49-60.
Rec #: 434
KEYWORDS: glaucous gull; *Larus hyperboreus*; liver; Greenland; North America; metals; Cd; Hg; temporal trends; spatial variations; age variations; bioaccumulation; biomagnification
ABSTRACT: Data for cadmium and mercury in Greenland marine biota (blue mussels, polar cod, shorthorn sculpin, glaucous gull and ringed seals) over a period of 20 years has been analyzed in order to assess temporal changes. Most of the comparisons were conducted between tissue samples collected in the mid-1980s and mid-1990s. Cadmium data from a few time series obtained at reference sites during monitoring of mining activities were also included. No overall temporal trends in cadmium or mercury concentrations were found within the 20-years period assessed. However, cadmium concentrations in ringed seals tended to increase in the period from late-1970s to the mid-1980s. From the mid-1980s to the mid-1990s cadmium concentrations in ringed seals decreased again, while mercury concentrations showed a tendency to increase in the same period. The observed changes may reflect natural fluctuations caused by factors such as a shift in feeding behavior, rather than changes in anthropogenic exposure.

Riget, F.; Dietz, R.; Johansen, P.; and Asmund, G. 2000. Lead, cadmium, mercury and selenium in Greenland marine biota and sediments during AMAP phase 1. Sci. Total Environ. 245(1-3):3-14.
Rec #: 435
KEYWORDS: metals; Pb; Cd; Hg; Se; Greenland; North America; glaucous gull; *Larus hyperboreus*; liver; age variations; bioaccumulation; biomagnification; spatial variations
ABSTRACT: Lead, cadmium, mercury and selenium levels in the Greenland marine environment from the first phase of the AMAP are presented. Samples were collected in 1994-

Annotated Bibliography (Cont.).

1995 covering four widely separated regions in Greenland. Samples included sediments, soft tissue of blue mussel; and liver of polar cod, shorthorn sculpin, glaucous gull, Iceland gull and ringed seal. Concentrations of lead were found to increase with the size of blue mussel, but not with the age of gulls or ringed seal. Both cadmium and mercury concentrations were found to increase with the size/age of all species. Selenium concentrations decreased with increasing size of blue mussel, but increased with the age of gulls and ringed seal. Element levels found are within the range of those found in previous studies in Greenland. Relative to global background levels, lead levels must be considered low, whereas levels of cadmium, mercury and selenium in Greenland marine biota are high. Significant differences in element levels in sediments and biota among regions in Greenland were seen in several cases. There was a tendency for the highest lead and mercury concentrations to be found in east Greenland, whereas the highest cadmium concentrations were found in central west Greenland. However, the geographical differences among the media did not show a consistent pattern.

Risebrough, R. W.; Reiche, P.; Herman, S. G.; Peakall, D. B.; and Kirven, M. N. 1968. Polychlorinated biphenyls in the global ecosystem. Nature. 220(5172):109-112.
Rec #: 529
KEYWORDS: organochlorines; PCBs; DDTs; dieldrin; brain; carcass; biomagnification; reproductive success; effects; eggshell thickness; toxicology; CA; USA; North America; Great Britain; Europe; peregrine falcon; *Falco peregrinus*
ABSTRACT: Results are given of analyses of peregrine falcons which died from a variety of causes shortly after being trapped for falconry. Significant amounts of polychlorinated biphenyls (PCB) were present in Arctic peregrines only a few months old, but higher residues were present in a second year Arctic bird, and exceptionally high residues were present in an adult trapped in California. In 2 birds the total lipid reserves were very low, and in both the brain concentrations of DDE and PCB were high, perhaps at toxic levels. Dieldrin concentrations were lower than in the peregrines analyzed in Britain, but DDE concentrations were approximately comparable. In 1968 no pairs of peregrines were observed to hatch or fledge more than a single young, compared to 2-4 young in the past. The thickness of fragments of a peregrine eggshell, with its membrane, collected in 1968 was 0.24 mm., a decrease of 34 % from the mean thickness of 0.34 mm. ± 0.015 mm. in 23 eggs collected in the same wilderness region before 1947. In California, numbers of breeding peregrines have been reduced by at least 80 % in recent years. Results of analyses of marine and terrestrial birds and of 3 species of freshwater fish for PCB and DDT content and concentrations show that regional fallout patterns exist. A significant fraction of chlorinated hydrocarbons must come from the atmosphere, and air transport best explains the presence of PCB in remote areas. Body concentrations of 40 ppm. of p,p'-DDE significantly increased the rate of estradiol degradation by the induced hepatic enzymes. On a weight basis the PCB preparation had an estradiol-degrading potential approximately 5-fold that of p,p'-DDE or tech. DDT. Both DDE and PCB have the capacity to produce sublethal physiological effects in birds. The reductions in eggshell thickness and eggshell weight increase the chances of egg breakage and water retention, which affects hatching success.

Risebrough, Robert W.; Menzel, Daniel B.; Martin, D. James; and Olcott, Harold S. 1967. DDT residues in Pacific sea birds: A persistent insecticide in marine food chains. Nature. 216:589-590.
Rec #: 542
KEYWORDS: organochlorines; DDTs; Pacific Ocean; CA; USA; North America; carcass;

Annotated Bibliography (Cont.).

muscle; brain; liver; fat; Cassin's auklet; *Ptychoramphus aleuticus*; western gull; *Larus occidentalis*; pelagic cormorant; *Phalacrocorax pelagicus*; Brandt's cormorant; *Phalacrocorax penicillatus*; brown pelican; *Pelecanus occidentalis*; common murre; *Uria aalge*; ancient murrelet; *Synthliboramphus antiquus*; red phalarope; *Phalaropus fulicarius*; rhinoceros auklet; *Cerorhinca monocerata*; northern fulmar; *Fulmarus glacialis*; black-legged kittiwake; *Rissa tridactyla*; sooty shearwater; *Puffinus griseus*; short-tailed shearwater; *Puffinus tenuirostris*; biomagnification

ABSTRACT: Collections of birds, fish, and invertebrates from Pacific Ocean localities were analyzed for DDT residues. Residue levels in the resident California sea birds, in nonresident (winter migrant) California sea birds, and in approximately 14 marine invertebrates from coastal localities between Monterey and Point Arena were tabulated; DDT residue levels in the resident California birds were considerably higher than those in the northern migrants. With 8.4-32 ppm DDT residues, the shearwaters, strictly pelagic migrants from the southern hemisphere contained as much as, or more, pesticide than the local birds. Fish from California coastal waters contained more residue (in general, total concentrations were 10-20 % of those in the birds) than other Pacific marine fish. The DDT residue levels in the marine invertebrates were lower by a factor of 10-50 than those in the fish analyzed.

Robinson, J.; Richardson, A.; Crabtree, A. N.; Coulson, J. C.; and Potts, G. R. 1967. Organochlorine residues in marine organisms. Nature. 214(5095):1307-1311.
Rec #: 396
KEYWORDS: organochlorines; dieldrin; DDTs; biomagnification; bioaccumulation; age variations; elimination; liver; egg; carcass; common eider; *Somateria mollissima*; herring gull; *Larus argentatus*; lesser black-backed gull; *Larus fuscus*; cormorant; *Phalacrocorax carbo*; shag; *Phalacrocorax aristotelis;* gannet; *Sula bassana*; sandwich tern; *Sterna sandvicensis;* common tern; *Sterna hirundo;* black-legged kittiwake; *Rissa tridactyla*; roseate tern; *Sterna dougallii*; razorbill; *Alca torda*; common murre; *Uria aalge;* Atlantic puffin; *Fratercula arctica*; Great Britain; Europe

ABSTRACT: Analytically significant amounts of p,p'-DDE (I) and dieldrin (II) were found in marine organisms. Residues of I and II tended to be greater in marine organisms of the higher trophic levels; it did not occur in all food chains. Significant positive correlation usually occurred between the concentration of both compounds and the trophic level. The simple food chain microplankton → mussels (*Mytilus*) → eider duck (*Somateria mollissima*) showed an appreciable concentration of II at each stage. The concentration of II in cod (*Gadus morhua*) was less than in the sand eel (*Ammodytes*). The concentration of II in planktonic crustacea was greater than in any fish examined. The concentrations of I and II followed the series Phadacrocoracidae > Alcidae > Laridae. II concentrations were not significantly different in the eggs of the cormorant (*Phalacrocorax carbo*) and shag (*P. aristotelis*). I concentrations were significantly greater in the cormorant. The sandwich tern (*Sterna sandvicensis*) had significantly greater amounts of I and II than the common tern (*S. hirundo*). II concentrations were independent of age in the shag and kittiwake (*Rissa tridactyla*). The concentration found in birds is probably in equilibrium between their intake and excretion of insecticide, and the latter is probably of more importance in determining the actual concentration present.

Rocque, Deborah A. and Winker, Kevin. 2004. Biomonitoring of contaminants in birds from two trophic levels in the North Pacific. Environ. Toxicol. Chem. 23(3):759-766.

Annotated Bibliography (Cont.).

Rec #: 516
KEYWORDS: metals; As; Cd; Cr; Cu; Hg; Ni; Se; Zn; organochlorines; dieldrin; HCB; DDTs; PCBs; chlordanes; stable isotopes; pelagic cormorant; *Phalacrocorax pelagicus*; red-faced cormorant; *Phalacrocorax urile*; rock sandpiper; *Calidris ptilocnemis*; liver; muscle; AK; USA; Pacific Ocean; North America; spatial variations; biomagnification
ABSTRACT: The presence and accumulation of persistent contaminants at high latitudes from long-range transport is an important environmental issue. Atmospheric transport has been identified as the source of pollutants in several arctic ecosystems and has the potential to severely impact high-latitude populations. Elevated levels of contaminants in Aleutian Island avifauna have been documented, but the great distance from potential industrial sources and the region's complex military history have confounded identification of contaminant origins. We sampled bird species across the natural longitudinal transect of the Aleutian Archipelago to test three contaminant source hypotheses. We detected patterns in some polychlorinated biphenyl congeners and mercury that indicate abandoned military installations as likely local point sources. Carbon isotopes were distinct among island groups, enabling us to rule out transfer through migratory prey species as a contaminant source. The long-range transport hypothesis was supported by significant west-to-east declines in contaminant concentrations for most detected organochlorines and some trace metals. Although relatively low at present, concentrations may increase in Aleutian fauna as Asian industrialization increases and emitted contaminants are atmospherically transported into the region, necessitating continued monitoring in this unique ecosystem.

Ruus, Anders; Sandvik, Morten; Ugland, Karl I.; and Skaare, Janneche U. 2002. Factors influencing activities of biotransformation enzymes, concentrations and compositional patterns of organochlorine contaminants in members of a marine food web. Aquat. Toxicol. 61(1-2):73-87.
Rec #: 360
KEYWORDS: liver; organochlorines; PCBs; DDTs; chlordanes; HCHs; HCB; mirex; toxicology; enzymes; EROD; CYP1A; herring gull; *Larus argentatus*; Norway; Europe; age variations; effects; bioaccumulation; biomagnification
ABSTRACT: The accumulation of polychlorinated biphenyls (PCBs; 34 congeners), ΣDDT (p,p'-DDT, o,p'-DDT, p,p'-DDD, o,p'-DDD and p,p'-DDE), chlordanes (ΣCHL; trans-chlordane, cis-chlordane, trans-nonachlor, cis-nonachlor and oxychlordane), hexachlorocyclohexanes (ΣHCH; α-, β- and γ-isomers), hexachlorobenzene (HCB) and mirex was investigated in members of a marine food web from the Hvaler and Torbjornskjaer archipelago, south-eastern Norway. The species studied were bullrout (*Myoxocephalus scorpius*), cod (*Gadus morhua*), herring gull (*Larus argentatus*) and harbor seal (*Phoca vitulina*). Furthermore, hepatic biotransformation enzyme activities (ethoxyresorufin-O-deethylase (EROD), pentoxyresorufin-O-depentylase (PROD) and glutathione S-transferase (GST)) were measured in all species. The objectives of the study were to investigate factors causing intraspecies variation in activities of biotransformation enzymes, as well as in concentrations and compositional patterns of the organochlorines (OCs). High correlations between EROD and PROD activities were found in all species, suggesting a single, common catalyst, CYP1A, and render the PROD assay questionable as biomarker for CYP2B inducers in marine wildlife. Furthermore, GST activities are shown to be dependent on biological factors, such as age (in harbor seal) and sex (in bullrout). In fish, the

Annotated Bibliography (Cont.).

OC concentrations vary between the sexes, likely due to differences in fat deposition strategies and possibly sex dimorphism. In seals, concentrations and compositional patterns of the OCs vary with age, owing to selective transfer from mother to pup in utero and mainly through lactation, but likely also due to age specific xenobiotic metabolizing capacity.

Ruus, Anders; Ugland, Karl Inne; and Skaare, Janneche Utne. 2002. Influence of trophic position on organochlorine concentrations and compositional patterns in a marine food web. Environ. Toxicol. Chem. 21(11):2356-2364.
Rec #: 357
KEYWORDS: biomagnification; liver; muscle; organochlorines; PCBs; DDTs; chlordanes; HCHs; HCB; mirex; stable isotopes; herring gull; *Larus argentatus*; Norway; Europe
ABSTRACT: The accumulation of polychlorinated biphenyls (PCBs), DDTs (p,p'-DDT [1,1,1-trichloro-2,2-bis(4-chlorophenyl)ethane], o,p'-DDT [1,1,1-trichloro-2-(2-chlorophenyl)-2-(4-chlorophenyl)ethane], p,p'-DDD [1,1-dichloro-2,2-bis(4-chlorophenyl)ethane], o,p'-DDD [1,1-dichloro-2-(2-chlorophenyl)-2-(4-chlorophenyl)ethane], and p,p'-DDE [1,1-dichloro-2,2-bis(4-chlorophenyl)ethene]), chlordanes (trans-chlordane, cis-chlordane, trans-nonachlor, cis-nonachlor and oxychlordane); hexachlorocyclohexanes (α-, β-, and γ-isomers), hexachlorobenzene, and mirex was investigated in a marine food web from southeastern Norway. The food web consisted of the polychaete Nereis diversicolor, lesser sandeel (*Ammodytes tobianus*), three species of gobys (*Gobiusculus flavescens, Pomatoschistus* sp., and *Gobius niger*), bullrout (*Myoxocephalus scorpius*), cod (*Gadus morhua*), herring gull (*Larus argentatus*), and harbor seal (*Phoca vitulina*). The results show that interspecies differences in organochlorine (OC) compositional patterns in the food web depend on several factors (allometric, biochemical, physiologic, and physicochemical) specific to both the chemicals and the organisms. The importance of dietary accumulation and metabolic capacity increases toward higher trophic levels, while the OC patterns are to a larger extent detected by the lipophilicity and water solubility of the compounds at lower trophic levels. Furthermore, stable nitrogen isotopes provided a continuous measure of trophic position, rendering us capable of quantifying the increases in the concentrations of sum PCB, sum dichorodiphenyltrichloroethane (DDT), and sum chlordane (CHL) and the percentages of highly chlorinated PCBs through the food web. The information provided may be important for future modeling of the fate of organochlorine contaminants in marine food webs.

Ryan, John J; DeWailly, Eric; Gilman, Andy; Laliberte, Claire; Ayotte, Pierre; and Rodrigue, Jean. 1997. Dioxin-like compounds in fishing people from the lower north shore of the St. Lawrence River, Quebec, Canada. Arch. Environ. Health. 52(4):309-316.
Rec #: 3
KEYWORDS: organochlorines; PCDDs; PCDFs; PCBs; TEQs; egg; herring gull; *Larus argentatus*; double-crested cormorant; *Phalacrocorax auritus*; common eider; *Somateria mollissima*; Quebec; Canada; North America; bioaccumulation
ABSTRACT: In this study, investigators assessed exposure to dioxin-like compounds in a fishing population that inhabits small coastal communities along the Lower North Shore of the St. Lawrence River, Quebec. This population relies heavily on wildlife foods for sustenance. Investigators analyzed chemically the most popular marine foods (i.e., fish, crustaceans, sea mammals, and sea-bird eggs), and they also obtained 25 human plasma samples from individuals in two villages along the river. The mean level of total polychlorinated biphenyls in this

Annotated Bibliography (Cont.).

population was approximately twice that found in the entire fishing cohort. Plasma levels of dioxin-like compounds, expressed as tetrachlorodibenzodioxin toxic equivalent, were approximately eight times higher than levels in urban residents. Most of the increase in tetrachlorodibenzodioxin toxic equivalents in the selected fish eaters resulted primarily from an elevation in polychlorited biphenyls. Concentrations of dioxin-like compounds from the Lower North Shore were low in fish and seals, but concentrations were elevated in the egg of sea birds. Given that from was also a significant statistical correlation in the entire population between human plasma levels and consumption of birds' eggs -- and not other traditional foods -- much of the increased human dose appeared to originate from this one food source. Because there appear to be increased, but uncertain, health risks from this elevated body burden, investigators advised the residents of the area to avoid consumption of wild birds'
eggs (i.e., a food source of minor nutritional importance).

Ryckman, D. P.; Weseloh, D. V.; Hamr, P.; Fox, G. A.; Collins, B.; Ewins, P. J.; and Norstrom, R. J. 1998. Spatial and temporal trends in organochlorine contamination and bill deformities in double-crested cormorants (*Phalacrocorax auritus*) from the Canadian Great Lakes. Environ. Monit. Assess. 53(1):169-195.
Rec #: 110
KEYWORDS: double-crested cormorant; *Phalacrocorax auritus*; Great Lakes; Canada; North America; organochlorines; HCB; DDTs; dieldrin; mirex; chlordanes; PCBs; abnormalities; eggshell thickness; egg; temporal trends; spatial variations; effects
ABSTRACT: The levels of organochlorine contaminants (OCs) in the eggs of double-crested cormorants (*Phalacrocorax auritus*) from the Canadian Great Lakes, Lake Nipigon and Lake-of-the- Woods were monitored between 1970 and 1995. PCBs and p,p'-DDE were present at the highest concentrations. Significant declines in OC concentrations on the Great Lakes were observed over this period for Lake Ontario, Lake Superior, Georgian Bay and North Channel but not Lake Erie where levels remained relatively stable. In the early 1970s, the greatest OC levels were generally observed in cormorant eggs from nesting sites in Georgian Bay and North Channel of Lake Huron. Between 1984 and 1995 mirex and PCB levels were consistently highest in samples from Lakes Ontario and Erie, respectively. Similar levels of PCDDs and PCDFs were observed from all regions of the Canadian Great Lakes in 1989. In general, OC levels in cormorant eggs between 1984-1995 were ranked as follows: Lake Erie>Lake Ontario>Lake Superior>Lake Huron. In 1995, eggshell thickness in Canadian Great Lakes cormorants, ranged from 0.423 to 0.440 mm and was on average only 2.3 % thinner than pre-DDT era values. Between 1988 and 1996, 31 cormorant chicks with bill defects were observed at 16 different colonies (21 % of all colonies surveyed) in Lakes Ontario and Superior, Georgian Bay and North Channel, and the main body of Lake Huron. No bill deformities were observed at reference sites in northwestern Ontario (Lake Nipigon and Lake-of-the-Woods). For the period 1988-1996, the prevalence of bill defects in cormorant chicks (0.0-2.8/10,000 chicks) did not differ significantly ($p > 0.05$) among most regions in the Canadian Great Lakes. Georgian Bay was the only region to show a significant decrease in the prevalence of bill defects between the periods 1979-1987 and 1988-1995.

Ryder, John P. 1974. Organochlorine and mercury residues in gull's eggs from Western Ontario. Can. Field. Nat. 88:349-352.
Rec #: 547

Annotated Bibliography (Cont.).

KEYWORDS: organochlorines; metals; Hg; egg; gull; Ontario; Canada; North America; eggshell thickness; effects; herring gull; *Larus argentatus*; ring-billed gull; *Larus delawarensis*
ABSTRACT: Organochlorine (DDE, dieldrin, PCBs) and mercury residues in eggs of Herring Gulls *(Larus argentatus)* and Ring-billed Gulls *(L. delawarensis)* from western Ontario are reported. Concentrations in wet weight (ppm) of DDE and PCBs approximate levels reported from aquatic birds in the Canadian prairie provinces. Mercury levels in both species are lower than those reported from heavily polluted waters.

Rüdel, Heinz; Lepper, Peter; Steinhanses, Jürgen; and Schröter-Kermani, Christa. 2003. Retrospective monitoring of organotin compounds in marine biota from 1985 to 1999: Results from the German Environmental Specimen Bank. Environ. Sci. Technol. 37(9):1731-1738.
Rec #: 479
KEYWORDS: organometallics; organotins; Germany; Europe; North Sea; herring gull; *Larus argentatus*; egg; temporal trends
ABSTRACT: In archived samples from the Environmental Specimen Bank in Germany, organotin compounds, including tributyltin (TBT) and triphenyltin (TPT), as well as their degradation products, were quantified. Biota samples from the North Sea and Baltic Sea areas were analyzed by gas chromatography/atomic emission detection coupling after extraction and Grignard or ethylborate derivatization. TBT and TPT were detected in nearly all samples. A decrease of TPT contamination was observed in bladder wrack, common mussels, and eelpout muscle tissues in the period 1985-1999. In this period, TPT concentrations in North Sea mussels decreased from 98 to 7 ng/g (as organotin cation concentration in wet tissue). Concentrations of TBT remained relatively const. with 17 ± 3 ng/g for mussels from a site with nearby marine traffic and 8 ± 2 ng/g for a more remote area. The results reflect that TBT is still used as a biocide in antifouling paints, whereas the use of TPT as a cotoxicant in such preparations ceased in the 1980s. The fact that the use of TBT in antifouling paints was banned in 1991 for small boats within the European Community seems not to have resulted in a decrease of TBT levels in marine biota.

Sagerup, Kjetil; Gabrielsen, Geir Wing; Skorping, Arne; and Skaare, Janneche Utne. 1998. Association between polychlorinated biphenyl (PCB) concentrations and intestinal nematodes in glaucous gulls, *Larus hyperboreus*, from Bear Island. Organohalogen Compounds. 39:449-451.
Rec #: 324
KEYWORDS: glaucous gull; *Larus hyperboreus*; parasites; PCBs; Svalbard; Norway; Europe; liver; intestine; effects
NOTES: FIGURES ONLY
ABSTRACT: The relationship between PCB levels and intestinal nematode infection intensities was investigated in glaucous gulls from Bear Island. The total PCB concentration (Σ9PCB) in liver samples were 15147-292439 ng/g lipid weight The PCB-153 congener is the most abundant accounting for 36 % of Σ9PCB. 12 Helminth parasite species were identified in the gastrointestinal tract of the gulls. The nematode *Paracuaria adunca* showed the highest prevalence and infected 63 % of the gulls. The trematode *Cryptocotyle lingua* showed the highest intensity with a maximum of 269 individuals in 1 host. The intensity of nematodes (sum of all nematodes) was positively correlated to Σ9PCB. The nematodes in the gulls penetrated tissues in the pro-ventricle and ventricle area.

Annotated Bibliography (Cont.).

Sagerup, Kjetil; Henriksen, Espen O.; Skaare, Janneche U.; and Gabrielsen, Geir W. 2002. Intraspecific variation in trophic feeding levels and organochlorine concentrations in glaucous gulls (*Larus hyperboreus*) from Bjørnøya, the Barents Sea. Ecotoxicology. 11(2):119-125.
Rec #: 380
KEYWORDS: glaucous gull; *Larus hyperboreus*; Barents Sea; Norway; Europe; liver; muscle; organochlorines; HCB; chlordanes; DDTs; mirex; PCBs; stable isotopes; biomagnification
ABSTRACT: Biomagnification contributes to high concentrations of persistent organochlorines (OC) in some Arctic vertebrates. Glaucous gulls (*Larus hyperboreus*) on Bjørnøya in the western Barents Sea were studied to compare the intraspecific variation in OC concentration with variation in trophic feeding levels, established from ratios of nitrogen isotopes. Liver tissue samples from 40 adult glaucous gulls were analyzed for hexachlorobenzene (HCB), oxychlordane, p,p'-dichlorodiphenyltrichloroethane (DDT) and its metabolite p,p'-DDE, mirex, and 9 congeners of polychlorinated biphenyl (PCB). The ratios of the heavier to lighter isotope of carbon ($^{13}C/^{12}C$) and nitrogen ($^{15}N/^{14}N$), expressed as $\delta^{13}C$ and $\delta^{15}N$, were measured in liver and muscle. Hepatic concentrations of the 9 PCB congeners (Σ 9 PCB) ranged from 16 µg/g lipid weight to 292 µg/g lipid weight. The $\delta^{15}N$ ranged from 14.0 % to 15.3 % in muscle. Seven of the 14 OC measured, Σ DDT, and Σ 9 PCB were positively correlated to $\delta^{15}N$ from muscle tissue. No correlations were found between OC and $\delta^{13}C$. The present results indicate that OC concentrations are partly dependent on the foraging strategy of the gull. The r^2 of the linear regressions suggests that up to 18 % of the variation in the OC concentrations could be explained by variation in food preference.

Sagerup, Kjetil; Henriksen, Espen O.; Skorping, Arne; Skaare, Janneche Utne; and Gabrielsen, Geir Wing. 2000. Intensity of parasitic nematodes increases with organochlorine levels in the glaucous gull. J. Appl. Ecol. 37(3):532-539.
Rec #: 433
KEYWORDS: organochlorines; DDTs; PCBs; HCB; chlordanes; DDTs; DDTs; mirex; glaucous gull; *Larus hyperboreus*; Barents Sea; Norway; Europe; liver; parasites; effects
ABSTRACT: Organochlorines probably suppress immune functions in birds and mammals, but few field assessments are available. If establishment and(or) survival of parasites is limited by host immunity, increased parasite intensities in animals with high organochlorine burdens, such as the glaucous gull Larus hyperboreus, would be expected. The authors collected 40 adult glaucous gulls on Bear Island in the western Barents Sea. Concentrations of 9 selected polychlorinated biphenyls (PCB; 28, 52, 99, 101, 118, 138, 153, 170, and 180) and 5 chlorinated pesticides (hexachlorobenzene, oxychlordane, DDE, DDT, and mirex) were measured in the liver. The abundance of 12 species of intestinal helminths, including 1 trematode, 6 cestodes, 4 nematodes, and 1 acanthocephalan, was detected. After controlling for nutritional condition, no single parasite species was significantly associated with concentrations of PCB or chlorinated pesticides. However, the intensity of all nematodes grouped together was positively correlated with 10 of the 14 organochlorine concentrations measured. The strongest correlations were with p,p'-DDT, mirex, Σ9PCB, and PCB congeners 28, 118, 153, 138, 170, and 180. Although correlative and collected in the absence of immunological data, these data do not refute the hypothesis that organochlorines might affect avian immune function.

Sanderson, J. Thomas; Norstrom, Ross J.; Elliott, John E.; Hart, Leslie E.; Cheng, Kimberly M.;

Annotated Bibliography (Cont.).

and Bellward, Gail D. 1994. Biological effects of polychlorinated dibenzo-p-dioxins, dibenzofurans, and biphenyls in double-crested cormorant chicks (*Phalacrocorax auritus*). J. Toxicol. Environ. Health. 41(2):247-265.
Rec #: 87
KEYWORDS: effects; hydrocarbons; PAHs; organochlorines; PCBs; PCDDs; PCDFs; TEQs; toxicology; EROD; cytochrome P_{450}; double-crested cormorant; *Phalacrocorax auritus*; great blue heron; *Ardea herodias;* British Columbia; Great Lakes; Canada; North America; egg
ABSTRACT: The present project assessed the effect of environmental contamination with polychlorinated dibenzo-p-dioxins (PCDDs), dibenzofurans (PCDFs), and biphenyls (PCBs) on hepatic microsomal ethoxyresorufin O-deethylase (EROD) activities and morphological parameters in matched double-crested cormorant (*Phalacrocorax auritus*) hatchings from egg clutches chosen for chemical analysis. Double-crested cormorant eggs were collected from five colonies across Canada, with differing levels of contamination. Levels of contamination expressed in sum of 2,3,7,8-tetrachlorodibenzo-p-dioxin-toxic equivalents (TCDD-toxic equivalents or TEQ, ng/kg egg) were: Saskatchewan, 250; Chain Islands, 672 Christy Islet, 276; Crofton, 131; and Lake Ontario, 1606. In the hatchlings, hepatic EROD activities (pmol/min/mg protein) were: Saskatchewan, 283; Chain Islands, 516; Christy Islet, 564; Crofton, 391; and Lake Ontario, 2250. Hepatic microsomal EROD activity (pmol/min/mg protein) regressed positively on TEQ ($r^2 = .69$; $p < .00005$; $n = 25$). Yolk weight (g) regressed negatively on TEQ ($r^2 = .44$; $p = .00005$). Wing length (mm) regressed negatively on PCB-169 ($r^2 = .28$; $p = .007$). Monospecific antibodies raised against rat cytochrome P_{450} 1A1 recognized a protein in the hepatic microsomes of the double-crested cormorant, and also in those of the great blue heron (*Ardea herodias*), using immunoblotting. The intensity of the stained band increased with increased EROD activity, supporting the assumption that ethoxyresorufin is a suitable substrate for avian cytochrome P_{450} 1A1. These results validate the use of avian hepatic microsomal EROD activity as an index of cytochrome P_{450} 1A1 induction by environmental levels of polychlorinated aromatic hydrocarbons and as a useful screening tool to determine the extent of exposure to such chemicals. Furthermore, the induction of cytochrome P_{450} 1A1 observed in the cormorant indicates that the Ah receptor-mediated process, by which TCDD and related chemicals exert many of their toxicities, has been activated.

Savinov, Vladimir M.; Gabrielsen, Geir W.; and Savinova, Tatiana N. 2003. Cadmium, zinc, copper, arsenic, selenium and mercury in seabirds from the Barents Sea: Levels, inter-specific and geographical differences. Sci. Total Environ. 306(1-3):133-158.
Rec #: 421
KEYWORDS: metals; Cd; Zn; Cu; As; Se; Hg; muscle; liver; thick-billed murre; *Uria lomvia;* common murre; *Uria aalge*; Atlantic puffin; *Fratercula arctica*; black guillemot; *Cepphus grylle*; little auk; *Alle alle*; razorbill; *Alca torda*; common eider; *Somateria mollissima*; king eider; *Somateria spectabilis*; glaucous gull; *Larus hyperboreus*; herring gull; *Larus argentatus*; black-legged kittiwake; *Rissa tridactyla*; northern fulmar; *Fulmarus glacialis*; Arctic tern; *Sterna paradisaea*; Barents Sea; Norway; Europe; spatial variations
ABSTRACT: Trace elements Cd, Zn, Cu, As, Se and Hg were analyzed in muscle and liver of Brünnich's guillemot, Common guillemot, Puffin, Black guillemot, Little auk, Razorbill, Common eider, King eider, Glaucous gull, Herring gull, Black-legged kittiwake, Northern fulmar and Arctic tern collected in 1991-1992 at the main breeding colonies in the Barents Sea. The highest levels of the most toxic elements Cd and Hg were found in birds nesting north of

Annotated Bibliography (Cont.).

Spitsbergen. Extremely high levels of As were detected in tissues of all seabird species collected at colonies in Chernaya Guba (Novaya Zemlya), where nuclear tests were carried out in the 1960s. In general, levels of all of the trace elements in the Barents Sea seabirds were similar or lower in comparison with those reported for the same seabird species from the other Arctic areas. Data on metallothionein concentrations in different seabird species need to be collected in order to understand the mechanism of bioaccumulation and possible toxic effects of trace elements in Arctic seabirds.

Savinov, Vladimir M.; Gabrielsen, Geir Wing; and Savinova, Tatiana N. 2000. Trace elements in seabirds from the Barents and Norwegian seas, 1991-1993. Norsk Polarinstitutt Internrapport Nr. 5. Troms: Norsk Polarinstitutt. 108 pp.
Rec #: 47
KEYWORDS: Norway; Baltic Sea; Barents Sea; Europe; black-legged kittiwake; *Rissa tridactyla*; herring gull; *Larus argentatus*; great black-backed gull; *Larus marinus*; glaucous gull; *Larus hyperboreus*; Arctic tern; *Sterna paradisaea*; little auk; *Alle alle*; thick-billed murre; *Uria lomvia*; common murre; *Uria aalge*; razorbill; *Alca torda*; black guillemot; *Cepphus grylle*; Atlantic puffin; *Fratercula arctica*; common eider; *Somateria mollissima;* king eider; *Somateria spectabilis*; long-tailed duck; *Clangula hyemalis*; northern fulmar; *Fulmarus glacialis*; metals; Cd; Zn; Cu; Mn; Cr; As; Se; Hg; liver; muscle; spatial variations; age variations; gender differences; bioaccumulation
ABSTRACT: The main objective of the project was to study the levels of trace elements (including heavy metals) in different seabird species breeding in the Barents Sea area. There are limited comparable data on environmental pollutants in seabirds from this area, and this report presents such information on trace element levels. The results are a contribution to the Arctic Monitoring and Assessment Programme (AMAP).

Savinova, T. N. 1990. Chemical pollution of the northern seas. Apatity, USSR: Kol. Nauch. Tsentr, Akad. Nauk SSSR. 144 pp.
Rec #: 543
KEYWORDS: review; oil; organochlorines; DDTs; PCBs; Russia; Asia; North Sea; Baltic Sea; Finland; Ireland; Great Britain; Europe; Pacific Ocean; CA; ME; VA; USA; Great Lakes; Bay of Fundy; Canada; North America; *Rissa tridactyla;* black-legged kittiwake; *Larus argentatus;* herring gull; *Larus fuscus;* lesser black-backed gull
ABSTRACT: The monograph summarizes ecotoxicological and biogeochemical materials on the northern seas pollution with chlorinated and petroleum hydrocarbons. There are given some data on the bioaccumulation of the above mentioned pollutants in hydrobios and sea birds. The problems of the effect of pollutants on the northern seas phytoplankton are also discussed. The factors determining the bioaccumulation of pollutants in the liver of fishes are examined. The monograph may present a certain interest for biologists, toxicologists, hygienists and the environment pollution control experts. Figures - 13, tables - 15, references - 424.

Savinova, T. N. 1992. The content of pollutants in seabirds from the Barents Sea: Results and investigation's perspectives. In: Theoretical Approaches to the Study of Ecosystems at Arctic and Subarctic Seas. Apatity, The Kola Scientific Centre Publishing House. 113-116.
Rec #: 544
KEYWORDS: organochlorines; DDTs; PCBs; metals; Cu; Ni; Co; Zn; Mn; herring gull; *Larus*

Annotated Bibliography (Cont.).

argentatus; liver; muscle; Barents Sea; Europe

Savinova, T. N. and Gabrielsen, G. W. 1994. Chlorinated hydrocarbons and heavy metals in birds of the Barents Sea. Preprint, Kol'Skij Nauchn. Tsentr Ran, Apatity (Russia). 36 pp.
Rec #: 65
KEYWORDS: organochlorines; PCBs; metals; bioaccumulation; Barents Sea; Norway; Europe; effects
NOTES: NOT AVAILABLE -SEE 61 for INFO
ABSTRACT: Data are presented on the content of residual organochlorine pesticides, polychlorinated biphenyls and heavy metals in the organs and tissues of 8 species of colonial seabirds collected in 1991 from the western coast of Spitsbergen, Bear I., northern Norway coast and Franz Josef Land.

Savinova, T. N.; Polder, A.; Gabrielsen, G. W.; and Skaare, J. U. 1995. Chlorinated hydrocarbons in seabirds from the Barents Sea area. Sci. Total Environ. 160/161:497-504.
Rec #: 66
KEYWORDS: organochlorines; HCHs; HCB; chlordanes; DDTs; PCBs; common eider; *Somateria mollissima*; black-legged kittiwake; *Rissa tridactyla*; glaucous gull; *Larus hyperboreus*; Barents Sea; Norway; Europe; liver; muscle; fat; brain; spatial variations; temporal trends
ABSTRACT: This study presents preliminary results on the levels of organochlorines (OCs) in three seabird species: common eider (*Somateria mollissima*), kittiwake (*Rissa tridactyla*) and glaucous gull (*Larus hyperboreus*) from the Barents Sea area, including Frans Josef Land, Ny-Llesund, Bjørnøya, and Hornøya, in July-August, 1991. Kittiwake was the only species collected at all four sampling areas, and all three species were sampled only at Ny-Llesund. Samples of liver, fat, muscle and brain tissue were analyzed for chlorinated pesticides, hexachlorobenzene (HCB), hexachlorocyclohexanes (alpha, beta, gamma -HCH), chlordanes (oxychlordane, trans-nonachlor and heptachlor epoxide), the DDT group (op', pp'-DDT, DDD and DDE) and the industrial chemical polychlorinated biphenyls (PCBs). In all species, the highest OC levels were found in fat tissue, followed by liver, muscle and brain. PCBs were the major OC contaminants, followed by the DDT group, the chlordane group, HCB and the HCHs. Only low levels of HCHs were found. The lowest mean level of capital sigma PCB - sum of concentrations of the 19 individual congeners (2.8 µg/kg wet weight (w.w.)) - was found in the liver of juvenile common eider from Frans Josef Land. The highest mean level of ΣPCB (15980 µg/kg w.w.) was found in the fat of adult glaucous gulls from Ny-Llesund. There were no significant geographical trends in OC contamination in the species investigated. However, the OC levels in kittiwake were highest at Bjørnøya. The OC contamination levels in seabirds from the Barents Sea region are in general lower in the 1990s than in past decades.

Savinova, Tatiana N.; Falk-Petersen, Stig; and Gabrielsen, Geir Wing. 1995. Chemical Pollution in the Arctic and Sub-Arctic Marine Ecosystems An Overview of Current Knowledge. NINA Fagrapport: Report / The Joint Norwegian-Russian Commission on Environmental Cooperation, the Seabird Expert Group, 1. 1995:3. Trondheim: Norsk Institutt for Naturforskning. 68 pp.
Rec #: 61
KEYWORDS: review; organochlorines; DDTs; PCBs; metals; Hg; Cd; As; Zn; Cr; Cu; Pb; Se; Ni; egg; muscle; liver; fat; brain; spleen; heart; bone; black-legged kittiwake; *Rissa tridactyla*;

Annotated Bibliography (Cont.).

herring gull; *Larus argentatus*; glaucous gull; *Larus hyperboreus;* great black-backed gull; *Larus marinus;* lesser black-backed gull; *Larus fuscus*; northern fulmar; *Fulmarus glacialis;* little auk; *Alle alle*; Atlantic puffin; *Fratercula arctica*; thick-billed murre; *Uria lomvia*; common murre; *Uria aalge;* black guillemot; *Cepphus grylle*; king eider; *Somateria spectabilis*; common eider; *Somateria mollissima*; long-tailed duck; *Clangula hyemalis*; razorbill; *Alca torda*; gannet; *Sula bassana;* North Sea; Pacific Ocean; MA; CA; USA; Canada; North America; Faroe Islands; Greenland; Atlantic Ocean; Great Lakes; Barents Sea; Baltic Sea; Sweden; Finland; Poland; Great Britain; Scotland; Norway; Europe; Russia; Asia; spatial variations; temporal trends
ABSTRACT: This report is part of a research project in the framework of the Norwegian-Russian Environmental Cooperation. The project was initiated in 1991 in order to elucidate the present status of environmental contaminants in the arctic marine ecosystem, with special focus on seabirds. This report is a part of the AMAP and is intended to review the levels and types of chlorinated hydrocarbons and heavy metal contaminants in the Arctic and Sub-Arctic marine waters, with reference to the situation in boreal waters.

Scharenberg, W. 1991. Prefledging terns (*Sterna paradisaea, Sterna hirundo*) as bioindicators for organochlorine residues in the German Wadden Sea. Arch. Environ. Contam. Toxicol. 21(1):102-105.
Rec #: 472
KEYWORDS: organochlorines; PCBs; HCHs; HCB; OCS; DDTs; Arctic tern; *Sterna paradisaea*; common tern; *Sterna hirundo*; Germany; Wadden Sea; Europe; carcass; age variations; bioaccumulation; biomagnification; effects
ABSTRACT: Residues of organochlorine chemicals (OC) including polychlorinated biphenyls were analyzed in carcasses of young Arctic terns (*Sterna paradisaea*) and common terns (*Sterna hirundo*) which were collected in 1988 in a breeding colony of the German Wadden Sea. The birds were separated into age-groups of 2-14 and 15-27 days old. Most of the chemicals showed a higher concentration in younger birds. In contrast, the absolute amount of the chemicals was greater or at least equal in older birds. If food contamination is as high as shown in this study, the accumulation of chemical hazards in growing terns can only be recognized by measurement of the absolute content, because the concentration is diluted by growth effects. In comparison to food samples from the same breeding colony, one can recognize an accumulation from one trophic level to the next. This study proves that prefledging terns can be accepted as bioindicators. Presumably the residues of OC alone did not cause the death of the terns, yet they were definitely important stress factors.

Scott, J. Michael; Wiens, John A.; and Claeys, Robert R. 1975. Organochlorine levels associated with a common murre die-off in Oregon. J. Wildl. Manage. 39(2):310-320.
Rec #: 338
KEYWORDS: common murre; *Uria aalge*; muscle; brain; organochlorines; DDTs; PCBs; toxicology; effects; OR; USA; North America
ABSTRACT: During the summer of 1969 an abnormally high mortality of common murres (Uria aalge) was observed on the Oregon coast. Autopsy of dead birds suggested that drowning was the proximate cause of death. The body weights of dead males collected during the die-off period were significantly lower than those of healthy birds collected at the same time or in August of the following year. Analyses for chlorinated hydrocarbons revealed an average p,p'-DDE concentration of 8.7 ppm wet weight in brain tissues of dead birds, whereas the level in

Annotated Bibliography (Cont.).

healthy birds collected at the same time was 1.1 ppm. The ratio of p,p'-DDE in brain and muscle tissues (p,p'-DDE brain/p,p'-DDE muscle) averaged 28 in dead birds collected during the die-off. Collections of living birds in the summer of 1970 when there was no die-off indicated an average p,p'-DDE level of 0.44 ppm in brain tissues, and brain-to-muscle ratios did not exceed 0.97. Brain levels of polychlorinated biphenyls (PCBs) averaged 4.0 ppm in the dead birds, 3.7 ppm in the live birds collected during the dieoff, and 1.1 ppm in the June 1970 sample. All p,p'-DDE and PCB levels were considerably less than reported lethal concentrations in other species of birds; however, under conditions of environmental stress, the levels recorded for dead birds in 1969 may have been sufficient to contribute directly or indirectly to mortality through effects on metabolism and behavior.

Seip, Knut Lehre; Sandersen, Erik; Mehlum, Fridtjof; and Ryssdal, Jostein. 1991. Damages to seabirds from oil spills: Comparing simulation results and vulnerability indexes. Ecol. Model. 53(1-2):29-59.
Rec #: 72
KEYWORDS: effects; oil; black-legged kittiwake; *Rissa tridactyla*; common murre; *Uria aalge*; common eider; *Somateria mollissima*
NOTES: SUPPLEMENTAL
ABSTRACT: The present model for estimating oil damage to seabirds is intended as a tool for a decision-support system in oil combat planning. The model systematizes the most well-known factors which determine seabirds' susceptibility to oil spills and the subsequent recovery of bird populations. The model is formulated as a simulation model describing the two-dimensional distribution pattern of seabird population on the coast and the two-dimensional spreading of an oil slick drifting towards land. Formulations for population dynamics include seabird age structure, recruitment, mortality, and migration. We have demonstrated the use of the model by applying it to a spill affecting three different seabird populations: kittiwake, *Rissa tridactyla*; common guillemot, *Uria aalge*; and common eider, *Somateria mollissima*. Model results are compared to the results of an index system ranging seabird species' vulnerability to oil spills.

Sellström, U.; Kierkegaard, A.; de Wit, C.; Jansson, B.; Bignert, A.; and Olsson, M. 1999. Temporal trend studies on polybrominated diphenyl ethers in guillemot egg from the Baltic Sea. In: Sellström, U. Determination of Some Polybrominated Flame Retardants in Biota, Sediment, and Sewage Sludge. Stockholm: Stockholm University, Ph.D. Dissertation. 10 pp.
Rec #: 347
KEYWORDS: PBDEs; egg; common murre; *Uria aalge*; Baltic Sea; Europe; temporal trends
ABSTRACT: Guillemot eggs from the Baltic Sea, sampled between 1969 and 1997 were analysed for tetra- and pentabromodiphenyl ethers (2,2',4,4'-tetraBDE (BDE-47), 2,2',4,4',5-pentaBDE (BDE-99) and 2,2'4,4',6-pentaBDE (BDE-100)). This temporal trend study indicates that the concentrations of the polybrominated diphenyl ether (PBDE) compounds increased from the 1970s to the 1980s, peaking around the mid to the late 1980s. These peaks are then followed by a rapid decrease in concentrations during the rest of the study period.

Sellström, Ulla; Bignert, Anders; Kierkegaard, Amelie; Häggberg, Lisbeth; de Wit, Cynthia A.; Olsson, Mats; and Jansson, Bo. 2003. Temporal trend studies on tetra- and pentabrominated diphenyl ethers and hexabromocyclododecane in guillemot egg from the Baltic Sea. Environ. Sci. Technol. 37(24):5496-5501.

Annotated Bibliography (Cont.).

Rec #: 494
KEYWORDS: PBDEs; common murre; *Uria aalge*; egg; Baltic Sea; Sweden; Europe; temporal trends
ABSTRACT: Guillemot eggs from the Baltic Sea, sampled from 1969 to 2001, were analyzed for tetra- and pentabromodiphenyl ethers (BDE-47, BDE-99, BDE-100) and hexabromocyclododecane (HBCD). This temporal trend study indicated polybrominated di-Ph ether compound concentrations increased from the 1970s to the 1980s and peaked in the mid- to late-1980s. These peaks were followed by a rapid decrease in concentrations during the rest of the study period, with concentrations of the major BDE congeners <100 ng/g lipid weight at the end of the period. This corresponded to <10 % of peak values. HBCD concentrations showed a different pattern over time. After a peak in the mid-1970s, followed by a decrease, concentrations increased in the later 1980s. In the recent 10-years period, no significant change occurred and annual mean concentrations were more or less stable at a higher level vs. the beginning of the study period.

Sellström, Ulla; Jansson, Bo; Kierkegaard, Amelie; de Wit, Cynthia; Odsjö, Tjelvar; and Olsson, Mats. 1993. Polybrominated diphenyl ethers (PBDE) in biological samples from the Swedish environment. Chemosphere. 26(9):1703-1718.
Rec #: 348
KEYWORDS: PBDEs; egg; common murre; *Uria aalge*; Baltic Sea; Sweden; Europe; temporal trends; spatial variations; biomagnification
ABSTRACT: Polybrominated diphenyl ethers, PBDE, are widespread contaminants in the Swedish environment and are present in both background and industrialised areas. This study presents results from analyses of a variety of species from different sampling sites in Sweden. The spatial trend along the Swedish coast is similar to that of polychlorinated biphenyls (PCB) and the DDTs. PBDE seem to bio-magnify in fish consumers like grey seal and guillemot (egg). The relative amounts of the investigated tetra- and penta-brominated PBDE congeners are different in different species and in different areas. The importance of a sampling strategy when doing time-trend studies is demonstrated for guillemot eggs.

Seys, J. and Meire, P. 1997. Oil pollution and seabirds. In: Jauniaux, Thierry; Bouquegneau, Jean Marie; Coignoul, Freddy (Eds.). Marine Mammals, Seabirds and Pollution of Marine Systems. Bulletin De La Societe Royale Des Sciences De Li. 61-66.
Rec #: 15
KEYWORDS: review; oil; effects; North Sea; Great Britain; Europe
NOTES: SUPPLEMENTAL
ABSTRACT: In many parts of the world oil deposits close to the earth's surface seep out. This natural input (0.25 million tons/year) is only a minor part of the total amount of oil entering the world seas (2.5 million tons/year) every year. Transport operations (including tanker- and non tanker accidents, cleaning of cargo compartments, deballasting of fuel tanks and bilge water and dry dock operations: 0.56 million tons/year), fixed installations (0.18 million tons/year), municipal and industrial waste (0.90 million tons/year), urban and river run-off (0.16 m tons/year) and atmospheric fall-out (0.30 million tons/year) are the other main sources of oil input. Although tanker accidents are the most spectacular way of oil input, the large majority of oils in the North Sea originates from "operational" (but illegal) discharges of mainly fuel oils from ships.

Annotated Bibliography (Cont.).

Crude oil (petroleum) consists mainly of hydrocarbons, sulfur and smaller amounts of fatty acids, nitrogen compounds, vanadium and nickel. The number and position of the molecules in the hydrocarbons of crude oil and its refining products, determine the characteristics in terms of biodegradability, state, volatility, toxicity   Water-soluble components of crude oil and refined products include a variety of compounds that are toxic to a wide spectrum of marine plants and animals. Weathered oil on a beach causes great damage through its physical properties.
Birds are directly affected since oil causes the fine structure of the feathers to clog, resulting in loss of thermal insulation, and water absorption (leading to hypothermia, illness, drowning). By preening their plumage they also ingest oil, causing all kind of pathological effects.
Oil is the major cause of death of most seabird species in the North Sea and Channel. Evidence for specific oiling incidents and for the level of chronic oil pollution can be obtained from beached bird surveys. This method is complementary to aerial surveys and a rather cheap way to monitor vast areas. The percentage of birds that are oil contaminated (oil-rate) is an integrated and useful measure to determine the oil pollution.
These oil-rates are particularly high in diving seabirds and species spending most of their lifespan at sea. Although some local examples of declining oil-rates have demonstrated the effect of policy measures, the overall North Sea oil-rates do not seem to have declined substantially the last few decades. The 1000-6000 seabirds washing ashore each year on the Belgian coast, have oil-rates of over 50 %.

Sheffield, Steven R.; Matter, John M.; Rattner, Barnett A.; and Guiney, Patrick D. 1998. Fish and wildlife species as sentinels of environmental endocrine disruptors. In: Kendall, Ronald J., Richard L. Dickerson, John P. Giesy, William P. Suk (Eds.). Principles and Processes for Evaluating Endocrine Disruption in Wildlife, Proceedings From Principles and Processes for Evaluating Endocrine Disruption in Wildlife; March 1996; Kiawah Island, S. C. Pensacola, FL: Society of Environmental Toxicology and Chemistry (SETAC). 515 Pp. 369-430.
Rec #: 252
KEYWORDS: review; North America; CA; WA; USA; Great Lakes; EDCs; organochlorines; DDTs; dieldrin; chlordanes; PCBs; PCDDs; hydrocarbons; PAHs; effects; toxicology; cytochrome P_{450}; AHH; EROD; retinol; porphyrin; eggshell thickness; reproductive success; western gull; *Larus occidentalis*; Japanese quail; *Coturnix coturnix*; double-crested cormorant; *Phalacrocorax auritus;* ring-billed gull; *Larus delawarensis;* bald eagle; *Haliaeetus leucocephalus;* herring gull; *Larus argentatus;* black-crowned night-heron; *Nycticorax nycticorax;* common tern; *Sterna hirundo*; Caspian tern; *Sterna caspia*; Forster's tern; *Sterna forsteri*; wedge-tailed shearwater; *Puffinus pacificus;* glaucous-winged gull; *Larus glaucescens*
NOTES: SUPPLEMENTAL; NO TABLES
ABSTRACT: Much like the caged canary used by miners last century to detect toxic gases in underground mines, fish and wildlife species today are still used to monitor environmental quality. Many different species have been promoted as sentinels, biomonitors, or bioindicators of environmental health hazards. The use of animal sentinel species and sentinel bioassays is essential to the idea of environmental biomonitoring as a means of determining possible health hazards from exposure to environmental contaminants. Among the most insidious and least characterized of contaminants in the environment are chemicals which act to disrupt normal endocrine function in organisms. Much attention has recently been paid to potential endocrine disruption in the environment, and one of the most important aspects involved in solving this

Annotated Bibliography (Cont.).

potentially major environmental problem is the use of fish and wildlife species as sentinels of exposure and effects of these chemicals. This paper examines the utility of fish and wildlife species and bioassays as sentinels of environmental endocrine disruption. Major vertebrate taxa, including mammals, birds, reptiles, amphibians, and fishes, are reviewed for use of species or species groups as sentinel species in general and as sentinels of environmental endocrine disruption. Overall, there have been many studies using fishes, birds, and mammals, and few studies using reptiles and amphibians. In addition, sentinel bioassay endpoints, including residue analysis, biochemical, physiological, immunological, behavioral, reproductive, developmental, and ecological endpoints are reviewed in general and specifically for endocrine disruption. Generally, a suite of different bioassay endpoints appears to be most effective in examining endocrine disruption in the environment. Finally, the study of sentinel species and sentinel bioassays is a rapidly emerging area in the field of wildlife toxicology, and many different directions or future research are discussed.

Shore, Richard F.; Wright, Julian; Horne, Janice A.; and Sparks, Timothy H. 1999. Polycyclic aromatic hydrocarbon (PAH) residues in the eggs of coastal-nesting birds from Britain. Mar. Pollut. Bull. 38(6):509-513.
Rec #: 188
KEYWORDS: Great Britain; Europe; hydrocarbons; PAHs; egg; herring gull; *Larus argentatus;* cormorant; *Phalacrocorax carbo;* shag; *Phalacrocorax aristotelis;* chough; *Pyrrhocorax pyrrhocorax*
ABSTRACT: As part of investigations carried out on the ecological consequences of the grounding of the Sea Empress off Milford Haven, Wales, in February 1996, the eggs of a number of coastal-breeding birds were analysed for PAHs to determine whether there was contamination from the spilled oil. Although no evidence was found of elevated hydrocarbon residues in eggs from the area in which the Sea Empress ran ashore, the study provided general data on PAH levels in eggs. The aim of this paper is to describe the background levels of PAH contamination in eggs of the herring gull *Larus argentatus*, the cormorant *Phalacrocorax carbo*, the shag *Phalacrocorax aristotelis* and the chough *Pyrrhocorax pyrrhocorax*, four species that nest on the British coast but differ markedly in the extent to which they forage on marine organisms. Thirty eggs were collected in the Spring of 1996 and analysed for PAHs. The results of the present study suggest that accumulation of background environmental levels of PAHs in eggs by coastal nesting birds in at least two areas of Britain is common but is unlikely to be sufficient to cause embryotoxic effects. However, it is possible that individuals nesting in more polluted areas and species which feed predominantly on prey that accumulate high PAH residues, such as bivalve molluscs, may have higher levels of PAH contamination in eggs. Furthermore, assessment of the potential embryotoxic effects of PAHs are based on single compound tests and it is known that simultaneous exposure to different PAHs can result in additive toxicity.

Shutt, J. L. 1993. Reproductive success of herring gulls nesting near a bleached kraft pulp mill. Organohalogen Compounds. 12:211-214.
Rec #: 549
KEYWORDS: organochlorines; DDTs; mirex; dieldrin; oxychlordane; PCDDs; PCDFs; PCBs; toxicology; EROD; retinol; porphyrin; herring gull; *Larus argentatus*; egg; carcass; effects; reproductive success; pulp mill; Great Lakes; North America
NOTES: FIGURES ONLY

Annotated Bibliography (Cont.).

ABSTRACT: Exposure to pulp mill effluents has been shown to have various negative impacts on aquatic biota. In order to determine the effect of pulp mill effluent on herring gulls (*Larus argentatus*), eggs were collected from effluent-contaminated and control sites in the Great Lakes for organochlorine analysis as well as EROD induction and total intracellular porphyrin accumulation. Incubating adults were also collected for organochlorine body burdens, hepatic EROD, hepatic retinol, retinyl palmitate, plasma retinol, and total highly carboxylated porphyrins. Organochlorine contaminant levels were low compared to other Great Lakes sites. EROD induction was not elevated above background levels. However, 1991 and 1992, all nests followed in the contaminated Jackfish Bay failed to produce any young of fledgling age. Eggs from Jackfish Bay and a control site hatched equally well. Body condition indices indicated lowered condition for birds nesting in Jackfish Bay compared to other sites in the Great Lakes. A possible change in the traditional diet of fish and insects may play a role in these effects.

Skaare, Janneche Utne. 1998. Environmental pollutants in top predators from the Norwegian coast and Arctic - Occurrence, levels and effects. Organohalogen Compounds. 39:397-401.
Rec #: 325
KEYWORDS: glaucous gull; *Larus hyperboreus*; Norway; Europe; liver; intestine; organochlorines; PCBs; DDTs; chlordanes; HCB; HCHs; toxicology; EROD; CYP1A; retinol; vitamin A; immunology; effects
NOTES: NO TABLES
ABSTRACT: A review with 41 references is given summarizing results on biologic/toxic effects of polychlorinated biphenyls, DDT, chlordanes, HCB, and HCH in marine top predator mammals and seabirds from the Norwegian coast and Arctic including effects on the immunosystem, reproduction, and biochemical and physiological parameters (i.e. thyroid hormones, retinol, EROD activity).

Skaare Janneche Utne; Bernhoft, Aksel; Derocher, Andrew; Gabrielsen, Geir Wing; Gøksyr, Anders; Henriksen, Espen; Larsen, Hans Jørgen; Lie, Elisabet; and Wiig, Øystein. 2000. Organochlorines in top predators at Svalbard--occurrence, levels and effects. Toxicol. Lett. 112-113:103-109.
Rec #: 321
KEYWORDS: glaucous gull; *Larus hyperboreus*; Norway; Svalbard; Europe; liver; intestine; PCBs; EROD; CYP1A; retinol; vitamin A; toxicology; effects
NOTES: NO TABLES
ABSTRACT: Alarmingly high polychlorinated biphenyl (PCB) levels have been found in the top predators such as glaucous gull (*Larus hyperboreus*) and polar bear (*Ursus maritimus*) at Svalbard [Gabrielsen, G.W., Skaare, J.U., Polder, A., Bakken, V., 1995. Chlorinated hydrocarbons in glaucous gull (*Larus hyperboreus*). Sci. Total Environ. 160/161, 337-346; Bernhoft, A., Skaare, J.U., Wiig, O., 1997. Organochlorines in polar bears (*Ursus maritimus*) at Svalbard. Environ. Pollut. 95, 159-175; Henriksen, E.O., Gabrielsen, G.W., Trudeau, S., Wolkers, H., Sagerup, K., Skaare, J.U., 1999. Organochlorines and possible biochemical effects in glaucous gull (*Larus hyperboreus*) from Bear Island, the Barents Sea. Arch. Environ. Contam. Toxicol. (in press).]. Studies of the possible toxic effects, particularly on the immune system and reproduction, of the very high PCB levels in these species are currently being investigated. Data obtained in the field (f.i. reproductive success in polar bears and intestinal nematodes in glaucous gulls), as well as levels of various biochemical and physiological parameters (f.i. thyroid

Annotated Bibliography (Cont.).

hormones, retinol, EROD activity, CYP1A, IgG), have been coupled with the PCB levels [Skaare, J.U., Wiig, O., Bernhoft, A., 1994. Klorerte organiske miljogifter; Nivaer og effekter i isbjorn. Norwegian Polar Institute Reportseries no. 86, 1-23 (in Norwegian); Bernhoft, A., Skaare, J.U., Wiig, O., 1997. Organochlorines in polar bears (*Ursus maritimus*) at Svalbard. Environ. Pollut. 95, 159-175; Bernhoft, A., Skaare, J.U., Wiig, O., Derocher, A.E., Larsen, H.J., 2000. Possible immunotoxic effects of organochlorines in polar bears (*Ursus maritimus*) at Svalbard (in press); Henriksen, E.O., Gabrielsen, G.W., Skaare, J.U., Skjegstad, N., Jensen, B.M., 1998a. Relationship between PCB levels, hepatic EROD activity and plasma retinol in glaucous gull, Larus hyperboreus. Marine Environ. Res. 46, 45-49; Henriksen, E.O., Gabrielsen, G.W., Trudeau, S., Wolkers, H., Sagerup, K., Skaare, J.U., 1999. Organochlorines and possible biochemical effects in glaucous gull (*Larus hyperboreus*) from Bear Island, the Barents Sea. Arch. Environ. Contam. Toxicol. (in press); Sagerup, K., Gabrielsen, G.W., Skorping, A., Skaare, J.U., 1998. Association between PCB concentrations and intestinal nematodes in glaucou gulls, Larus hyperboreus, from Bear Island. Organohalogen compounds 39, 449-451; Skaare, J.U., Wiig, O., Bernhoft, A., 1994. Klorerte organiske miljogifter; Nivaer og effekter i isbjorn. Norwegian Polar Institute Reportseries no. 86, 1-23. (in Norwegian)].

Somers, J. D.; Goski, B. C.; Barbeau, J. M.; and Barrett, M. W. 1993. Accumulation of organochlorine contaminants in double-crested cormorants. Environ. Pollut. 80(1):17-23.
Rec #: 404
KEYWORDS: organochlorines; HCB; HCHs; chlordanes; dieldrin; endrin; mirex; DDTs; PCBs; double-crested cormorant; *Phalacrocorax auritus*; egg; fat; feather; Alberta; Saskatchewan; Canada; North America; effects; reproductive success; eggshell thickness; bioaccumulation; spatial variations
ABSTRACT: Cormorant eggs and lipid samples from juvenile cormorants were analyzed for 14 organochlorine contaminants. Low concentrations (geometric mean <0.05 µg/g) of HCB, lindane, oxychlordane, heptachlor epoxide, dieldrin, endrin, mirex, DDD, and DDT in eggs primarily reflected the wintering-ground origin of organochlorine contaminants. Overall geometric mean concentrations of DDE and PCBs in cormorant eggs were 3.90 and 2.22 µg/g egg, respectfully, and would not affect reproduction or eggshell thickness. Eggshells averaged 0.44 mm in thickness and no correlation ($r^2 = 0.17$) with log-transformed DDE residues in cormorant eggs was evident. Only DDE and PCBs were detected in lipid samples from 5- to 8-wk-old cormorants (geometric mean approximately 1.0 µg/g lipid for each compound). The PCB:DDE ratios in cormorant lipid from some individual colonies were 2-3.5 times greater than the ratio in eggs from the same colony, suggesting an accumulation of PCBs related to local diet. Juvenile cormorants might serve as regional indicators of chemical residue contamination in Alberta, and provide a temporal perspective on changes in contaminant burdens in aquatic ecosystems.

Spear, Philip A.; Bourbonnais, Diane H.; Norstrom, Ross J.; and Moon, Thomas W. 1990. Yolk retinoids (Vitamin A) in eggs of the herring gull and correlations with polychlorinated dibenzo-p-dioxins and dibenzofurans. Environ. Toxicol. Chem. 9(8):1053-1061.
Rec #: 230
KEYWORDS: organochlorines; PCDDs; PCDFs; hydrocarbons; PAHs; TEQs; bioaccumulation; yolk; egg; retinol; vitamin A; toxicology; herring gull; *Larus argentatus*; Great

Annotated Bibliography (Cont.).

Lakes; North America

ABSTRACT: Little is known of the combined effects associated with chronic, low-level exposure of wildlife to the polyhalogenated dibenzo-p-dioxins (PCDD), dibenzofurans (PCDF), certain biphenyls and other related compounds. To examine possible effects upon egg yolk retinoids, herring gull (*Larus argentatus*) eggs were collected at early (i.e., days 2-12) and late (i.e., approximately day 20) phases of incubation. Analysis of egg yolks by reversed-phase high-performance liquid chromatography revealed compounds that comigrated with all-trans-retinol and all-trans-retinyl palmitate standards. The retinol concentration and the molar ratio of retinol to retinyl palmitate changed significantly between the early and late phases of incubation. Within the 2- to 12-d period of incubation, however, retinoid values were constant. Gull eggs were collected from two breeding colonies on the Great Lakes in 1986 and from five colonies in 1987. Significant correlations existed between the molar ratio of retinoids and 2,3,7,8-TCDD concentration, toxic equivalents of PCDDs and PCDFs and the sum of PCDD and PCDF concentrations.

Speich, Steven M.; Calambokidas, John; Shea, David W.; Peard, John; Witter, Michael; and Fry, D. Michael. 1992. Eggshell thinning and organochlorine contaminants in western Washington waterbirds. Colonial Waterbirds. 15(1):103-112.
Rec #: 532
KEYWORDS: organochlorines; DDTs; PCBs; egg; eggshell thickness; double-crested cormorant; *Phalacrocorax auritus*; pelagic cormorant; *Phalacrocorax pelagicus*; pigeon guillemot; *Cepphus columba*; great blue heron; *Ardea herodias*; glaucous-winged gull; *Larus glaucescens*; WA; USA; North America; temporal trends; spatial variations; effects
ABSTRACT: Within the Puget Sound, Washington, marine environment there are urban-industrial areas with a known variety of pollutants present, including PCBs and DDT/DDE. In 1984 we studied five species if seabirds for effects of pollutants at urban-industrial sites and in more remote areas of western Washington. No significant thinning was observed in the eggs of Double-crested and Pelagic Cormorants (*Phalacrocorax auritus* and *P. pelagicus*), all taken in remote areas. The same was true of Pigeon Guillemot (*Cepphus columba*) eggs taken in Puget Sound. Significant average eggshell thinning (to 13 %) was observed in eggs of Great Blue Herons (*Ardea herodias*) from heronries near agricultural areas. Samples from near urban-industrial areas showed less average thinning (5-7 %). The generally low average concentrations of total DDT (0.35-2.22 μg/g, wet weight) is heron eggs probably reflect local contamination from agricultural areas and wind drift from elsewhere. Higher average concentrations of PCBs its heron eggs were in those from Seattle (15.58 μg/g) and Tacoma (5.46 μg/g), the most industrially developed parts of Puget Sound. The largest average amounts of thinning (7-9 %) in Glaucous-winged gull (*Larus glaucescens*) eggs were for colonies in or near the urban-industrial areas of Puget Sound. Average shell thinning in eggs from the remote area colony was less (2 %). Concentrations of total DDT in eggs (0.49-1.19 μg/g) do not account for the amount of thinning observed. The levels of the contaminants and the degree of eggshell thinning observed in the study species are all currently below levels associated with reproductive impairment in other studies. There is no current or historical evidence of significant pollution related impairment of reproduction in the study species in western Washington.

Stalling, D. L.; Norstrom, R. J.; Smith, L. M.; and Simon, M. 1985. Patterns of PCDD, PCDF,

Annotated Bibliography (Cont.).

and PCB contamination in Great Lakes fish and birds and their characterization by principal components analysis. Chemosphere. 14(6-7):627-643.
Rec #: 224
KEYWORDS: organochlorines; PCBs; PCDDs; PCDFs; Great Lakes; WI; USA; North America; double-crested cormorant; *Phalacrocorax auritus*; herring gull; *Larus argentatus*; Forster's tern; *Sterna forsteri*; egg; carcass; spatial variations
ABSTRACT: Contamination of the Great Lakes with polychlorinated dibenzo-p-dioxins (PCDDs), dibenzofurans (PCDFs), and biphenyls (PCBs) has created concern because of the adverse effects of these chemicals on fish and wildlife. Concerns about these residues are based on the observations that PCDDs and PCDFs in fish and birds are composed primarily of the highly toxic 2,3,7,8-tetrachlorodibenzo-p-dioxin (TCDD) and 2,3,7,8-tetrachloro-dibenzofuran (TCDF) and other 2,3,7,8-chlorine substitute penta- and hexa-chlorodibenzo-p-dioxins and -furans. The concentrations of 2,3,7,8-TCDD in fish and birds are greatest in regions where chlorinated organic chemicals are manufactured or near hazardous waste sites.

Stepanova, Ludmila I.; Glaser, Vadim M.; Savinova, Tatiana I.; Kotelevtsev, Sergey V.; and Savva, Demetris. 1999. Accumulation of mutagenic xenobiotics in fresh water (Lake Baikal) and marine (Hornøya Island) ecosystems. Ecotoxicology. 8(2):83-96.
Rec #: 457
KEYWORDS: Norway; Europe; Russia; Asia; mallard; *Anas platyrhynchos*; pintail; *Anas acuta*; shoveler; *Anas clypeata*; tufted duck; *Aythya fuligula;* pochard; *Aythya ferina;* grey heron; *Ardea cinerea;* herring gull; *Larus argentatus*; common gull; *Larus canus*; black-headed gull; *Larus ridibundus*; common tern; *Sterna hirundo*; horned grebe; *Podiceps auritus;* tree sparrow; *Passer montanus;* Daurian redstart; *Phoenicurus auroreus;* common pigeon; *Columba livia*; black-legged kittiwake; *Rissa tridactyla*; egg; liver; toxicology; biomagnification; effects; enzymes; metabolism
NOTES: SUPPLEMENTAL
ABSTRACT: extracts from tissues of a wide range of aquatic organisms (plants, plankton, decapods, molluscs, fish, Baikal seals, fish-eating birds and their eggs) from Lake Baikal and from the Selenga River estuary, and tissue extracts of birds breeding on Hornøya Island (northern Norway) were assayed for mutagenicity using the Ames Salmonella/microsome test. Cytochrome P_{450} activity and enzymes of phase II of detoxification were also studied in the liver of fish and birds. Evidence was found for accumulation of mutagens in the food chain. Relationship of bioaccumulation to levels of enzyme activity possessing both detoxification and activation functions is discussed in the cases of fish and birds. Accumulation of mutagens depended on the activity and level of induction of enzymes providing detoxification and metabolic activation in liver of fish and birds.

Stewart, F. M.; Phillips, R. A.; Catry, P.; and Furness, R. W. 1997. Influence of species, age, and diet on mercury concentrations in Shetland seabirds. Mar. Ecol. Prog. Ser. 151:237-244.
Rec #: 295
KEYWORDS: metals; Hg; parasitic jaeger; *Stercorarius parasiticus*; great skua; *Catharacta skua*; Arctic tern; *Sterna paradisaea*; black-legged kittiwake; *Rissa tridactyla*; common murre; *Uria aalge;* Shetland; Great Britain; Europe; feather; age variations; bioaccumulation; biomagnification
ABSTRACT: Chick down, chick feathers and feathers from adults of 5 seabird species (Arctic

Annotated Bibliography (Cont.).

skua *Stercorarius parasiticus*, great skua *Catharacta skua*, Arctic tern *Sterna paradisaea*, kittiwake *Rissa tridactyla* and common guillemot *Uria aalge*) were analysed for mercury. Individual female Arctic and great skuas' body feather mercury concentrations correlated with concentrations in their chicks' down, but not feathers (Arctic skua: r = 0.64; great skua: r = 0.66). This demonstrated that mercury in chick down originated from the egg, and that mercury in the egg and in the adult females' plumage could have the same dietary source. Inter-specific differences in mercury concentrations were found for all age classes sampled, and these could be explained partly in terms of dietary specialization, although physiological variations may also be important. All 3 age classes of great skua showed a direct increase in mercury with increasing proportion of bird meat in the diet of individual pairs. in kittiwake, Arctic skua, and great skua, adults had higher mercury concentrations than chicks and chick down had higher concentrations than chick feathers. However, in 2 species (Arctic terns and guillemots) chick down had higher concentrations than adult feathers. Chick down could be sampled for mercury content as an alternative to using eggs in national biomonitoring schemes. Feathered chicks could be sampled to determine mercury availability around the breeding colony between hatching and fledging.

Stewart, F. M.; Thompson, D. R.; Furness, R. W.; and Harrison, N. 1994. Seasonal variation in heavy metal levels in tissues of common guillemots, *Uria aalge*, from northwest Scotland. Arch. Environ. Contam. Toxicol. 27(2):168-175.
Rec #: 59
KEYWORDS: effects; toxicology; seasonal variations; biomagnification; common murre; *Uria aalge*; metals; Hg; Cd; Zn; Cu; liver; kidney; muscle; Scotland; Europe
ABSTRACT: Mercury, cadmium, zinc, and copper concentrations were analyzed in three samples of common guillemot (in April, June and November). Levels measured were uniformly low, and not enough to have any toxic effects. Adult guillemots had significantly more cadmium in their livers and kidneys than juveniles, with juvenile levels ranging from 25 % to 89 % of adult levels. Mercury concentrations in liver and kidney were also higher in adults. Juvenile levels represented from 80 % to 94 % of adults, but there were no age differences in feather and muscle mercury. Mercury levels declined throughout the year in internal tissues from April through June to November. There was a strong seasonal fluctuation in cadmium levels in liver and kidney, rising significantly between April and June and declining again from June to November. These changes were apparent in both adult and juvenile birds. The influences of seasonal processes (namely breeding and moult) and seasonal dietary differences as causative factors in the changes in metal burdens are discussed. These findings have implications for the use of seabirds as monitors of heavy metals in the marine environment.

Stickel, Lucille F. and Dieter, Michael P. 1979. Ecological and physiological /toxicological effects of petroleum on aquatic birds: A summary of research activities FY 76 through FY 78. Biol. Ser. Program Fish. Wildl. Serv. (U.S.). 14 pp.
Rec #: 71
KEYWORDS: effects; oil; reproductive success; toxicology; gender differences; bioaccumulation; elimination; abnormalities; mallard; *Anas platyrhynchos*; common eider; *Somateria mollissima;* great black-backed gull; *Larus marinus*; laughing gull; *Larus atricilla*; Louisiana heron; *Egretta tricolor*; sandwich tern; *Sterna sandvicensis*
NOTES: SUPPLEMENTAL
ABSTRACT: The physiological and ecological effects of oil on waterbirds were examined in a

Annotated Bibliography (Cont.).

series of laboratory and field experiments. Chemical methodology was developed in support of these studies. Research conducted from 1 July 1975 to 30 September 1978 by the US Fish and Wildlife Service about the effects of petroleum on aquatic birds is summarized. The following assessments were made: effects of oiling on hatchability of eggs; effects of oil ingestion on physiological condition and survival of birds; effects of oil ingestion on reproduction in birds; accumulation and loss of oil by birds; and development of analytical methods for identification and quantification of oil breakdown products in tissues and eggs of ducks.

Stout, Jordan H.; Trust, Kimberly A.; Cochrane, Jean F.; Suydam, Robert S.; and Quakenbush, Lori T. 2002. Environmental contaminants in four eider species from Alaska and arctic Russia. Environ. Pollut. 119(2):215-226.
Rec #: 343
KEYWORDS: AK; USA; North America; Russia; Asia; metals; As; B; Cd; Cr; Cu; Fe; Hg; Mg; Mn; Mo; Se; Sr; Zn; organochlorines; HCB; PCBs; dieldrin; chlordanes; DDTs; toxaphene; benzenes; common eider; *Somateria mollissima*; king eider; *Somateria spectabilis*; spectacled eider; *Somateria fischeri*; Steller's eider; *Polysticta stelleri*; kidney; liver; gender differences
ABSTRACT: Population declines in four species of eider; common (*Somateria mollissima*), king (*Somateria spectabilis*), spectacled (*Somateria fischeri*) and Steller's (*Polysticta stelleri*), have raised concerns about exposure to contaminants. Livers and kidney tissues were collected from eiders in Alaska and Russia for organic and elemental analyses. Results showed that organochlorine and many elemental levels were below toxic thresholds; however, in many cases, cadmium, copper, lead and selenium appeared high relative to other waterfowl and may warrant concern. With the exception of lead, local anthropogenic sources for these elements are not known. Although adverse physiological responses have not been documented in eiders, these four elements cannot be ruled out as contaminants of potential concern for some eider species.

Stow, Craig A. 1995. Great Lakes herring gull egg PCB concentrations indicate approximate steady-state conditions. Environ. Sci. Technol. 29(11):2893-2897.
Rec #: 95
KEYWORDS: Great Lakes; North America; organochlorines; PCBs; egg; herring gull; *Larus argentatus*; temporal trends
NOTES: FIGURES ONLY
ABSTRACT: PCB concentrations in Great Lakes herring gull eggs exhibited decreases following the imposition of regulations banning PCB manufacture. However, gull egg concentrations in Lakes Superior, Michigan, Huron, and Ontario now appear to have stabilized or slightly increased. Concentrations in Lake Erie gull eggs still appear to be decreasing, though the variability in these data may limit the ability to differentiate trend changes until more data are available. This pattern is consistent with previously reported findings of stabilizing PCB concentrations in Great Lakes fishes and suggests that future improvements will be slow and difficult to discern.

Strauch, J. G. Jr. 1980. Birds. In: Environmental Assessment of the Alaskan Continental Shelf. Northeast Gulf of Alaska Interim Synthesis Report. Science Applications, Inc., Boulder, CO 80301, USA.
Rec #: 147
KEYWORDS: oil; effects; AK; USA; North America

Annotated Bibliography (Cont.).

NOTES: SUPPLEMENTAL
ABSTRACT: The distribution of suitable breeding areas and the seasonal variation in the food supply are probably the major selective forces on the breeding biology of marine birds. Two kinds of hazards to bird populations in the Kodiak area can result from petroleum development: contamination of the environment by oil and disturbance by humans. The effects of oil pollution on birds may be direct fouling of the plumage by floating oil or indirect due to being ingested during feeding or preening. Another indirect effect is contamination of their prey and of the food source of their prey.

Struger, J.; Elliott, J. E.; and Weseloh, D. V. 1987. Metals and essential elements in herring gulls from the Great Lakes, 1983. J. Great Lakes Res. 13(1):43-55.
Rec #: 85
KEYWORDS: bioaccumulation; age variations; spatial variations; metals; effects; herring gull; *Larus argentatus*; Great Lakes; North America; liver; kidney; feather; bone; Ag; Al; Ba; Ca; Cd; Co; Cr; Cu; Fe; Hg; K; Mg; Mn; Mo; Na; Ni; P; Pb; Sr; Th; Ti; V; Zn; Zr
ABSTRACT: Adult and prefledged herring gulls (*Larus argentatus*) were collected from one location each in Lakes Ontario, Erie, Huron, and Superior. Composite samples of liver, kidney, and feather were analyzed for 24 elements and composite samples of bone for 22 elements. After consideration of quality assurance results, concentrations of 16 elements (Al, Ca, Cd, Cr, Cu, Fe, Hg, K, Mg, Mn, Na, P, Pb, Sr, Ti, Zn) in liver, kidney, and feather were accepted for presentation while 6 elements were accepted from bone (Ca, Cd, Hg, P, Pb, Zn). Only lead, cadmium, and mercury values were of toxicological interest. Data on other trace elements are presented as baseline values among locations for each tissue and age class. Concentrations of Cd, Pb, and Hg were higher in adults than in prefledged young. Metal levels varied among different tissues with Cd highest in kidney (2.16 µg/g; Hamilton Harbour, Lake Ontario), Pb highest in bone (30.0 µg/g; Double Island, Lake Huron), and Hg highest in feather (6.11 µg/g; Middle Island, Lake Erie).

Struger, J.; Weseloh, D. V.; Mineau, P.; and Hallett, D. J. 1983. Levels and trends of organochlorines in herring gulls in the Great Lakes, 1974-1981. Proceedings of the 26th Conference on Great Lakes Research. May 23-27, 1983, State University of New York at Oswego. 39 pp.
Rec #: 171
KEYWORDS: organochlorines; DDTs; HCB; mirex; PCBs; dieldrin; egg; herring gull; *Larus argentatus*; North America; Great Lakes; temporal trends
NOTES: ABSTRACT ONLY - NOT AVAILABLE
ABSTRACT: Levels of major organochlorine residues found in the eggs of herring gulls (*Larus argentatus*) are reported for two colonies from each of Lakes Superior, Huron, Erie and Ontario from 1974 to 1981. Levels of DDE, DDT, HCB, Mirex and PCBs were greatest in Lake Ontario. Dieldrin levels were the highest in Lake Superior. Significant declines occurred in levels of DDT, Mirex, PCBs, DDE, Dieldrin and HCB for many colonies between 1974 and 1981. However, between 1980 and 1981, there were significant increases in levels of PCBs, Mirex, DDE and HCB on some colonies, especially Snake Island in Lake Ontario. During the same period, none of these compounds decreased significantly on any of the monitor colonies. These figures may indicate that the declining residue levels in herring gulls during the last six years have begun to level off.

Annotated Bibliography (Cont.).

Sydeman, William J. and Jarman, Walter M. 1998. Trace metals in seabirds, Steller sea lion, and forage fish and zooplankton from Central California. Mar. Pollut. Bull. 36(10):828-832.
Rec #: 339
KEYWORDS: biomagnification; metals; Al; Cr; Fe; Cu; Zn; As; Se; Ag; Cd; Hg; egg; common murre; *Uria aalge*; Brandt's cormorant; *Phalacrocorax penicillatus*; rhinoceros auklet; *Cerorhinca monocerata*; pigeon guillemot; *Cepphus columba;* CA; USA; North America
ABSTRACT: We studied concentrations of the trace metals, aluminum (Al), chromium (Cr), iron (Fe), copper (Cu), zinc (Zn), arsenic (As), selenium (Se), silver (Ag), cadmium (Cd), and mercury (Hg), in krill (*Euphausia pacifica* and *Thysanoessa spinifera*), two species of fish (short-bellied rockfish *Sebastes jordani* and northern anchovy *Engraulis mordax*), four species of marine bird (Common Murre *Uria aalge*, Brandt's Cormorant *Phalacrocorax penicillatus*, Rhinoceros Auklet *Cerorhinca monocerata* , and Pigeon Guillemot *Cepphus columba*) and a pinniped, (Steller sea lion *Eumetopias jubatus*) from the Gulf of the Farallones, central California in 1993. Geometric mean levels of some trace elements in Steller sea lions were elevated (Cu, 91.0 mg/kg dry weight; Hg, 19.0 mg/kg; Se, 4.1 mg/kg). Levels of Hg in Pigeon Guillemot (3.5 mg/kg) were also elevated. Mercury increased whereas Pb decreased with increasing trophic level in the Gulf of the Farallones food web. Selenium levels were highest for krill and sea lions, and intermediate for fish and birds occupying mid trophic levels. Results indicate little to relatively high trace metal contamination of upper trophic level marine wildlife in the central California coastal marine ecosystem.

Szaro, Robert C.; Coon, Nancy C.; and Kolbe, Elizabeth. 1979. Pesticide and PCB of common eider, herring gull, and great black-backed gull eggs. Bull. Environ. Contam. Toxicol. 22(3):394-399.
Rec #: 535
KEYWORDS: organochlorines; PCBs; DDTs; chlordanes; toxaphene; HCB; mirex; ME; VA; USA; North America; egg; common eider; *Somateria mollissima*; herring gull; *Larus argentatus*; great black-backed gull; *Larus marinus*; eggshell thickness; reproductive success; effects; biomagnification
ABSTRACT: All but 1 common eider egg collected on the southeastern coast of Maine contained detectable residues of DDE (I) [72-55-9]. Polychlorinated biphenyls (PCBs) were detected in all eggs. No other organochlorine pesticides were detected. There were no significant differences in the residue levels of DDT [50-29-3] and its metabolites or PCBs in the herring gull eggs collected in Maine and Northern Virginia. However, all were significantly greater than the residues in common eider eggs. Great black-backed gull eggs contained significantly more I and PCBs than herring gull eggs from either Maine or Virginia. The former birds also showed low levels of several other pesticides. The differences in pesticide burdens in the 3 species appear to be related to their feeding habits. None of the residues were high enough to cause any apparent reproductive problems.

Särkkä, Jukka; Hattula, Marja-Liisa; Janatuinen, Jorma; Paasivirta, Jaakko; and Palokangas, Risto. 1978 . Chlorinated hydrocarbons and mercury in birds of lake Paijanne, Finland - 1972-1974. Pestic. Monit. J. 12:26-34.
Rec #: 530
KEYWORDS: metals; Hg; organochlorines; PCBs; DDTs; HCHs; dieldrin; muscle; liver; Finland; Europe; spatial variations; biomagnification; temporal trends; gender differences; Arctic

Annotated Bibliography (Cont.).

loon; *Gavia arctica*
ABSTRACT: The levels of Hg, polychlorinated biphenyls (PCBs), DDT (I) [50-29-3] and its analogs, lindane [58-89-9], and dieldrin [60-57-1] were examined in aquatic birds nesting on the shores of Lake Paijanne, the 2nd largest lake in Finland, which is polluted by a wood-processing industry and urban sewages. The primary food of the 10 species examined was fish. In muscle of approximately 350 individuals, the highest average residues were PCBs; in livers, Hg was the highest. Lindane was found in some individuals; dieldrin appeared in none. The differences among levels in 1972, 1973, and 1974 were not significant. Some regional differences were found, particularly for Hg. Some PCB contamination was observed near the town of Jyvaskyla. I was distributed evenly. A stronger correlation existed between residues of PCBs and I than between residues of any other compounds. In some gulls, males had higher average residues than females. The I:PCB ratio generally corresponded to that of the North Atlantic Ocean, but the difference among species was great. Higher Hg, PCB, and I values existed in livers than in muscles. Black-throated divers had the highest Hg residues; in herring gulls, PCBs and I were highest. The levels generally correspond to those found in other studies.

Tanaka, Hiroyuki. 1996. PCBs and drug-metabolizing enzymes in seabirds. Kankyo Kagaku. 6(4):559-565.
Rec #: 453
KEYWORDS: organochlorines; PCBs; laysan albatross; *Diomedea immutabilis*; northern fulmar; *Fulmarus glacialis*; sooty shearwater; *Puffinus griseus*; short-tailed shearwater; *Puffinus tenuirostris*; pomarine jaeger; *Stercorarius pomarinus*; black-legged kittiwake; *Rissa tridactyla*; red-legged kittiwake; *Rissa brevirostris*; muscle; liver; toxicology; EROD; AHH; enzymes; cytochrome P_{450}; Pacific Ocean; Asia; effects
ABSTRACT: Concentrations of PCBs and activities of hepatic drug-metabolizing enzymes were measured in seven species of seabirds including Laysan Albatross (*Diomedea immutabilis*), Northern Fulmar (*Fulmarus glacialis*), Sooty Shearwater (*Puffinus griseus*), Short-tailed Shearwater (*P. tenuirostris*), Pomarine Jaeger (*Stercorarius pomarinus*), Black-legged Kittiwake (*Larus tridactyla*), and Red-legged Kittiwake (*L. brevirostris*). PCB concentrations in pectoral muscle ranged from 0.010 to 2.8 $\mu g\ g^{-1}$. Twenty six PCB components, mainly IUPAC number 156, 118, 138, 180, and 99 were detected. The combined concentration of these five compounds accounted for more than 70 % of total PCB concentration. Cytochrome P_{450} contents ranged from 0.21 to 0.53 nmol mg^{-1} protein. The ranges of activities (nmol product formed $min^{-1}\ mg^{-1}$ protein) of three drug-metabolizing enzymes, 7-ethoxyresorufin O-deethylase (EROD), benzo[a]pyrene hydroxylase (AHH) and aldrin epoxidase (ALDE) in liver microsomes were 0.08-1.20, 0.01-0.36 and 0.11-1.63, respectively The relation between these activities and PCB concentrations were analyzed by multiple regression analysis.

Tasker, Mark L. and Becker, Peter H. 1992. Influences of human activities on seabird populations in the North Sea. Neth. J. Aquat. Ecol. 26(1):59-73.
Rec #: 121
KEYWORDS: review; effects; North Sea; Europe; egg; red-throated loon; *Gavia stellata*; northern fulmar; *Fulmarus glacialis*; Manx shearwater; *Puffinus puffinus*; storm-petrel; *Hydrobates pelagicus*; Leach's storm-petrel; *Oceanodroma leucorhoa*; gannet; *Sula bassana*; cormorant; *Phalacrocorax carbo*; shag; *Phalacrocorax aristotelis;* parasitic jaeger; *Stercorarius parasiticus*; great skua; *Catharacta skua*; Mediterranean gull; *Larus melanocephalus*; black-

Annotated Bibliography (Cont.).

headed gull; *Larus ridibundus*; common gull; *Larus canus*; lesser black-backed gull; *Larus fuscus*; herring gull; *Larus argentatus*; great black-backed gull; *Larus marinus*; black-legged kittiwake; *Rissa tridactyla*; gull-billed tern; *Sterna nilotica*; sandwich tern; *Sterna sandvicensis*; common tern; *Sterna hirundo*; Arctic tern; *Sterna paradisaea*; roseate tern; *Sterna dougallii*; little tern; *Sterna albifrons*; common murre; *Uria aalge*; razorbill; *Alca torda*; black guillemot; *Cepphus grylle*; Atlantic puffin; *Fratercula arctica*; reproductive success; metals; Hg; organochlorines; HCB; OCS; HCHs; DDTs; PCBs
NOTES: NO TABLES
ABSTRACT: This paper describes the North Sea's seabird resource, which is of great international importance. It reviews the significance of human activities on seabirds; these are persecution, fishing, eutrophication, chemical pollution, introduction of alien predators, rising sea level, and oil pollution. The paper speculates on future trends. Some suggestions to reduce human pressures, to use seabirds as monitors of the marine environment and to conserve their populations more effectively are made.

Teeple, Stanley M. 1977. Reproductive success of herring gulls nesting on Brothers Island, Lake Ontario, in 1973. Can. Field Nat. 91(2):148-157.
Rec #: 521
KEYWORDS: organochlorines; DDTs; herring gull; *Larus argentatus*; Great Lakes; North America; reproductive success
ABSTRACT: Breeding success and causes of breeding failure were assessed in 1973 in a Herring Gull colony of 34 pairs in eastern Lake Ontario. Breeding synchrony was normal, 13 pairs laid repeat clutches, but 77 % of all eggs laid failed to hatch. The number of chicks fledged per pair averaged at least 0.06 but not more than 0.18, an exceptionally low result. Geometric mean concentrations of DDE and PCBs in 15 eggs that failed to hatch were 134 and 420 ppm dry weight. Concentrations of dieldrin, p,p'DDD, p,p'DDT, heptachlor epoxide, β-benzene hexachloride, hexachlorobenzene, and mercury were each less than 6 ppm. Arithmetic mean shell thickness of 13 of those 15 eggs was 0.339 mm, and mean thickness index of 11 of those 13 was 1.60; both are low values. Pathological examinations and analyses for organochlorine pesticides and PCBs in brains were conducted on 12 chicks that died. For 11 of the 12, no clear cause of death could be determined. A general association was established between high organochlorine levels and the low breeding success.

Ternes, W. and Russel, H. A. 1986. Distribution of the heavy metals lead, cadmium, and mercury in birds and bird eggs in relation to nutrition and location. Fortschritte in Der Atomspektrometrischen Spurenanalytik. 2:531-542.
Rec #: 459
KEYWORDS: metals; Pb; Cd; Hg; liver; egg; Germany; Europe; black-legged kittiwake; *Rissa tridactyla*; common murre; *Uria aalge*; common eider; *Somateria mollissima*; bioaccumulation; biomagnification; age variations; seasonal variations
NOTES: FIGURES ONLY
ABSTRACT: The content of Cd, Pb, and Hg was detected in the liver and eggs of 15 species of birds collected in 1980-83 in the coastal region of German Federal Republic. Pb and Cd were not accumulated in eggs; thus eggs are not suitable as bioindicators of environmental pollution. The contamination of the birds with the metals is discussed in relation to their feeding habits, age, and season. The birds may be used as bioindicators of environmental pollution with the

Annotated Bibliography (Cont.).

metals.

Thompson, D. R.; Bearhop, S.; Speakman, J. R.; and Furness, R. W. 1998. Feathers as a means of monitoring mercury in sea birds: Insights from stable isotope analysis. Environ. Pollut. 101(2):193-200.
Rec #: 355
KEYWORDS: great skua; *Catharacta skua;* northern fulmar; *Fulmarus glacialis*; feather; metals; Hg; stable isotopes; biomagnification; Shetland; Great Britain; Europe
ABSTRACT: Mercury concentrations, together with nitrogen and carbon stable isotope signatures, were detected in body feather samples from northern fulmars *Fulmarus glacialis* and great skuas *Catharacta skua*, and in different flight feathers from great skuas. There were no significant relationships between trophic status, as defined using isotope analysis, and mercury concentration in the same feather type, in either species. Mercury concentrations in body feather samples were markedly different between fulmars and skuas, reflecting differences in diet, but there was no corresponding difference in trophic status as measured through nitrogen stable isotope signatures. We conclude that mercury concentrations and stable isotope values in feathers are uncoupled, mercury concentrations apparently reflecting the body pool of accumulated mercury at the time of feather growth while stable isotope values reflect the diet at the time of feather growth. There were significant positive correlations between the different flight feathers of great skuas for all three parameters measured. These were strongest between primary 10 and secondary 8, suggesting that these two feathers are replaced at the same time in the molt sequence in great skuas. Stable isotope analysis of different feathers may provide a means of investigating molt patterns in birds.

Thompson, D. R.; Becker, P. H.; and Furness, R. W. 1993. Long-term changes in mercury concentrations in herring gulls, *Larus argentatus* and common terns, *Sterna hirundo*, from the German North Sea coast. J. Appl. Ecol. 30(2):316-320.
Rec #: 60
KEYWORDS: herring gull; *Larus argentatus*; common tern; *Sterna hirundo*; bioaccumulation; metals; Hg; temporal trends; Germany; North Sea; Europe; feather; temporal trends; spatial variations
ABSTRACT: Mercury concentrations in body feathers of herring gulls, *Larus argentatus Pontoppidan*, from the German North Sea coast showed no significant variation among seasons or regions. Concentrations were higher in adults than in juveniles and two times higher after 1940 than in earlier years. Among adult herring gulls, concentrations increased abruptly to a peak of 12 $\mu g\ g^{-1}$ during the 1940s, presumed to be due to high discharges of mercury during the Second World War. Concentrations dropped in the 1950s then increased to a second peak in the 1970s (10 $\mu g\ g^{-1}$) before falling in the late 1980s. This pattern fits well with known discharges of mercury into the rivers Elbe and Rhine, indicating that herring gull feather samples provide a measure of mercury pollution of the German North Sea coastal ecosystem, pollution that appears to be due predominantly to river inputs rather than atmospheric inputs. Over the same period, mercury concentrations in body feathers of common terns *Sterna hirundo L.* from the German North Sea coast show a 380 % increase in adults and 140 % increase in young. As with herring gulls, museum samples did not provide evidence of geographical variation at a local level within Germany evident from other studies using eggs or chick body feathers. We predict that seabird feather sampling will show a progressive reduction in mercury contamination of the German

Annotated Bibliography (Cont.).

North Sea coastal ecosystem as a result of recent and anticipated measures to reduce river-borne pollution.

Thompson, D. R. and Furness, R. W. 1989. Comparison of the levels of total and organic mercury in seabird feathers. Mar. Pollut. Bull. 20(11):577-579.
Rec #: 49
KEYWORDS: feather; metals; Hg; organometallics; organic Hg; northern fulmar; *Fulmarus glacialis;* shag; *Phalacrocorax aristotelis;* great skua; *Catharacta skua*; parasitic jaeger; *Stercorarius parasiticus*; black-legged kittiwake; *Rissa tridactyla*; razorbill; *Alca torda*; common murre; *Uria aalge*; Atlantic puffin; *Fratercula arctica*; Scotland; Europe
ABSTRACT: In this paper we present data for total and organic mercury concentrations in seabird feathers from a range of species. We also discuss the implications for studies using feather samples from museum collections to investigate historical trends in mercury contamination.

Thompson, D. R.; Furness, R. W.; and Barrett, R. T. 1992. Mercury concentrations in seabirds from colonies in the Northeast Atlantic. Arch. Environ. Contam. Toxicol. 23(3):383-389.
Rec #: 34
KEYWORDS: metals; Hg; bioaccumulation; spatial variations; common murre; *Uria aalge*; black-legged kittiwake; *Rissa tridactyla*; northern fulmar; *Fulmarus glacialis*; razorbill; *Alca torda*; thick-billed murre; *Uria lomvia*; Atlantic puffin; *Fratercula arctica*; shag; *Phalacrocorax aristotelis*; feather; effects; Iceland; Scotland; Norway; Europe
ABSTRACT: Total mercury concentrations were determined in samples of body feathers from a range of common seabird species breeding at Låtrabjarg, northwest Iceland, St. Kilda, Foula and the Firth of Forth, Scotland and Bleiksøy, Syltefjord, and Hornøy, Norway. Seabirds from Låtrabjarg generally exhibited the highest mercury concentrations, with a trend of decreasing mercury concentrations in a southwest to northeast direction in seabirds at the other colonies; seabirds at Hornøy were generally found to have the lowest mercury concentrations. Some species at the Firth of Forth exhibited relatively elevated mercury concentrations compared to those at Foula and Norwegian sites. Inter-colony differences in diet were thought to be relatively small for most species and unlikely to account for the range of mercury concentrations measured in the seabirds (Låtrabjarg: lowest arithmetic mean mercury concentration in common guillemots *Uria aalge*, 1.6 µg/g, s.d. = 0.6, n = 45; highest arithmetic mean mercury concentration in kittiwakes *Rissa tridactyla*, 5.5 µg/g, s.d. = 1.7, n = 36).

Thompson, D. R.; Furness, R. W.; and Walsh, P. M. 1992. Historical changes in mercury concentrations in the marine ecosystem of the north and north-east Atlantic Ocean as indicated by seabird feathers. J. Appl. Ecol. 29(1):79-84.
Rec #: 51
KEYWORDS: metals; Hg; Scotland; Great Britain; Europe; feather; northern fulmar; *Fulmarus glacialis*; Manx shearwater; *Puffinus puffinus*; Atlantic Ocean; gannet; *Sula bassana*; great skua; *Catharacta skua*; Atlantic puffin; *Fratercula arctica*; temporal trends; biomagnification
ABSTRACT: Body feather samples from preserved study skins held in museum collections and from contemporary specimens of a range of species of adult seabirds from the north and north-east Atlantic were analysed for mercury in order to assess historical trends in contamination of the marine environment with this toxic heavy metal. *Fulmars Fulmarus glacialis* (L.) from both

Annotated Bibliography (Cont.).

St. Kilda and Shetland/Orkney exhibited significant declines in mercury concentrations associated with pronounced dietary change over the study period. Other species analysed showed a general increase in mercury burdens, particularly towards the south and west of Britain. Increased pluvial deposition of atmospheric mercury over specific oceanic areas, associated with jet-streams and the pollution of the northern hemisphere by mercury, is suggested as being a likely cause of the observed trends. The use of time-series of feather samples, coupled with the extraction method for methyl mercury used in this study, offer great potential for elucidating changes in the mercury contamination of other environments.

Thompson, David R.; Stewart, Fiona M.; and Furness, Robert W. 1990. Using seabirds to monitor mercury in marine environments: The validity of conversion ratios for tissue comparisons. Mar. Pollut. Bull. 21(7):339-342.
Rec #: 52
KEYWORDS: metals; Hg; bioaccumulation; organometallics; organic Hg; liver; muscle; feather; Scotland; Europe; wandering albatross; *Diomedea exulans*; yellow-nosed albatross; *Diomedea chlororhynchos*; sooty albatross; *Phoebetria fusca*; Atlantic petrel; *Pterodroma incerta*; soft-plumaged petrel; *Pterodroma mollis*; Tristan skua; *Catharacta hamiltoni*; northern fulmar; *Fulmarus glacialis*; great skua; *Catharacta skua*; lesser black-backed gull; *Larus fuscus*; herring gull; *Larus argentatus;* common murre; *Uria aalge*
ABSTRACT: The "7:3:1 rule" for converting mercury concentrations between feather, liver and muscle tissues was evaluated by measuring total and methyl mercury levels in liver, muscle and body feathers of a range of seabird species. Mean mercury concentrations were used to calculate feather:liver (both total and methyl) and feather:muscle ratios, the results obtained being compared with the predicted values. Feather:liver ratios were found to approximate to 2.3 when liver methyl mercury concentrations were considered, but elevated inorganic mercury concentrations in the livers of some species resulted in greatly reduced feather:liver ratios for total mercury. Feather:muscle ratios varied from 3.8 to 15.3. Factors likely to affect the value of feather:liver and feather:muscle mercury concentration ratios, such as the predominant form of mercury present in the liver tissue, sampling date relative to the stage of the moult sequence and types of feather used for analysis, are discussed and the authors emphasize that the 7:3:1 conversion ratio should be treated with caution.

Tillitt, Donald E.; Ankley, Gerald T.; and Giesy, John P. 1989. Planar chlorinated hydrocarbons (PCHs) in colonial fish-eating water bird eggs from the Great Lakes. Mar. Environ. Res. 28(1-4):505-508.
Rec #: 406
KEYWORDS: egg; organochlorines; PCBs; PCDDs; PCDFs; toxicology; EROD; cytochrome P_{450}; TEQs; effects; egg; double-crested cormorant; *Phalacrocorax auritus*; Caspian tern; *Sterna caspia*; Great Lakes; North America; dosing study; spatial variations; reproductive success
NOTES: FIGURES ONLY
ABSTRACT: Reproductive impairment was studied in double-crested cormorants (*Phalacrocorax auritus*) and Caspian terns (*Hydroprogne caspia*) of the Great Lakes of North America with respect to the levels of planar chlorinated hydrocarbons (PCHs), which include polychlorinated biphenyls (PCBs), dibenzo-p-dioxins (PCDDs), and dibenzofurans (PCDFs), in the bird eggs. The relative potency of the egg extracts was assessed by their ability to induce cytochrome P_{450}-dependent ethoxyresorufin O-deethylase (EROD) in H4IIE rat hepatoma cells.

Annotated Bibliography (Cont.).

The magnitude of the response was compared with EROD induction in cell cultures by TCDD. The highest TCDD-equivalents were found in water bird egg composites from areas with greater PCH concentrations and more severe reproductive effects. Significant concentrations of PCHs were detected in all sites tested; the range of TCDD-equivalents in the water bird eggs was 49-415 pg/g, uncorrected for external efficiencies. The evidence is strong for at least a partial role of PCHs as causal agents in the reproductive impairment of fish-eating water birds from the Great Lakes of North America.

Tillitt, Donald E.; Ankley, Gerald T.; Giesy, John P.; Ludwig, James P.; Kurita-Matsuba, Hiroko; Weseloh, D. Vaughn; Ross, Peter S.; Bishop, Christine A.; Sileo, Lou; Stromborg, Ken L.; Larson, Jill; and Kubiak, Timothy J. 1992. Polychlorinated biphenyl residues and egg mortality in double-crested cormorants from the Great Lakes. Environ. Toxicol. Chem. 11(9):1281-1288.
Rec #: 405
KEYWORDS: egg; organochlorines; PCBs; PCDDs; toxicology; TEQs; reproductive success; effects; double-crested cormorant; *Phalacrocorax auritus*; Great Lakes; North America
ABSTRACT: The authors evaluated the overall potency of PCB-containing extracts from double-crested cormorant (*Phalacrocorax auritus*) eggs with an in vitro bioassay system, the H4IIE rat hepatoma cell bioassay. Results from the H4IIE bioassay were strongly correlated with the hatching success of eggs in the colonies, whereas conventional methods of PCB analysis correlated poorly with hatching success of eggs from the same colonies. These observations suggest that even though concentrations of total PCB residues have declined in almost all compartments of the environment, their effects are still being observed. The significance of this observation is that the adverse symptoms presently observed in certain Great Lakes fish-eating waterbird populations do not appear to be caused by some as yet unidentified industrial chemical or chemicals and seem not to be the result of pesticides, but rather to the dioxin-like activity of PCBs. Evidence is presented to suggest that the relative enrichment of the potency of PCBs in the environment may play a role in the persistence of the observed adverse symptoms.

Tittlemier, Sheryl A.; Fisk, Aaron T.; Hobson, Keith A.; and Norstrom, Ross J. 2002. Examination of the bioaccumulation of halogenated dimethyl bipyrroles in an Arctic marine food web using stable nitrogen isotope analysis. Environ. Pollut. 116(1):85-93.
Rec #: 431
KEYWORDS: little auk; *Alle alle*; black guillemot; *Cepphus grylle*; black-legged kittiwake; *Rissa tridactyla*; glaucous gull; *Larus hyperboreus*; liver; Northwater Polynya; Canada; Greenland; Atlantic Ocean; North America; organohalogens; HDBPs; stable isotopes; biomagnification
ABSTRACT: Except for ringed seals, halogenated di-Me bipyrrole (HDBP) congeners biomagnified in a sample aquatic food web of invertebrate>fish>seabird. concentrations of four possibly naturally produced organohalogens - 1,1'-dimethyl-3,3',4-tribromo-4,5,5'-trichloro-2,2'-bipyrrole (DBP-Br_3Cl_3), 1,1'-dimethyl-3,3',4,4'-tetrabromo-5,5'-dichloro-2,2'-bipyrrole (DBP-Br_4Cl_2), 1,1'-dimethyl-3,3',4,4',5-pentabromo-5'-chloro-2,2'-bipyrrole (DBP-Br_5Cl) and 1,1'-dimethyl-3,3',4,4',5,5'-hexabromo-2,2'-bipyrrole (DBP-Br_6) - were quantitated and the extent of their magnification through an entire Arctic marine food web [measured as integrated trophic magnification factors (TMFs)] were calculated The food web consisted of three zooplankton species (*Calanus hyperboreus, Mysis oculata,* and *Sagitta* sp.), one fish species [Arctic cod

Annotated Bibliography (Cont.).

(*Boreogadus saida*)], four seabird species [dovekie (*Alle alle*), black guillemot (*Cepphus grylle*), black-legged kittiwake (*Rissa tridactyla*), and glaucous gull (*Larus hyperboreus*)], and one marine mammal species [ringed seal (*Phoca hispida*)]. Trophic levels in the food web were calculated from ratios of stable isotopes of nitrogen ($^{15}N/^{14}N$). All halogenated di-Me bipyrrole (HDBP) congeners were found to significantly (P<0.02) biomagnify, or increase in concentration with trophic level in the invertebrate - fish - seabird food web. DBP-Br$_4$Cl$_2$ (TMF=14.6) was found to biomagnify to a greater extent than DBP-Br$_3$Cl$_3$ (TMF=5.2), DBP-Br$_5$Cl (TMF=6.9), or DBP-Br$_6$ (TMF=7.0), even though the K$_{ow}$ of DBP-Br$_4$Cl$_2$ was predicted to be lower than those of DBP-Br$_5$Cl and DBP-Br$_6$. None of the four HDBP congeners in ringed seals followed the general trend of increasing concentration with trophic level, which was possibly due to an ability of the seals to metabolize HDBPs.

Tittlemier, Sheryl A.; Norstrom, Ross J.; and Elliott, John E. 1998. Levels of a possible naturally-produced bromo/chloro compound in bird eggs from the Great Lakes, Atlantic, and Pacific coastal areas. Organohalogen Compounds. 39:117-120.
Rec #: 386
KEYWORDS: organochlorines; C$_{10}$H$_6$N$_2$Br$_4$Cl$_2$; egg; liver; Great Lakes; Atlantic Ocean; Pacific Ocean; Canada; North America; Leach's storm petrel; *Oceanodroma leucorhoa*; rhinoceros auklet; *Cerorhinca monocerata*; glaucous-winged gull; *Larus glaucescens;* black-footed albatross; *Diomedea nigripes*; herring gull; *Larus argentatus*; Atlantic puffin; *Fratercula arctica*; bald eagle; *Haliaeetus leucocephalus*; spatial variations; biomagnification
ABSTRACT: The unknown halogenated compound C$_{10}$H$_6$N$_2$Br$_4$Cl$_2$ (I) was isolated from bald eagle liver and quantified using GC-FID. Absolute and relative (to PCB 153) concentrations of I were determined in bird eggs. The concentration of I in the eagle liver extract was 160 ng/µL. Levels of I were only detected in marine egg samples and the absolute and relative concentrations of I were higher in the Pacific samples (140-1.8 ppb) than in the Atlantic samples (4.8-0.6 ppb). This geographical trend was also apparent for Leach's storm petrels (the Pacific samples had approximately 25 times higher absolute and relative concentrations of I than the Atlantic samples) and the glaucous-winged gulls (5-11 times higher than the herring gulls on the Atlantic coast). The offshore surface feeders from both the Pacific and Atlantic coasts had the highest absolute and relative concentration of I compared to both the offshore subsurface feeders and inshore omnivores. The 2 observed trends suggest that the presence of I is strictly a marine phenomenon and that it predominantly occurs in the surface layer. The absence of I from the Great Lakes also implies that atmosphere transport dose not play a role in the distribution of the compound.

Tittlemier, Sheryl A.; Simon, Mary; Jarman, Walter M.; Elliott, John E.; and Norstrom, Ross J. 1999. Identification of a Novel C$_{10}$H$_6$N$_2$Br$_4$Cl$_2$ Heterocyclic Compound in Seabird Eggs. A Bioaccumulating Marine Natural Product? Environ. Sci. Technol. 33(1):26-33.
Rec #: 429
KEYWORDS: egg; Leach's storm-petrel; *Oceanodroma leucorhoa*; rhinoceros auklet; *Cerorhinca monocerata*; glaucous-winged gull; *Larus glaucescens*; black-footed albatross; *Diomedea nigripes*; Atlantic puffin; *Fratercula arctica*; herring gull; *Larus argentatus*; organochlorines; C$_{10}$H$_6$N$_2$Br$_4$Cl$_2$; bioaccumulation; Pacific Ocean; Atlantic Ocean; British Columbia; Nova Scotia; Newfoundland; Bay of Fundy; Canada; North America; spatial variations

Annotated Bibliography (Cont.).

ABSTRACT: A brominated and chlorinated compound, $C_{10}H_6N_2Br_4Cl_2$, bioaccumulating in seabird eggs was identified and characterized by low- and high-resolution electron impact ionization (EI), electron capture negative ionization (ECNI), and ammonia positive chemical ionization (PCI) mass spectrometry. This compound is the major congener of a series of 4 hexahalogenated species. The major congener was detected in egg samples from Leach's storm-petrel, Rhinoceros auklet, glaucous-winged gull, and black-footed albatross from the Pacific coast area; Leach's storm-petrel, Atlantic puffin, and herring gull from the Atlantic coast; and herring gull from the Great Lakes using GC-ECNI-MS. The concentrations of $C_{10}H_6N_2Br_4Cl_2$ in the Pacific Ocean samples were 1.8-140 ng/g (wet weight), and were significantly higher than the Atlantic Ocean samples ($p = 0.037$). The Pacific Ocean samples contained levels of $C_{10}H_6N_2Br_4Cl_2$ approximately 1.5-2.5 times higher than in the Atlantic Ocean samples of the same or ecologically similar species. The compound was not detected in any of the samples from the Great Lakes. The Pacific Ocean offshore surface feeders had the highest concentrations (34-140 ng/g) when compared to the other samples (0.61-5.6 ng/g). Its strictly marine occurrence and relatively high nitrogen content indicate that $C_{10}H_6N_2Br_4Cl_2$ probably is a marine natural product, found at highest concentrations in the Pacific Ocean surface feeding birds. A possible structure of $C_{10}H_6N_2Br_4Cl_2$ is 1,1'-dimethyl-tetrabromodichloro-2,2'-bipyrrole.

Trivelpiece, Wayne Z.; Butler, Ronald G.; Miller, David S.; and Peakall, David B. 1984. Reduced survival of chicks of oil-based adult Leach's storm-petrels. Condor. 86(1):81-82.
Rec #: 136
KEYWORDS: oil; toxicology; dosing study; effects; reproductive success; feather; Leach's storm-petrel; *Oceanodroma leucorhoa*
NOTES: SUPPLEMENTAL
ABSTRACT: Birds with petroleum-coated plumage may ingest and/or inhale substantial amounts of the substance while preening, small amounts of ingested oil have resulted in marked physiological changes in nestling larids and alcids. Similar oil-induced physiological aberrations in adult birds may well impair reproductive success. In the present study, the authors examined the effect of sub-lethal oil ingestion by adult Leach's Storm-Petrels (*Oceanodroma leucorhoa*) on the survival and growth of their chicks.

Turle, R.; Norstrom, R. J.; and Collins, B. 1991. Comparison of PCB quantitation methods: Re-analysis of archived specimens of herring gull eggs from the Great Lakes. Chemosphere. 22(1-2):201-213.
Rec #: 220
KEYWORDS: organochlorines; PCBs; egg; herring gull; *Larus argentatus*; North America; Great Lakes; temporal trends
ABSTRACT: Eggs have been collected from colonies of Herring Gulls (*Larus argentatus*) throughout the Great Lakes since 1970, and analyzed as a means of assessing trends in levels of organochlorine contaminants. PCBs were originally determined by packed-column GC-ECD as Aroclor 1254/1260 1:1 equivalents. Since 1986, forty-one individual congeners have been determined and summed to obtain Sigma PCB. In order to relate the new and old data and remove discontinuities due to method improvements, representative samples of a pooled gulls eggs from the 1970s and 1980s were retrieved from a specimen bank and reanalysed for individual PCB congeners. Conversion factors from Aroclor 1254/1260 1:1 Sigma PCB were 0.450 for Lake Superior, 0.484 for Lake Huron, 0.444 for Lake Erie and 0.461 for Lake Ontario.

Annotated Bibliography (Cont.).

Turle, Richard and Collins, Brian. 1992. Validation of the use of pooled samples for monitoring of contaminants in wildlife. Chemosphere. 25:463-469.
Rec #: 228
KEYWORDS: herring gull; *Larus argentatus*; egg; North America; Great Lakes; organochlorines; DDTs; dieldrin; chlordanes; mirex; HCB; benzenes; OCS; HCHs
ABSTRACT: The high cost of sample analysis for organic contaminants requires strategies such as pooling of samples. Eggs collected as part of the Great Lakes Herring Gull (*Larus argentatus*) monitoring have been pooled since 1986. Individual eggs from colonies from each lake have also been analysed. Any differences between the means of individual analyses and those of pooled samples are generally less than the normal analytical variation.

Vander Pol, Stacy S. 2002. Persistent organic pollutants (POPs) in Alaskan murre (*Uria* spp.) eggs. Charleston, SC, USA: College of Charleston, Masters Thesis. 125 pp.
Rec #: 478
KEYWORDS: organochlorines; PCBs; HCB; chlordanes; DDTs; HCHs; mirex; dieldrin; egg; common murre; *Uria aalge*; thick-billed murre; *Uria lomvia*; AK; USA; North America; Gulf of Alaska; Bering Sea
ABSTRACT: Several studies have examined murre eggs in the Baltic and Eastern Canada for persistent organic pollutants (POPs), but only one published in 1982 investigated colonies in Alaska where the eggs are consumed as part of some Alaskan Native diets. The current study examines POPs in murre eggs collected in 1999 and 2000 from seven Alaskan colonies. Nine to 11 eggs from each colony were individually cryohomogenized, extracted by pressurized fluid extraction (PFE) techniques, cleaned up with size exclusion and aminopropylsilane LC methods and analyzed by GC-ECD. Results revealed significant geographical differences among the colonies, with higher concentrations of most compounds in Gulf of Alaska murre eggs compared to Bering Sea eggs. Common and thick-billed murres from the same location also showed significant differences. There was a temporal decline in several organochlorine pesticides from the 1970's to 1999-2000. Unanalyzed portions of the cryohomogenized egg tissue will remain in the NIST Charleston specimen bank as part of the Seabird Tissue Archival and Monitoring Project (STAMP) for future analyses.

Vander Pol, Stacy S.; Becker, Paul R.; Kucklick, John R.; Pugh, Rebecca S.; Roseneau, David G.; and Simac, Kristin S. 2004. Persistent organic pollutants in Alaskan murre (*Uria* spp.) eggs: Geographical, species, and temporal comparisons. Environ. Sci. Technol. 38(5):1305-1312.
Rec #: 473
KEYWORDS: organochlorines; PCBs; DDTs; HCB; chlordanes; dieldrin; mirex; egg; common murre; *Uria aalge*; thick-billed murre; *Uria lomvia*; AK; USA; North America; Pacific Ocean; Gulf of Alaska; Bering Sea; spatial variations; temporal trends
ABSTRACT: Concentrations of persistent organic pollutants (POP) in eggs of common and thick-billed murres (*Uria aalge* and *U. lomvia*) from 5 Alaskan nesting colonies were dominated by 4,4'-DDE, total polychlorinated biphenyl (ΣPCB; 46 congeners comprised mainly of PCB congeners 153, 118, 138, 99, 151), hexachlorobenzene (HCB), β-hexachlorocyclohexane (β-HCH), and chlordane compounds (ΣCHL). 4,4'-DDE, cis-nonachlor, and heptachlor epoxide concentrations were lower than those reported for some of the same colonies in the 1970s; HCB concentrations were similar. Generally, significantly higher concentrations were observed in

Annotated Bibliography (Cont.).

eggs from Gulf of Alaska colonies vs. those from Bering Sea colonies except for HCB (higher in the Bering Sea) and β-HCH (no significant difference between the 2 regions). Thick-billed murre eggs contained higher 4,4'-DDE and ΣPCB concentrations; common murre eggs had higher HCB concentrations Possible factors contributing to POP patterns in eggs from these murre colonies are discussed.

Vander Pol, Stacy S.; Christopher, Steven J.; Roseneau, David G.; Becker, Paul R.; Day, Russell D.; Kucklick, John R.; Pugh, Rebecca S.; York, Geoff W.; and Simac, Kristin S. 2003. Seabird Tissue Archival and Monitoring Project: Egg collections and analytical results for 1999 - 2002. Gaithersburg, MD: National Institute of Standards and Technology. NISTIR 7029. 79 pp.
Rec #: 495
KEYWORDS: organochlorines; PCBs; DDTs; HCB; chlordanes; dieldrin; mirex; metals; Hg; egg; common murre; *Uria aalge*; thick-billed murre; *Uria lomvia*; black-legged kittiwake; *Rissa tridactyla*; AK; USA; North America; Gulf of Alaska; Bering Sea; spatial variations; temporal trends
ABSTRACT: The Seabird Tissue Archival and Monitoring Project (STAMP) was implemented in 1999 as a long-term collaborative Alaska-wide effort by the U.S. Fish and Wildlife Service's Alaska Maritime National Wildlife Refuge (USFWS/AMNWR), the U.S. Geological Survey's Biological Resources Division (USGS/BRD), and the National Institute of Standards and Technology (NIST) to monitor long-term trends in environmental quality by banking and analyzing the contents of colonial seabird eggs for contaminants (e.g., chlorinated pesticides, PCBs, mercury). Through 2002, a total of 222 common and thick-billed murre (*Uria aalge* and *U. lomvia*) and black-legged kittiwake (*Rissa tridactyla*) egg clutches were collected from 12 different locations in the Gulf of Alaska and Bering Sea.
Concentrations of persistent organic pollutants (POP) in eggs of common and thick-billed murres from 5 Alaskan nesting colonies were dominated by 4,4'-DDE, total polychlorinated biphenyl (ΣPCB; 46 congeners comprised mainly of PCB congeners 153, 118, 138, 99, 151), hexachlorobenzene (HCB), β-hexachlorocyclohexane (β-HCH), and chlordane compounds (ΣCHL). 4,4'-DDE, cis-nonachlor, and heptachlor epoxide concentrations were lower than those reported for some of the same colonies in the 1970s; HCB concentrations were similar. Generally, significantly higher concentrations were observed in eggs from Gulf of Alaska colonies vs. those from Bering Sea colonies except for HCB (higher in the Bering Sea) and β-HCH (no significant difference between the 2 regions). Thick-billed murre eggs contained higher 4,4'-DDE and ΣPCB concentrations; common murre eggs had higher HCB concentrations Possible factors contributing to POP patterns in eggs from these murre colonies are discussed.
An analytical method using isotope dilution cold vapor inductively coupled plasma mass spectrometry (ID-CV-ICPMS) was developed for the detection of total Hg in the eggs of seabirds. Components including error magnification, verification of method accuracy and assignment of analytical uncertainty are presented in the context of collecting Hg data for single sample aliquots. 41 Egg samples collected from common murre colonies on Little Diomede and Saint George Islands in the Bering Sea and East Amatuli and Saint Lazaria Islands in the Gulf of Alaska yielded Hg mass fraction values ranging from approximately 0.010 $\mu g\ g^{-1}$ to 0.360 $\mu g\ g^{-1}$. Relative expanded uncertainties for the individual detections ranged from 1.2 % to 4.4 %. A one-way analysis of variance including pairwise comparisons across the colonies showed that Hg levels in eggs collected from the Gulf of Alaska colonies were significantly higher than their

Annotated Bibliography (Cont.).

counterparts in the Bering Sea. Hg data from each colony were normally distributed, suggesting a ubiquitous regional deposition of Hg and corresponding incorporation into local food webs.

Vermeer, K. and Peakall, D. B. 1977. Toxic chemicals in Canadian fish-eating birds. Mar. Pollut. Bull. 8(9):205-210.
Rec #: 533
KEYWORDS: organochlorines; DDTs; PCBs; double-crested cormorant; *Phalacrocorax auritus*; herring gull; *Larus argentatus*; common tern; *Sterna hirundo*; egg; Great Lakes; Ontario; Canada; North America; eggshell thickness; reproductive success; effects; spatial variations; biomagnification
NOTES: FIGURES ONLY
ABSTRACT: Cross-country comparison of DDE and PCB residue levels in cormorant, gull and tern eggs in Canada reveal that bird populations at the Great Lakes are most contaminated with those pollutants. DDE levels have been correlated with reproductive failure in Double-crested Cormorants in the Great Lakes with eggshell thinning as a major factor. Low reproductive success in Herring Gull colonies at Lake Ontario is associated with high chlorinated hydrocarbon levels in eggs. Fish-eating birds in the Wabigoon River system, northwestern Ontario, are among the most known mercury contaminated birds. It is suggested that the effects of mercury on the reproduction of fish-eating birds should be further examined there.
Fish-eating birds occupy the highest levels of the food web and magnification of toxic chemicals through prey organisms in this web makes those birds vulnerable to the effects of environmental contaminants. Since fish-eating birds are present everywhere in Canada's freshwater and marine habitats and occupy various niches there, they may serve as pollution indicators in various food chains of our aquatic environment. Colonial birds are especially valuable indicators as pollution effects on total bird populations can be studied. Baseline information on fish-eating bird populations should now be collected everywhere in Canada for measuring present and future effects of environmental pollutants, as well as other man-made disturbances on their populations.

Vermeer, Kees and Reynolds, Lincoln M. 1970. Organochlorine residues in aquatic birds in the Canadian provinces. Can. Field. Nat. 84:117-130.
Rec #: 534
KEYWORDS: organochlorines; DDTs; PCBs; dieldrin; chlordanes; HCHs; herring gull; *Larus argentatus;* mallard; *Anas platyrhynchos;* California gull; *Larus californicus*; Franklin's Gull; *Larus pipixcan;* common tern; *Sterna hirundo*; double-crested cormorant; *Phalacrocorax auritus*; white pelican; *Pelecanus erythrorhynchos*; great blue heron; *Ardea herodias*; black-crowned night-heron; *Nycticorax nycticorax*; western grebe; *Aechmophorus occidentalis*; horned grebe; *Podiceps auritus*; eared grebe; *Podiceps caspicus*; American avocet; *Recurvirostra americana*; coot; *Fulica atra*; Canada goose; *Branta canadensis*; pintail; *Anas acuta*; gadwall; *Anas strepera*; American wigeon; *Anas americana*; blue-winged teal; *Anas discors*; lesser scaup; *Aythya affinis;* Canada; North America; fat; ovary; liver; brain; egg; spatial variations; seasonal variations; effects; reproductive success
ABSTRACT: A survey was conducted of organochlorine residues in 21 aquatic bird species at 31 locations in Alberta, Saskatchewan and Manitoba. As DDT and DDD residue levels from analyses without PCB separation proved to be unreliable, they were omitted from the results. DDE and dieldrin levels were higher in eggs of lairds and fish-eating birds than in those of geese and ducks, presumably reflecting different trophic levels between those two groups of birds.

Annotated Bibliography (Cont.).

Interspecific differences of DDE, dieldrin, and HE, and β BHC residues observed in lairds and fish-eating birds at the same breeding localities may reflect interspecific differences of feeding habits. DDE, dieldrin and PCB levels may be predicted in tissues of California Gull females when known in their eggs. Residue levels in eggs closely resembled those in the livers of females at the time of laying. Shell thickness was significantly and inversely correlated with the concentration of DDE in 40 Great Blue Heron eggs from Alberta, but no significant correlation was found between the concentration of PCBs and shell thickness in those eggs.

Verreault Jonathan; Skaare Janneche Utne; Jenssen, Bjørn Munro; and Gabrielsen, Geir Wing. 2004. Effects of organochlorine pollutants on thyroid hormone levels in Arctic breeding glaucous gulls (*Larus hyperboreus*). Environ. Health Perspect. 112(5):532-537.
Rec #: 518
KEYWORDS: glaucous gull; *Larus hyperboreus;* blood; organochlorines; HCB; chlordanes; DDTs; PCBs; immunology; thyroid hormones; Barents Sea; Norway; Europe; effects; gender differences; spatial variations
ABSTRACT: Studies on glaucous gulls (*Larus hyperboreus*) breeding in the Barents Sea have reported that high blood levels of halogenated organic contaminants in this species might cause reproductive, behavioral, and developmental stress. However, potential endocrine system modulation caused by contaminant exposure has yet not been reported in this Arctic apical predator. In this present study we aimed to investigate whether the current levels of a selection of organochlorines (OCs) were associated with altered circulating levels of thyroid hormones (THs) in free-ranging adult glaucous gulls breeding at Bear Island in the Barents Sea. Blood concentrations of 14 polychlorinated biphenyls, hexachlorobenzene (HCB), oxychlordane, and *p,p*′-dichlorodiphenyldichloroethylene (*p,p*′-DDE) were quantified, in addition to free and total thyroxine (T4) and triiodothyronine (T3), in plasma of 66 glaucous gulls in the spring of 2001. Negative correlations were found between plasma levels of T4 and T4:T3 ratio, and blood levels of OCs in male glaucous gulls. Despite their relatively low contribution to the total OC fraction, HCB and oxychlordane were the most prominent compounds in terms of their negative effect on the variation of the T4:T3 ratio. Moreover, lower T4 levels and T4:T3 ratios were measured in glaucous gulls breeding in a colony exposed to high levels of OCs, compared with a less exposed colony. Levels of T3 were elevated in the high-OC–exposed colony. This may indicate that the glaucous gull is susceptible to changes to TH homeostasis mediated by exposure to halogenated organic contaminants.

Vorkamp, Katrin; Christensen, Jan H.; Glasius, Marianne; and Riget, Frank F. 2004. Persistent halogenated compounds in black guillemots (*Cepphus grylle*) from Greenland--Levels, compound patterns and spatial trends. Mar. Pollut. Bull. 48(1-2):111-121.
Rec #: 476
KEYWORDS: organochlorines; PCBs; chlordanes; toxaphene; PBDEs; DDTs; HCHs; HCB; liver; egg; black guillemot; *Cepphus grylle*; Greenland; North America; spatial variations; age variations; bioaccumulation
ABSTRACT: Twenty-seven black guillemot eggs and 39 livers were analyzed for polychlorinated biphenyls (PCBs), chlorinated pesticides including chlordane-related compounds and toxaphene, and polybrominated diphenylethers (PBDEs). The samples were collected at Qeqertarsuaq (Godhavn, West Greenland) and Ittoqqortoormiit (Scoresbysund, East Greenland). The concentrations of halogenated organic compounds in samples from East Greenland were

Annotated Bibliography (Cont.).

somewhat higher than the corresponding concentrations from West Greenland. Differences in compound patterns were found between West and East Greenland, with higher percentages of the heavier PCB molecules, p,p'-DDE and α-HCH in the samples from Ittoqqortoormiit. Similarly, different levels and different compositions were observed for eggs and livers. The eggs had generally higher concentrations of all compounds as well as higher percentages of CHB-50, CHB-62 and α-HCH than liver samples from the same area. Dividing the liver samples into age groups revealed increasing concentrations with age.

Walker, C. H.; Craven, A. C. C.; and Kurukgy, M. 1975. Metabolism of organochlorine compounds by microsomal enzymes of the shag (*Phalacrocorax aristotelis*). Environ. Physiol. Biochem. 5(1):58-64.
Rec #: 391
KEYWORDS: organochlorines; aldrin; dieldrin; liver; metabolism; Scotland; Europe; shag; *Phalacrocorax aristotelis*; toxicology; enzymes
NOTES: SUPPLEMENTAL
ABSTRACT: The activities of microsomal enzymes of adult male shags (a fish-eating bird) towards the organochlorine substrates aldrin (I) [309-00-2], HCE (II) [21858-40-2], and HEOM (III) [56892-37-6] were compared with those of microsomal enzymes of the adult male Wistar rat. Liver homogenates showed similar epoxide hydrase [9048-63-9] activity to kidney homogenates in the shag, but in the rat the liver preparation was much more active than the kidney preparation. Liver microsomes of the shag showed >8 % of the epoxide hydrase activity and >14 % of the hydroxylating capacity of liver microsomes from the rat. The relatively low activity of these enzymes is probably the main reason why the shag has been found to contain relatively high levels of dieldrin in ecological studies.

Walsh, P. M. 1990. The use of seabirds as monitors of heavy metals in the marine environment. In: Furness, Robert W. and Rainbow, Philip S. (Eds.). Heavy Metals in the Marine Environment. Boca Raton, FL: CRC Press. 183-204.
Rec #: 200
KEYWORDS: review; effects; liver; kidney; muscle; brain; bone; feather; metals; Hg; Cd; Pb; Cu; Zn; penguins; albatross; fulmars; *Fulmarus*; tubenoses; *Pagodroma;* petrels; *Pterodroma;* broad-billed prion; *Pachyptila vittata;* shearwaters; *Puffinus; Calonectris;* storm petrels; brown pelican; *Pelecanus occidentalis*; gannet; *Sula bassana*; cormorants; *Phalacrocorax*; common eider*; Somateria mollissima*; skuas; *Catharacta*; gulls; *Larus; Rissa*; terns; *Sterna*; Netherlands; Europe; SC; USA; North America; spatial variations; temporal trends; seasonal variations; age variations; bioaccumulation; elimination; body burden
NOTES: NO TABLES
ABSTRACT: The review aims to draw attention to factors which need to be taken into account in designing programmes to monitor geographical or temporal variations in marine levels of heavy metals using seabird tissues. The following topics are discussed: metabolic regulation of metals; choice of species -- taxonomic and ecological factors; variation of metal concentrations between tissues; physiological and seasonal influences; age-related variation; correlations between metals, and evidence for detoxification mechanisms; dose-response relationships; and, current evidence for geographical variation in exposure of seabirds.

Wayland, M.; Garcia-Fernandez, A. J.; Neugebauer, E.; and Gilchrist, H. G. 2001.

Annotated Bibliography (Cont.).

Concentrations of cadmium, mercury, and selenium in blood, liver, and kidney of common eider ducks from the Canadian Arctic. Environ. Monit. Assess. 71(3):255-267.
Rec #: 466
KEYWORDS: metals; Hg; Se; Cd; organometallics; organic Hg; liver; kidney; blood; common eider; *Somateria mollissima*; Canada; North America
ABSTRACT: We detected concentrations of selected trace elements in livers, kidneys, and blood samples from common eiders (*Somateria mollissima borealis*) from the eastern Canadian arctic during 1997 and 1998. Concentrations of total Hg and organic Hg were generally low in the livers of these birds (less than 6 and 4 µg g-1 dry weight, respectively). Se ranged between II-47 $µg\ g^{-1}$ in livers. Renal Cd concentrations were among the highest ever published for this species (range: 47-281 $µg\ g^{-1}$). The regressions of log-transformed concentrations of these trace elements in blood samples on those in liver or kidney were significant (all P-values < 0.05) and positive. However, except for organic Hg ($R^2 = 0.83$), the co-efficients of detection were low to moderate (range of R^2: 0.26-0.52), suggesting poor to moderate predictive capability. Furthermore, the relationships between total Hg in blood and liver changed between 1997 and 1998, suggesting that it would not be possible to predict consistently, concentrations of Hg in blood from those in liver based on samples taken in one year. Blood samples can be used to detect concentrations of these trace elements in common eiders (and probably other sea duck species as well). The use of blood samples is especially warranted when it is undesirable to kill the animal such as when working with rare or endangered sea duck species or when the objective is to relate trace element exposure to annual survival rates. However, the predictive equations developed here should not be used to predict expected concentrations in one type of tissue from those in the other.

Wayland, M.; Gilchrist, H. G.; Dickson, D. L.; Bollinger, T.; James, C.; Carreno, R. A.; and Keating, J. 2001. Trace elements in king eiders and common eiders in the Canadian Arctic. Arch. Environ. Contam. Toxicol. 41(4):491-500.
Rec #: 465
KEYWORDS: metals; Se; Hg; Cu; Zn; Cd; organometallics; organic Hg; liver; kidney; common eider; *Somateria mollissima*; king eider; *Somateria spectabilis*; liver; kidney; Nunavut; Northwest Territories; Canada; North America; toxicology; parasites; effects; histopathology
ABSTRACT: The authors detected the concentrations of selected trace elements in tissues of king and common eiders at 3 locations in the Canadian Arctic. Renal and hepatic cadmium concentrations in king eiders at a location in the eastern Arctic were among the highest ever recorded in eider ducks: there, they were higher in king eiders than in common eiders. Cadmium concentrations were lower in king eiders from the western Arctic than in those from the east. In the western Arctic, cadmium concentrations did not differ between species. Hepatic mercury and zinc were higher in king eiders than in common eiders. Zinc and selenium were higher in eiders from the western Arctic than in those from the eastern Arctic. Trace element concentrations in these 2 duck species were below published toxicity thresholds. Positive correlations in trace element concentrations in both species were found between total and organic hepatic mercury, renal and hepatic cadmium as well as hepatic zinc, copper, mercury, and cadmium. Body mass of common but not king eiders and spleen mass of both species were negatively correlated with mercury concentrations. In common eiders, the number of nematode parasites was positively correlated with total and organic mercury. Histopathological evidence

Annotated Bibliography (Cont.).

of kidney or liver lesions that are typical of trace metal poisoning was not found. The authors did not find evidence to support the hypothesis that trace metal exposure may be contributing to adverse effects on the health of individuals of these species.

Wayland, M.; Gilchrist, H. G.; Marchant, T.; Keating, J.; and Smits, J. E. 2002. Immune function, stress response, and body condition in Arctic-breeding common eiders in relation to cadmium, mercury, and selenium concentrations. Environ. Res. 90(1):47-60.
Rec #: 463
KEYWORDS: metals; Cd; Hg; Se; common eider; *Somateria mollissima*; liver; kidney; Nunavut; Canada; North America; biomarkers; immunology; effects; gender differences
ABSTRACT: We examined relationships between trace metal concentrations in tissues of common eider ducks (cadmium, mercury, and selenium) and selected biomarkers of health (stress response, immune function, and body condition). This study was conducted at an eider nesting colony in the Canadian arctic in 1998 and 1999. Capture-induced stress, measured as the rise in corticosterone concentrations following capture, was positively related (P=0.03) to renal cadmium concentration in 1998 when incubating eiders were sampled, but not in 1999 when prenesting eiders were sampled. Stress response was inversely related (P=0.02) to selenium concentrations in 1999. Following capture and blood sampling in 1999, eiders were placed in a flight pen on-site for eight days in order to examine immune function. Cell-mediated immunity, measured as the skin-swelling response to an intradermal injection of phytohemagglutinin-P, (PHA-P), was positively related (P=0.003) to hepatic selenium. The heterophil:lymphocyte ratio was inversely related (P=0.08) to hepatic selenium. In 1998, selenium was positively related to body mass (P=0.01), abdominal fat mass (P=0.07), kidney mass (P=0.03), and liver mass (P=0.07). In 1999, hepatic mercury was negatively related to abdominal fat mass (P=0.01), spleen mass (P=0.07) and body mass at capture (P=0.09) in prenesting eiders.

Wayland, Mark; Hobson, Keith A.; and Sirois, Jacques. 2000. Environmental contaminants in colonial waterbirds from Great Slave Lake, NWT: Spatial, temporal and food-chain considerations. Arctic. 53(3):221-233.
Rec #: 349
KEYWORDS: herring gull; *Larus argentatus*; common gull; *Larus canus*; black tern; *Chlidonias niger;* Caspian tern; *Sterna caspia;* egg; liver; Northwest Territories; Canada; North America; metals; Hg; Se; organochlorines; DDTs; chlordanes; PCBs; stable isotopes; biomagnification; temporal trends; spatial variations; seasonal variations
ABSTRACT: Great Slave Lake in the Northwest Territories, Canada, differs regionally in trophic status and local and regional inputs of contaminants. Spatial and temporal trends in contaminant levels in bioindicator species such as colonial waterbirds could offer insights into the potential for contaminant bioaccumulation in Great Slave Lake. Persistent chlorinated hydrocarbon contaminants, mercury (Hg), and selenium (Se) were examined in herring gull (*Larus argentatus*) eggs and livers collected from various locations on Great Slave Lake in 1995. Eggs were collected in May and June, and livers in May and August. Also, the relationship between contaminants and trophic level, as inferred from stable-nitrogen isotope analysis (delta 15N), was examined in four colonial waterbird species: herring gull, mew gull (*L. canus*), Caspian tern (*Sterna caspia*), and black tern (*Chlidonias niger*). Finally, the co-accumulation of mercury and selenium was examined in eggs of these birds. There were no differences in chlorinated hydrocarbon concentrations among four sampling sites (colonies). Concentrations

Annotated Bibliography (Cont.).

did not differ between herring gull adults collected in early May and those collected in early August. Chlorinated hydrocarbon concentrations in eggs of herring gull, mew gull, Caspian tern, and black tern were related to their trophic positions as inferred from their delta 15N values in their lipid-free egg yolks. Concentrations in these colonial waterbirds were much higher than those in fish from Great Slave Lake, but lower than those in their conspecifics from the Great Lakes. It is probable that a relatively large proportion of the chlorinated hydrocarbon contaminant load in colonial waterbird eggs on Great Slave Lake results from exposure to and storage of such contaminants at more heavily contaminated wintering and staging areas. This possibility limits the usefulness of colonial waterbirds as indicators of chlorinated hydrocarbon bioaccumulation in Great Slave Lake. Selenium and mercury concentrations in herring gull eggs differed significantly among the four breeding colonies, and concentrations in adults declined between May and August. Selenium and mercury were positively correlated in eggs of all species.

Wayland, Mark; Smits, Judit J. E. G.; Gilchrist, H. Grant; Marchant, Tracy; and Keating, Jonathan. 2003. Biomarker responses in nesting, common eiders in the Canadian Arctic in relation to tissue cadmium, mercury and selenium concentrations. Ecotoxicology. 12(1-4):225-237.
Rec #: 462
KEYWORDS: metals; Cd; Hg; Se; common eider; *Somateria mollissima*; liver; kidney; Nunavut; Canada; North America; biomarkers; toxicology; retinol; vitamin A; immunology; effects
ABSTRACT: Populations of many North American sea ducks are declining. Biomarkers may offer valuable insights regarding the health and fitness of sea ducks in relation to contaminant burdens. In this study the authors examined body condition, immune function, corticosterone stress response, liver glycogen levels and vitamin A status in relation to tissue concentrations of Hg, Se and Cd in female common eiders during the nesting period. The study was conducted in the eastern Canadian arctic during July, 2000. Hepatic Hg, Se and renal Cd concentrations ranged 1.5-9.8, 6.5-47.5 and 74-389 µg/g, dry weight, respectively Hg concentrations were negatively related to dissection body mass, heart mass and fat mass. Cd concentrations were negatively related to mass at capture and dissection mass after controlling for the Hg concentration-dissection mass relationship. Cell-mediated immunity was assessed by the skin swelling reaction to an injection of phytohemagglutinin-P, and was unrelated to metal concentrations. After adjusting the corticosterone concentration to account for the time between capture and sampling, there was a negative relationship between the residual corticosterone concentration and Se. Liver glycogen concentrations were not significantly related to metal concentrations. Hg concentrations were positively related to those of hepatic retinol and retinyl palmitate and the ratio of the retinol to retinyl palmitate in liver. They were negatively related to the ratio of plasma to liver retinol. Our findings do not indicate that exposure to metals may have adversely affected the health of these birds. They do, however, suggest that more research is required to elucidate mechanisms by which exposure to these metals could impact body condition.

Wenzel, C. and Adelung, D. 1996. The suitability of oiled guillemots (*Uria aalge*) as monitoring organisms for geographical comparisons of trace element contaminants. Arch. Environ. Contam. Toxicol. 31(3):368-377.

Annotated Bibliography (Cont.).

Rec #: 238
KEYWORDS: metals; Cd; Hg; Se; Cu; Zn; Germany; Brittany; Europe; common murre; *Uria aalge*; liver; kidney; feather; muscle; lung; fat; spatial variations; bioaccumulation; age variations
ABSTRACT: The influence of the nutritional state (condition factor), age, and sex of Common Guillemots (*Uria aalge*) on trace element levels was examined to investigate the validity of geographical comparisons of metal accumulations in dying and dead oiled seabirds. A quotient of liver mass to kidney mass was calculated as a condition factor. Condition factors ranged from 0.85-4.74. Sex did not alter the distribution of any of the elements analyzed in immature or adult birds. Cadmium concentrations in soft tissues were strongly influenced by the age of the birds, with adult birds containing significantly higher amounts (0.025-88.28 ppm) than immature ones (0.004-17.9 ppm). When simultaneously considering age and nutritional condition of the birds, selenium levels were highest in liver samples in immature guillemots (7.37-41.27 ppm) as compared to adult birds (7.22-36.84 ppm). Mercury, copper, and zinc were independent of age. The condition factor had no effect on cadmium levels in birds. In contrast, copper and zinc concentrations in kidney, liver and feathers were negatively correlated with the nutritional condition. In the case of mercury a similar relationship was found only in feathers. On the basis of these results guillemots collected in the German Bight showed significantly higher mercury values (0.81-20.87 ppm) in soft tissues than birds collected in Brittany (0.85-17.95 ppm). In general, cadmium levels were higher in the liver of immature birds from the German Bight than in those from Brittany. With regard to copper and zinc only kidney samples had higher values in the German Bight than in Brittany. Selenium levels were lower in feathers of birds from the German Bight (0.09-2.20 ppm) than in those from Brittany (0.67-6.64 ppm). The results indicate that beached guillemots can be used to monitor geographical differences in the contamination with certain metals provided that birds of the same age and/or nutritional condition are compared.

Wenzel, C. and Gabrielsen, G. W. 1995. Trace element accumulation in three seabird species from Hornøya, Norway. Arch. Environ. Contam. Toxicol. 29(2):198-206.
Rec #: 45
KEYWORDS: Norway; Europe; bioaccumulation; metals; toxicology; Hg; Se; Cu; Cd; Zn; feather; liver; kidney; muscle; gonad; lung; feather; common murre; *Uria aalge;* thick-billed murre; *Uria lomvia;* black-legged kittiwake; *Rissa tridactyla*; effects
ABSTRACT: Soft tissues and body feathers of Common Guillemots (*Uria aalge*) Brünnich's Guillemots (*Uria lomvia*), and Kittiwakes (*Rissa tridactyla*) collected at Hornøya (northern Norway) were analyzed for total mercury, selenium, cadmium, zinc, and copper. Kittiwakes revealed highest cadmium and mercury concentrations in most tissues compared to both guillemot species, whereas interspecific differences in concentrations of essential elements were less obvious. The results are discussed in relation to feeding habits and migration patterns of the different seabird species. Age-dependent accumulation of selenium, mercury, and cadmium was clearly recognizable when comparing trace element contents in fully-fledged Kittiwakes to adult specimens. There was no evidence that the analyzed birds suffered from acute toxicities of heavy metals.

Wenzel, Christine; Adelung, Dieter; and Theede, Hans. 1996. Distribution and age-related changes of trace elements in kittiwake *Rissa tridactyla* nestlings from an isolated colony in the

Annotated Bibliography (Cont.).

German Bight, North Sea. Sci. Total Environ. 193(1):13-26.
Rec #: 80
KEYWORDS: black-legged kittiwake; *Rissa tridactyla*; North Sea; Germany; Europe; liver; kidney; feather; brain; metals; Hg; Se; Cd; Cu; Zn; age variations; bioaccumulation; biomagnification
ABSTRACT: Tissue distribution of five trace elements (Se, Hg, Cd, Zn, Cu) was investigated in soft tissues and feathers of Kittiwake (*Rissa tridactyla*) nestlings from the Island of Helgoland, North Sea. The tissue distribution of metals was similar in all age classes. Feathers and down contained highest mercury and zinc concentrations, whereas elevated levels of selenium and cadmium were found in the kidney. The concentrations of cadmium, copper and zinc in liver and kidney were low in hatchlings and increased with age of the nestlings, indicating the importance of the ingestion of contaminated food during chick growth. In the case of mercury, concentrations were high shortly after hatching and decreased when the chicks grew older. This indicates, that egg contamination was more important in chicks than contaminated food items. The total liver burden of all trace elements increased throughout chick development. Significant positive correlations between essential and non-essential elements were found in liver and/or kidney of all age classes. However, only the correlations between zinc and copper levels maintained throughout chick growth. The number of interactions between elements were increasing with progressing chick age. Correlations between cadmium and the essential elements copper and zinc appeared to depend on contaminant levels of cadmium, whereas mercury was correlated to essential elements in low and highly contaminated chicks. It was demonstrated that particularly older chicks (greater than or equal to 6 days old) were reliable bioindicators of mercury and cadmium contaminations around Helgoland Island.

Weseloh, D. V. and Struger, J. 1986. Organochloride contaminants and mercury in herring gull eggs from the Great Lakes, 1983-84. IAGLR-86 Program. International Association for Great Lakes Research 29th Conference, May 26-29, 1986. 51.
Rec #: 248
KEYWORDS: organochlorines; DDTs; chlordanes; HCHs; HCB; PCBs; dieldrin; endrin; benzenes; mirex; metals; Hg; egg; herring gull; *Larus argentatus*; North America; Great Lakes
NOTES: ABSTRACT ONLY - NOT AVAILABLE
ABSTRACT: Herring Gull (*Larus argentatus*) eggs from 13 colony sites from throughout the 5 Great Lakes and connecting channels were analyzed for 18 organic contaminants and mercury in 1983/84. The highest levels of eight different compounds: DDE (6.4 mg/kg wet weight), dieldrin (0.64), heptachlor epoxide (0.30), DDT (0.06), alpha-chlordane (0.04), oxychlordane (0.43), endrin (0.04) and beta-HCH (0.005), occurred in gull eggs at Big Sister Island, Green Bay, Lake Michigan. The highest levels of tetra-(0.35) and pentachlorobenzene (0.07) were found in eggs from Saginaw Bay, Lake Huron. PCBs (85.0) and HCB (0.21) were greatest in eggs from Fighting Island, Detroit River. Colonies in Lake Ontario had the highest levels of mirex (2.0), beta-HCH (0.04) and mercury (.033 mg/kg wet weight).

Weseloh, D. V. Chip; Ewins, Peter J.; Struger, John; Mineau, Pierre; and Norstrom, Ross J. 1994. Geographical distribution of organochlorine contaminants and reproductive parameters in herring gulls on Lake Superior in 1983. Environ. Monit. Assess. 29(3):229-251.
Rec #: 235
KEYWORDS: effects; egg; herring gull; *Larus argentatus*; organochlorines; DDTs; HCB;

Annotated Bibliography (Cont.).

mirex; PCBs; chlordanes; dieldrin; PCDDs; Great Lakes; North America; spatial variations; eggshell thickness; reproductive success; temporal trends

ABSTRACT: As part of the Great Lakes International Surveillance Plan,1978-1983, egg contaminant levels and reproductive output were determined for Herring Gull colonies on Lake Superior in 1983. Since 1974, the Herring Gull has been widely used in the Great Lakes as a spatial and temporal monitor of organochlorine (OC) contaminant levels and associated biological effects. Most eggs contained a wide range of OCs, the main compounds being DDE, polychlorinated biphenyls (PCBs), dieldrin, heptachlor epoxide, oxychlordane, hexachlorobenzene and mirex. Levels of an additional ten OCs and five polychlorinated dibenzo-p-dioxin (PCDD) congeners were also determined for some sites. Overall, levels varied significantly among colonies, but there was no obvious relationship to spatial distribution of contaminants in sediments or fish species. OC levels in eggs had declined by up to 84 % since 1974. Eggshells were only 8 % thinner than before the introduction of DDT, and shell thinning was not a cause of breeding failure. Average reproductive output varied from 0.15 to 1.57 young per apparently occupied nest in 1983: at 56 % of colonies the value was below that thought necessary to maintain stable populations. The main causes of failure were egg disappearance and cannibalism of chicks. Despite this, the population appeared to have been increasing at about 4 % per annum. Reduced availability of forage fish during the early 1980s was the most likely reason for the poor reproductive output in 1983.

Weseloh, D. V. Chip; Hamr, Premek; Bishop, Christine A.; and Norstrom, Ross J. 1995. Organochlorine contaminant levels in waterbird species from Hamilton Harbor, Lake Ontario: An IJC area of concern. J. Great Lakes Res. 21(1):121-137.
Rec #: 446
KEYWORDS: double-crested cormorant; *Phalacrocorax auritus*; herring gull; *Larus argentatus*; black-crowned night-heron; *Nycticorax nycticorax*; ring-billed gull; *Larus delawarensis*; common tern; *Sterna hirundo*; Caspian tern; *Sterna caspia*; Canada goose; *Branta canadensis*; bufflehead; *Bucephala albeola*; gadwall; *Anas strepera;* greater scaup; *Aythya marila*; lesser scaup; *Aythya affinis*; mallard; *Anas platyrhynchos*; egg; Great Lakes; Canada; North America; organochlorines; chlordanes; benzenes; HCB; DDTs; dieldrin; HCHs; mirex; OCS; PCBs; PCDDs; PCDFs; metals; Hg; temporal trends; spatial variations; biomagnification
ABSTRACT: The levels of organochlorine (OC) contaminants in eggs and tissues of waterbird species nesting in Hamilton Harbor, Ontario, an International Joint Commission Area of Concern, were monitored between 1981 and 1992. PCBs, DDE, and mirex were present at the highest concentrations of the 29 organochlorines and one trace metal measured. Most contaminants in the various species tested showed a declining temporal pattern during that period. Double-crested cormorant eggs generally had higher concentrations of contaminants than those in herring gull eggs which in turn were higher than those in eggs of black-crowned night-herons, Caspian terns and common terns. The levels of contaminants in the eggs reflect the dietary preferences of the species with the fish-eating birds containing consistently higher accumulations of contaminants when compared to the herbivorous Canada goose for which the levels were consistently much lower for all compounds. The concentrations of contaminants detected are among some of the highest in the Great Lakes but when compared to other sites on Lake Ontario, the levels in Hamilton Harbor are generally equal or lower. Elevated levels of contaminants such as PCBs, mirex, and DDE were also detected in liver and muscle tissues of migrant waterfowl species from the harbor. Despite habitat degradation and continued presence

Annotated Bibliography (Cont.).

of contaminants, the harbor supports a large number and wide variety of waterbird species. Except for black-crowned night-herons, the nesting populations of colonial waterbirds have increased between 21 and 1061 % since the last survey in 1987.

Weseloh, D. Vaughn; Hughes, Kimberly D.; Ewins, Peter J.; Best, Dave; Kubiak, Timothy; and Shieldcastle, Mark C. 2002. Herring gulls and great black-backed gulls as indicators of contaminants in bald eagles in Lake Ontario, Canada. Environ. Toxicol. Chem. 21(5):1015-1025.

Rec #: 424
KEYWORDS: bald eagle; *Haliaeetus leucocephalus;* great black-backed gull; *Larus marinus*; herring gull; *Larus argentatus*; organochlorines; DDTs; PCBs; mirex; chlordanes; dieldrin; HCB; egg; Great Lakes; Canada; North America; effects; reproductive success
ABSTRACT: In 2000, a pair of bald eagles (*Haliaeetus leucocephalus*) nested successfully along the shoreline of Lake Ontario in North America for the first time since 1957; however, it is a continuing question whether bald eagles will be able to reproduce successfully as they return to nest on Lake Ontario. Great black-backed gulls (*Larus marinus*) and herring gulls (*L. argentatus*) were selected as surrogate species to predict pollutant concentrations in eggs of bald eagles nesting on Lake Ontario. Due to suspected overlap in the diets of great black-backed gulls and bald eagles (i.e., fish, gull chicks, waterfowl), the 2 species probably occupy a similar trophic level in the Lake Ontario food web, thus, may have similar pollutant concentrations. Fresh great black-backed gull and herring gull eggs were collected from 3 sites in eastern Lake Ontario in 1993 and 1994 and analyzed for contaminants. Average p,p'-DDE, total polychlorinated biphenyls (PCB), and dieldrin concentrations in great black-backed gull eggs were 12.85, 26.27, and 0.27 µg/g, respectively The mean ratio of contaminant concentrations in great black-backed gull eggs to concentrations in herring gull eggs for these 3 pollutants was 2.09 (range of means, 1.73-2.38). Predicted pollutant concentrations in bald eagle eggs in Lake Ontario would be expected to be similar to mean concentrations reported for great black-backed gull eggs. As a comparison, contaminant concentrations in bald eagle eggs collected from other Great Lakes nesting sites were compared to mean concentrations reported for herring gull eggs collected from nearby sites in 1986 to 1995. The mean ratio of contaminant concentrations in bald eagle eggs to those in herring gull eggs from these sites for DDE, total PCB, and dieldrin was 2.40 (range of means, 1.73-3.28). These ratios are very similar to those reported using great black-backed gull eggs, illustrating the apparent similarity in trophic status shared by the 2 top predator species at these Great Lakes sites. Predicted contaminant concentrations in bald eagle eggs at Lake Ontario were similar to concentrations reported for bald eagles breeding at other Great Lakes sites, suggesting bald eagles may be able to breed on the shores of Lake Ontario. However, it is unclear what level of breeding success should be expected, given that productivity at other similarly polluted Great Lakes sites may be below that required to sustain a successful breeding population. The absence of an inland bald eagle population from which bald eagles may begin to colonize Lake Ontario shorelines may be delaying initiation of nesting site selection; other factors, e.g., habitat and prey availability, would likely not limit reproductive success.

Whitehead, P. E.; Harfenist, A.; Elliott, J. E.; and Norstrom, R. J. 1992. Levels of polychlorinated dibenzo-p-dioxins and polychlorinated dibenzofurans in waterbirds of Howe Sound, British Columbia. In: Levings, C. D.; Turner, R. B.; Ricketts, B. (Eds.) Proceedings of

Annotated Bibliography (Cont.).

the Howe Sound Environmental Science Workshop., Can. Tech. Rep. Fish. Aquat. Sci. No. 1879. 229-238.
Rec #: 183
KEYWORDS: egg; liver; muscle; double-crested cormorant; *Phalacrocorax auritus;* common merganser; *Mergus merganser;* western grebe; *Aechmophorus occidentalis*; surf scoter; *Melanitta perspicillata;* harlequin duck; *Histrionicus histrionicus;* long-tailed duck; *Clangula hyemalis*; common goldeneye; *Bucephala clangula*; Canada; British Columbia; North America; organochlorines; PCDDs; PCDFs; PCBs; biomagnification
ABSTRACT: In 1988 - 1990, eggs from Double-crested Cormorants (*Phalacrocorax auritus*) and livers and breast muscles of six species of diving ducks were collected from Howe Sound, B.C., as part of a Canadian Wildlife Service program to monitor contaminants in coastal habitats. PCB contamination of cormorant eggs was significantly lower in 1990 than in 1988. The levels of PCDDs and PCDFs in diving ducks varied considerably among species and between liver and muscle of the same species. 2,3,7,8-TCDF was the only contaminant found in all samples analyzed. Fish-eating species were the most contaminated and, for a given species, the liver contained higher concentrations of dioxins and furans than did the breast muscle. A health advisory regarding the consumption of livers from Western Grebes (*Aechmophorus occidentalis*) and Common Mergansers (*Mergus mergus*) from Howe Sound has been issued by Health and Welfare Canada.

Wideqvist, Ulla; Jansson, Bo; Reutergårdh, Lars; Olsson, Mats; Odsjö, Tjelvar; and Uvemo, Ulla-Britt. 1993. Temporal trends of PCC in guillemot eggs from the Baltic. Chemosphere. 27(10):1987-2001.
Rec #: 297
KEYWORDS: organochlorines; toxaphene; common murre; *Uria aalge*; Sweden; Baltic Sea; Europe; egg; temporal trends
ABSTRACT: Pooled samples of homogenized Guillemot eggs from 1974, 1978, 1982 and 1987 and ten individual eggs for each of the two years, 1976 and 1989 have been analyzed for PCC with GC-ECD and GC-MS in the ECNI mode. Regression analysis on PCC concentrations versus sampling year, using the results obtained in the GC-ECD analyses based on pooled egg samples and the mean values of the egg samples from 1976 and 1989 respectively, revealed a significant ($p<0.01$) negative trend. Student t-test applied on the two annual egg samples from 1976 and 1989 showed significant ($p<0.01$) lower concentrations in 1989, the mean concentration decreasing from 14 mg/kg lipid weight to 6 mg/kg. Concentrations obtained in the GC-MS analyses were higher, especially for the early years and the results confirmed the downward trend.

Wiens, J. A; Ford, G.; Heinemann, D.; and Pietruszka, C. 1980. Simulation Modeling of Marine Bird Population Energetics, Food Consumption, and Sensitivity to Perturbation. Environmental Assessment of the Alaskan Continental Shelf. Annual Reports of Principal Investigators for the Year Ending March 1980. NOAA/OMPA, Boulder, CO. NOAA-OMPA-AR-80-1. 1:1-93.
Rec #: 240
KEYWORDS: review; oil; effects; black-legged kittiwake; *Rissa tridactyla*; glaucous-winged gull; *Larus glaucescens*; common murre; *Uria aalge*; sooty shearwater; *Puffinus griseus*; tufted puffin; *Fratercula cirrhata*; AK; USA; North America; energetics; reproductive success
NOTES: SUPPLMENTAL

Annotated Bibliography (Cont.).

ABSTRACT: The sensitivity and response of breeding marine birds in the Kodiak area to environmental perturbations such as might accompany petroleum developments were examined. This approach combines the use and analysis of field observations on the distributional patterns of the birds and their life history and reproductive attributes with analytic and simulation modeling of population foraging distributions and behavior and population demography. Throughout, patterns of response are organized within the framework of energy flow patterns. For purposes of modeling and potential impacts of offshore development and providing a distributional baseline for management decisions, these data are synthesized into distributional maps for each of the five major species found in this region: Black-legged Kittiwakes (*Rissa tridactyla*), Glaucous-winged Gulls (*Larus glaucescens*), murres (principally Common Murres, *Uria aalge*), Sooty Shearwaters (*Puffinus griseus*), and Tufted Puffins (*Lunda cirrhata*).

Wiens, J. A.; Ford, R. G.; and Heinemann, D. 1984. Information Needs and Priorities for Assessing the Sensitivity of Marine Birds to Oil Spills. Biol. Conserv. 28(1):21-49.
Rec #: 239
KEYWORDS: review; oil; effects; thick-billed murre; *Uria lomvia*; black-legged kittiwake; *Rissa tridactyla*; tufted puffin; *Fratercula cirrhata;* AK; USA; North America; reproductive success; energetics
NOTES: SUPPLEMENTAL
ABSTRACT: Experience in developing models to predict the potential impacts of oil spills on colonially breeding marine birds has revealed some major gaps in the information available on these systems. The authors consider the availability of data for a variety of parameters of seabird biology that are required in modeling efforts, and assign provisional priorities to information needs. In order to develop means of predicting the impacts of oil spills on seabirds, they suggest that colony- or site-specific information on the timing of reproduction and colony occupancy, chick growth rates and body weights, several metabolic parameters, flight speed, and food load size is of relatively low overall priority. They suggest that studies of seabird biology should give highest priority to obtaining information on population sizes, the probability of adult death upon encountering a spill, age-specific fecundity and survivorship, the time required in foraging trips, the lag time in the response of birds to an oil spill, foraging rate as a function of resource density, and changes in the availability of resources to the birds as a consequence of oil spills.

Wiens, John A. 1996. Oil, seabirds, and science. The effects of the Exxon Valdez oil spill. Bioscience. 46(8):587-597.
Rec #: 245
KEYWORDS: oil; AK; USA; North America; effects; common murre; *Uria aalge;* thick-billed murre; *Uria lomvia*; pigeon guillemot; *Cepphus columba*; marbled murrelet; *Brachyramphus marmoratus*; black oystercatcher; *Haematopus bachmani*; harlequin duck; *Histrionicus histrionicus*; scoters; *Melanitta*; loons; *Gavia*; common gull; *Larus canus*; Arctic tern; *Sterna paradisaea*; black-legged kittiwake; *Rissa tridactyla*; shag; *Phalacrocorax aristotelis*; black guillemot; *Cepphus grylle*; reproductive success
NOTES: SUPPLEMENTAL
ABSTRACT: When the supertanker Exxon Valdez ran aground on Bligh Reef in Prince William Sound, Alaska, on the morning of 24 March 1989, it aroused widespread concern about possible environmental devastation. Within hours, some 41 million liters of crude oil were released into the marine ecosystem, making this spill the largest in US history. Eventually, oil

Annotated Bibliography (Cont.).

was found more than 900 km from the spill site, and roughly 2100 km of shoreline were contaminated with oil. Reports of mortality of marine birds were immediate, and images of oiled seabirds figured prominently in media coverage of the spill. Within a few months, more than 30,000 oiled carcasses had been retrieved from the water and beaches in the spill area, and estimates of overall mortality were substantially greater. Studies of the effects of the Exxon Valdez oil spill on seabirds were initiated shortly following the spill by researchers working for the State of Alaska and several federal agencies (the "Trustees") or supported by Exxon. By mid-summer 1989, however, litigation became a priority, and as a result the two groups conducted their studies separately, each with little knowledge of what the other group was doing. Reports of many of these studies have now been made public, and enough information is available to develop a general assessment of how the spill affected seabirds and whether the initial concerns were justified. The studies also provide some insights into how the scientific process itself may be affected by such well-publicized environmental accidents and into the relationships among preconceptions, advocacy, and science.

Wiens, John A. 1995. Recovery of seabirds following the Exxon Valdez oil spill: An overview. In: Wells, P.G.; Butler, J.N.; Hughes, J.S. (Eds.). Exxon Valdez Oil Spill: Fate and Effects in Alaskan Waters. Philadelphia, PA: American Society for Testing and Materials. 854-893.
Rec #: 241
KEYWORDS: review; AK; USA; North America; oil; effects; common murre; *Uria aalge*; thick-billed murre; *Uria lomvia*; northern fulmar; *Fulmarus glacialis*; black-legged kittiwake; *Rissa tridactyla*; shag; *Phalacrocorax aristotelis*; Atlantic puffin; *Fratercula arctica*; little auk; *Alle alle*; reproductive success
NOTES: SUPPLEMENTAL
ABSTRACT: Assessing oil-spill effects requires rigorous definitions of "impact" and "recovery." Impact is defined as a statistically significant difference between samples exposed to oil and reference samples. Recovery is then the disappearance through time of such a statistical difference. Both impact and recovery must be assessed in relation to the background of natural variation that characterizes marine environments. There are three primary avenues of potential spill impacts on seabirds: on population size and structure, on reproduction, and on habitat occupancy and use. Detecting oil-spill effect involves comparisons of (1) observations taken following the spill with prespill data; (2) data gathered following the spill from oiled areas ("treatments") and unoiled areas ("controls") surveyed at the same time; or (3) measurements taken from sites along a gradient of oiling magnitude. The strengths and weaknesses of these approaches are discussed. In many situations, the third approach may be most useful. Following the Exxon Valdez oil spill in March 1989, over 35 000 dead birds were retrieved. Model analyses suggested that actual seabird mortality could have been in the hundreds of thousands, prompting concerns about severe and persistent impacts on populations of several species, especially murres (Uria spp.). Recovery for some populations was projected to take decades. The findings of several studies conducted following the oil spill, however, indicate that these concerns may not be justified. These studies examined colony attendance and reproduction of murres as well as habitat utilization for the prevalent species in Prince William Sound and along the Kenai Peninsula. Surveys of attendance by birds at murre breeding colonies in 1991 indicated no overall differences from prespill attendance levels when colonies were grouped by the degree of oiling in the vicinity. At a large colony in the Barren Islands, where damage was described as especially severe, counts of murres were generally similar to historical estimates made in the late

Annotated Bibliography (Cont.).

1970s.

Wiens, John A.; Crist, Thomas O.; Day, Robert H.; Murphy, Stephen M.; and Hayward, Gregory D. 1996. Effects of the Exxon Valdez oil spill on marine bird communities in Prince William Sound, Alaska. Ecol. Appl. 6(3):828-841.
Rec #: 249
KEYWORDS: AK; USA; North America; oil; effects; Arctic tern; *Sterna paradisaea*; Bonaparte's gull; *Larus philadelphia*; pigeon guillemot; *Cepphus columba;* double-crested cormorant; *Phalacrocorax auritus*; pelagic cormorant; *Phalacrocorax pelagicus*; red-faced cormorant; *Phalacrocorax urile*; horned grebe; *Podiceps auritus*; red-necked grebe; *Podiceps grisegena*; yellow-billed loon; *Gavia adamsii*; red-throated loon; *Gavia stellata*; Pacific loon; *Gavia pacifica*; common loon; *Gavia immer;* common murre; *Uria aalge*; tufted puffin; *Fratercula cirrhata*; horned puffin; *Fratercula corniculata*; black-legged kittiwake; *Rissa tridactyla*; fork-tailed storm petrel; *Oceanodroma furcata*; parakeet auklet; A*ethia psittacula*; rhinoceros auklet; *Cerorhinca monocerata*; ancient murrelet; *Synthliboramphus antiquus*; long-tailed duck; *Clangula hyemalis*; red-necked phalarope; *Phalaropus lobatus*; Kittlitz's murrelet; *Brachyramphus brevirostris*; marbled murrelet; *Brachyramphus marmoratus*; parasitic jaeger; Stercorarius parasiticus; pomarine jaeger; *Stercorarius pomarinus*; long-tailed jaeger; *Stercorarius longicaudus*; red-breasted merganser; *Mergus serrator*; common raven; *Corvus corax*; black-billed magpie; *Pica pica;* sharp-shinned hawk; *Accipiter striatus*; Steller's jay; *Cyanocitta stelleri*; northwestern crow; *Corvus caurinus*; least sandpiper; *Calidris minutilla*; harlequin duck; *Histrionicus histrionicus*; black scoter; *Melanitta nigra*; surf scoter; *Melanitta perspicillata*; white-winged scoter; *Melanitta fusca*; Barrow's goldeneye; *Bucephala islandica*; bufflehead; *Bucephala albeola*; greater scaup; *Aythya marila*; bufflehead; *Bucephala albeola*; peregrine falcon; *Falco peregrinus*; belted kingfisher; *Ceryle alcyon*; common merganser; *Mergus merganser;* bald eagle; *Haliaeetus leucocephalus*; great blue heron; *Ardea herodias*; glaucous-winged gull; *Larus glaucescens*; glaucous gull; *Larus hyperboreus*; common gull; *Larus canus*; lesser yellowlegs; *Tringa flavipes*; greater yellowlegs; *Tringa melanoleuca*; black oystercatcher; *Haematopus bachmani*; herring gull; *Larus argentatus*; pintail; *Anas acuta*; American wigeon; *Anas americana*; shoveler; *Anas clypeata*; ruddy duck; *Oxyura jamaicensis*; ring-necked duck; *Aythya collaris*; hudsonian godwit; *Limosa haemastica*; sandhill crane; *Grus canadensis*; dunlin; C*alidris alpina*; whimbrel; *Numenius phaeopus*; wandering tattler; *Heteroscelus incanus*; spotted sandpiper; *Actitis macularia*; black turnstone; *Arenaria melanocephala*; western sandpiper; *Calidris mauri*; rock sandpiper; *Calidris ptilocnemis*; surfbird; *Aphriza virgata*; gadwall; *Anas strepera*; mallard; *Anas platyrhynchos*; green-winged teal; *Anas crecca*; Canada goose; *Branta canadensis*
NOTES: SUPPLEMENTAL
ABSTRACT: The supertanker Exxon Valdez ran aground on 24 March 1989, spilling 41×10^6 L of oil into Prince William Sound, Alaska. To examine effects of this oil spill on the marine bird community, we analyzed data from 11 survey cruises between June 1989 and August 1991. Cruises were conducted in 10 study bays differing in the magnitude of initial oiling. We gauged bird responses to the spill in terms of habitat use, measured by frequency of bay occupancy and species abundances as functions of initial bay oiling. We focused on community-level measures to obtain a broader perspective than can be obtained from studies directed toward individual species of concern. Effects of the oil spill on community measures were most apparent shortly after the spill but diminished rapidly. Species richness was significantly lower in 1989 than at the

Annotated Bibliography (Cont.).

same season 1-2 years later, especially in heavily oiled bays. Species diversity (log-series alpha) was also significantly reduced in more heavily oiled bays in early summer 1989 and 1990, but impacts evident in midsummer and fall 1989 were absent 1 year later, and there were no significant relationships between diversity and bay oiling after midsummer 1990. Species occurrence in bays was more restricted immediately following the spill than 1-2 years later, and widespread species were less abundant in early summer and fall 1989 than at the same seasons 1 year later. This latter pattern was reversed in the midsummer surveys, perhaps because spill clean-up activities attracted large numbers of nonbreeding gulls. Our analyses indicated that the Exxon Valdez oil spill had significant initial impacts on marine bird community structure, although they were not evenly distributed among ecological guilds. Even during the first survey, many species were present in the most heavily oiled bays. Although a few species continued to show spill impacts in late 1991, none of the community measures indicated continuing negative oiling effects. This suggests that, at the community level, recovery was well underway, consistent with observations that seabird habitat had apparently returned to normal in all but a few localized areas by mid-1991. Seabird communities appear to have considerable resiliency to such severe but relatively short-term perturbations, possibly because birds move over a regional scale. It may, therefore, be important to consider regional processes in evaluating recovery following environmental accidents.

Williams, J. M.; Tasker, M. L.; Carter, I. C.; and Webb, A. 1995. A method of assessing seabird vulnerability to surface pollutants. Ibis. 137(Suppl. 1):S147-S152.
Rec #: 14
KEYWORDS: North Sea; Europe; oil; effects
NOTES: SUPLEMENTAL
ABSTRACT: Substantial information on the offshore distribution of seabirds exists for the North Sea and nearby areas. This information is extensive but is not easily applicable to management problems, such as oil pollution incidents. A quantitative oil vulnerability index has been developed, based on four easily scored factors, to assess the vulnerability of seabird species to surface pollution. The index has been applied to calculate area vulnerability scores for 15' latitude times 30' longitude rectangles in the North Sea by combining the species vulnerability scores with information on seabird densities. Maps of vulnerable concentrations of seabirds for each month of the year have been published with interpretative text. These maps can be used as a management tool, for example, in assessing potential risk or in the event of an oil pollution incident.

Williams, L. L.; Giesy, J. P.; Verbrugge, D. A.; Jurzysta, S.; and Stromborg, K. 1995. Polychlorinated biphenyls and 2,3,7,8-tetrachlorodibenzo-p-dioxin equivalents in eggs of double-crested cormorants from a colony near Green Bay, Wisconsin, USA. Arch. Environ. Contam. Toxicol. 29(3):327-333.
Rec #: 32
KEYWORDS: organochlorines; PCBs; PCDDs; toxicology; TEQs; egg; WI; USA; North America; Great Lakes; double-crested cormorant; *Phalacrocorax auritus*; abnormalities; effects
ABSTRACT: Great Lakes colonial waterbirds have experienced poor reproduction and a greater incidence of birth defects than those in remote areas. An egg was collected from each of 1,000 marked cormorant nests at Spider Island (Lake Michigan). Nine pools comprised of three eggs were randomly selected for instrumental quantification of polychlorinated biphenyls (PCB)

Annotated Bibliography (Cont.).

congeners, calculation of 2,3,7,8- tetrachlorodibenzo-p-dioxin equivalents (TEQ) and measurement of equivalents by bioassay (TCDD-EQ). PCB analysis of the nine samples was semi-automated with high performance liquid chromatography (HPLC) columns including a porous graphitic carbon column. TEQs were calculated from concentrations of PCB congeners and bioassay-derived toxic equivalency factors (TEFs) and TCDD-EQ were measured directly with an H4IIE bioassay. Total PCB concentrations ranged from 9.7 to 38 µg/g wet weight (ww). Mean concentrations of PCB 77, 126, and 169 were 2, 7, and 1 ng/g ww. The mean TEQs and TCDD-EQ were 150 and 350 pg/g ww, respectively. Thus PCB congeners contributed less than 50 % of the total TCDD-EQs as measured by the bioassay.

Wilson, Heather M.; Petersen, Margaret R.; and Troy, Declan. 2004. Concentrations of metals and trace elements in blood of spectacled and king eiders in northern Alaska, USA. Environ. Toxicol. Chem. 23(2):408-414.
Rec #: 470
KEYWORDS: metals; As; Ba; Cd; Pb; Hg; Se; blood; king eider; *Somateria spectabilis*; spectacled eider; *Somateria fischeri*; AK; USA; North America; toxicology; gender differences; seasonal variations
ABSTRACT: In 1996, we measured concentrations of arsenic, barium, cadmium, lead, mercury, and selenium in blood of adult king (*Somateria spectabilis*) and spectacled (*Somateria fischeri*) eiders and duckling spectacled eiders from northern Alaska, USA. Concentrations of selenium exceeded background levels in all adults sampled and 9 of 12 ducklings. Mercury was detected in all adult spectacled eiders and 5 of 12 ducklings. Lead concentrations were above the clinical toxicity threshold in one duckling (0.64 ppm) and two adult female spectacled eiders (0.54 and 4.30 ppm). Concentrations of cadmium and mercury varied between species; barium, cadmium, mercury, and selenium varied between sexes. In female spectacled eiders, mercury concentrations increased during the breeding season and barium and selenium levels decreased through the breeding season. Selenium declined at 2.3 ± 0.9 % per day and levels were lower in spectacled eiders arriving to the breeding grounds in northern Alaska than in western Alaska. The variation in selenium levels between breeding areas may be explained by differences in timing and routes of spring migration. Most trace elements for which we tested were not at levels currently considered toxic to marine birds. However, the presence of mercury and elevated lead in ducklings and adult female spectacled eiders suggests these metals are available on the breeding grounds.

Wilson, James G. and Earley, John J. 1986. Pesticide and PCB levels in the eggs of shag *Phalacrocorax aristotelis* and cormorant *P. carbo* from Ireland. Environ. Pollut. (B. Chem. Phys.). 12(1):15-26.
Rec #: 393
KEYWORDS: organochlorines; PCBs; DDTs; HCHs; dieldrin; endrin; chlordanes; shag; *Phalacrocorax aristotelis*; cormorant; *Phalacrocorax carbo*; egg; Ireland; Great Britain; Europe; eggshell thickness; effects; spatial variations
ABSTRACT: Shag and cormorant eggs were collected from 3 sites off the east, south-east and south coasts of Ireland and the pesticide and PCB levels determined by gas chromatography. Of the pesticides, p,p'-DDT [50-29-3], p,p'-DDE [72-55-9], o,p'-DDE [3424-82-6], lindane [58-89-9], dieldrin [60-57-1], endrin [72-20-8], α-BHC [319-84-6], α-chlordane [5103-71-9],

Annotated Bibliography (Cont.).

oxychlordane [27304-13-8], heptachlor [76-44-8], heptachlor-epoxide [1024-57-3], and quintofene [82-68-8] were detected, whereas o,p'-DDT [789-02-6], o,p'-DDD [53-19-0], aldrin [309-00-2], endosulphan-I [959-98-8], endosulphan-II [33213-65-9], endosulfan sulfate [1031-07-8], methoxychlor [72-43-5], β-BHC [319-85-7], and γ-chlordane [5566-34-7], were not found. PCB levels were an order of magnitude greater than those of the pesticides, but levels of all substances were, in general, rather low, thus organochlorine contamination is not at present a serious problem in the Irish marine environment. In general, the highest levels were found at the site off the east coast, and there was a significant inter-site difference in total pesticides, while the difference in PCBs was very close to significance. These levels did not cause either lethal or sublethal (egg-shell thinning) effects.

Wong, M. 1985. Environmental residues in Canadian game birds. Toxic Chemical Division Report, Canadian Wildlife Service. 293 pp.
Rec #: 546
KEYWORDS: Canada; North America
NOTES: UNABLE TO OBTAIN – See #s 419 and 422 for info.

Wong, Michael P.; Braune, Birgit M.; and Marshall, W. Keith. 1989. The use of wing parts for monitoring environmental residues. Technical Report Series No. 63. Ottawa, Ontario, Canada: Canadian Wildlife Service Ottawa Region. 173 pp.
Rec #: 422
KEYWORDS: review; age variations; gender differences; temporal trends; spatial variations; elimination; common merganser; *Mergus merganser*; red-breasted merganser; *Mergus serrator*; hooded merganser; *Lophodytes cucullatus*; American black duck; *Anas rubripes*; gadwall; *Anas strepera*; Eurasian wigeon; *Anas penelope*; American wigeon; *Anas americana*; green-winged teal; *Anas crecca*; blue-winged teal; *Anas discors*; shoveler; *Anas clypeata*; pintail; *Anas acuta*; wood duck; *Aix sponsa*; redhead duck; *Aythya americana*; canvasback duck; *Aythya valisineria*; greater scaup; *Aythya marila*; lesser scaup; *Aythya affinis*; ring-necked duck; *Aythya collaris*; common goldeneye; *Bucephala clangula*; Barrow's goldeneye; *Bucephala islandica*; bufflehead; *Bucephala albeola*; long-tailed duck; *Clangula hyemalis*; harlequin duck; *Histrionicus histrionicus*; common eider; *Somateria mollissima*; king eider; *Somateria spectabilis*; black scoter; *Melanitta nigra*; white-winged scoter; *Melanitta fusca*; surf scoter; *Melanitta perspicillata*; ruddy duck; *Oxyura jamaicensis*; snow goose; *Chen caerulescens*; Ross' goose; *Chen rossii*; white-fronted goose; *Anser albifrons*; Canada goose; *Branta canadensis*; brant; *Branta bernicla*; Newfoundland; Nova Scotia; New Brunswick; Quebec; Ontario; Manitoba; Saskatchewan; Alberta; British Columbia; Northwest Territories; Yukon; Canada; PA; NY; LA; ME; MI; NH; NJ; GA; NC; SC; CT; MA; MN; OH; VT; WI; FL; IA; OK; MT; CA; OR; UT; MO; CO; IL; TX; USA; Great Lakes; North America; Great Britain; Baltic Sea; Europe; skin; kidney; muscle; gland; pancreas; brain; gonad; liver; feather; organochlorines; DDTs; dieldrin; HCHs; PCBs; chlordanes; HCB; mirex; toxaphene; metals; Hg; Pb; Zn; Cd; Se; Fe; Mn; Cu; Co; Ni; Cr; Ag; As; Sr; V; Al; B; Mg; Li; S; Ti; Br; P
ABSTRACT: This report reviews the literature on the use of wing parts for surveying toxic chemical residues. The possibility of using wing parts of harvested game birds, which are submitted as part of an annual survey, to monitor environmental residues is evaluated. The studies reviewed employed whole defeathered wings, wing muscles, wing bones and feathers as media for chemical analysis. The focus of attention has centered on persistent environmental

Annotated Bibliography (Cont.).

residues (i.e. whole defeathered wings: organochlorines, wing muscles: mercury, wing bones: lead from spent shot, wing feathers: mercury). Other than studies of DDT residues in whole defeathered wings and lead residues in wing bones, little attention has been focused on the relationship of contaminants in the wing parts to other body tissues.

Yamashita, Nobuyoshi; Tanabe, Shinsuke; Ludwig, James P.; Kurita, Hiroko; Ludwig, Matthew E.; and Tatsukawa, Ryo. 1993. Embryonic abnormalities and organochlorine contamination in double-crested cormorants (*Phalacrocorax auritus*) and Caspian terns (*Hydroprogne caspia*) from the upper Great Lakes in 1988. Environ. Pollut. 79(2):163-173.
Rec #: 251
KEYWORDS: effects; abnormalities; toxicology; TEQs; organochlorines; PCBs; PCDDs; PCDFs; DDTs; chlordanes; HCHs; HCB; egg; North America; Great Lakes; double-crested cormorant; *Phalacrocorax auritus*; Caspian tern; *Sterna caspia*
ABSTRACT: Persistent organochlorine contaminants including polychlorinated dibenzo-p-dioxins (PCDDs), polychlorinated dibenzofurans (PCDFs) and polychlorinated biphenyls (PCBs) were determined in eggs with normal and deformed embryos collected in 1988 from different colonies during an epizootiological survey of double-crested cormorants (*Phalacrocorax auritus*) and Caspian terns (*Hydroprogne caspia*) from the upper Great Lakes. The toxic effects of these contaminants were also estimated in Caspian tern eggs, where elevated levels of coplanar PCBs, PCDDs and PCDFs were observed in concordance with increased rate of anomalies in eggs during a breeding season in the Great Lakes.

Young, David R.; Heesen, Theadore C.; Esra, Gerald N.; and Howard, Edwin B. 1979. DDE-contaminated fish off Los Angeles are suspected cause in deaths of captive marine birds. Bull. Environ. Contam. Toxicol. 21(4-5):584-590.
Rec #: 417
KEYWORDS: Brandt's cormorant; *Phalacrocorax penicillatus*; Guanay cormorant; *Phalacrocorax bougainvillii*; California gull; *Larus californicus*; organochlorines; DDTs; PCBs; metals; Hg; brain; liver; CA; USA; North America; effects; biomagnification
ABSTRACT: Brain and liver tissues of Brandt's cormorants (*Phalacrocorax penicillatus*), Guanay cormorants (*P. bougainvillii*), and Calif. gulls (*Larus californicus*) that died at the Los Angeles Zoo contained total DDT [50-29-3] levels 2 orders of magnitude above the mean value for wild specimens. P,p'-DDE (I) [72-55-9] constituted >90 % of the total DDT. Queenfish used to feed the birds had relatively high I levels. The zoo birds also had polychlorinated biphenyl (PCB) accumulations of 140-500 ppm in the liver and 49-66 ppm in the brain. However, PCB and Hg levels in the zoo birds were about equal to that in wild birds and I was believed to be the primary cause of death.

Zitko, V. 1972. Absence of chlorinated dibenzodioxins and dibenzofurans from aquatic animals. Bull. Environ. Contam. Toxicol. 7(2):105-10.
Rec #: 409
KEYWORDS: organochlorines; DDTs; PCBs; PCDDs; PCDFs; egg; Bay of Fundy; Canada; North America; double-crested cormorant; *Phalacrocorax auritus*; herring gull; *Larus argentatus*
ABSTRACT: Chlorinated dibenzodioxins and dibenzofurans were absent from the muscle and liver of white shark (*Carcharodon carcharias*), the eggs of double-crested cormorants

Annotated Bibliography (Cont.).

(*Phalacrocorax auritus*) and herring gulls (*Larus argentatus*), the muscle of eel (*Anguilla rostrata*) and chain pickerel (*Esox niger*), in Canada and in commercial samples of herring oil and groundfish-herring fishmeal.

Zitko, V. 1973. Determination of phthalates in biological samples. Int. J. Environ. Anal. Chem. 2(3):241-252.
Rec #: 450
KEYWORDS: phthalates; double-crested cormorant; *Phalacrocorax auritus*; herring gull; *Larus argentatus*; egg; Canada; North America
ABSTRACT: Phthalates, extracted from biological samples with hexane are partially separated from lipids by chromatography on alumina to yield fractions in which the common phthalate plasticizers can be quantitated by gas chromatography. An additional cleanup is achieved by the extraction of phthalates from hexane into DMF. Phthalates can then be confirmed by measurement of fluorescence in concentrated H_2SO_4. Analyses of spiked samples are reported. Dibutyl phthalate [84-74-2] was detected in eggs of double-crested cormorants (*Phalacrocorax auritus*) and herring gulls (*Larus argentatus*) at 11-19 µg/g lipid. Di-2-ethylhexyl phthalate [117-81-7] was detected in hatchery-reared juvenile Atlantic salmon (*Saamo salar*) at 13-16 µg/g lipid, and in the blubber of a common seal pup (*Phoca vitulina*) at 11 µ g/g lipid.

Zitko, V. 1976. Levels of chlorinated hydrocarbons in eggs of double-crested cormorants from 1971 to 1975. Bull. Environ. Contam. Toxicol. 16(4):399-405.
Rec #: 408
KEYWORDS: organochlorines; PCBs; HCB; DDTs; dieldrin; mirex; egg; double-crested cormorant; *Phalacrocorax auritus*; Bay of Fundy; Canada; North America; temporal trends
ABSTRACT: Levels of Aroclor 1254 [11097-69-1], and DDE (I) [72-55-9] measured in eggs of double-crested cormorants, (*Phalacrocorax auritus*) decreased between 1971 and 1973 with no further change between 1973 and 1975. The levels of dieldrin [60-57-1], DDD [72-54-8], DDT [50-29-3], Mirex [2385-85-5], and hexachlorobenzene [118-74-1] were several orders of magnitude lower than those of Aroclor 1254 and I and may show the same trend, except that the 1975 concentrations were generally higher than those in 1973 and 1974. The observed levels of chlorinated hydrocarbons in eggs of cormorants had no apparent effect on the cormorant population.

Zitko, V. 1974. Trends of PCB and DDT in fish and aquatic birds. Proceedings International Conference on Transport of Persistent Chemicals in Aquatic Ecosystems. Ottawa, Canada, May 1-3, 1974. 61-64.
Rec #: 449
KEYWORDS: egg; double-crested cormorant; *Phalacrocorax auritus*; herring gull; *Larus argentatus*; Bay of Fundy; Nova Scotia; Canada; North America; organochlorines; PCBs; DDTs; temporal trends
ABSTRACT: Muscle analysis of herring (*Clupea harengus*) from the Bay of Chaleur showed an increase in Aroclor 1254 (I) [11097-69-1] from 0.29 to 0.39 µg/g, and a decrease in DDE (II) [72-55-9], DDD III [72-54-8], and DDT (IV) [50-29-3] from 0.21 to 0.072, 0.055 to 0.042, and 0.104 to 0.028 µg/g, respectively, over a period of 5 years, while herring from Nova Scotia showed a decrease in I, II, III, and IV, over a 2 year period. Double-crested cormorant

Annotated Bibliography (Cont.).

(*Phalacrocorax auritus*) eggs showed a decrease in I, II, III, and IV, of 14.3 to 5.57, 9.7 to 2.89, 0.113 to 0.053, and 0.167 to 0.073, respectively, over a 2 to 3 year period. Herring gull (*Larus argentatus*) eggs showed a decrease in I, II, III, and IV of 5.29 to 3.77, 2.48 to 2.10, 0.04 to <0.01, and 0.148 to 0.039, respectively, over a 2 to 3 year period.

Zitko, V. and Choi, P. M. K. 1972. PCB and p,p'-DDE in eggs of cormorants, gulls, and ducks from the Bay of Fundy, Canada. Bull. Environ. Contam. Toxicol. 7(1):63-64.
Rec #: 410
KEYWORDS: organochlorines; DDTs; PCBs; egg; New Brunswick; Bay of Fundy; Canada; North America; double-crested cormorant; *Phalacrocorax auritus*; herring gull; *Larus argentatus*; American black duck; *Anas rubripes*
ABSTRACT: Relatively high levels of polychlorinated biphenyls and p,p'-DDE were found in eggs of double-crested cormorants (*Phalacrocorax auritus*), herring gulls (*Larus argentatus*), and black ducks (*Anas rubripes*) from Fatpot Island and Hospital Island in Canada collected in May 1971.

Zitko, V.; Hutzinger, O.; Jamieson, W. D.; and Choi, P. M. K. 1972. Polychlorinated terphenyls in the environment. Bull. Environ. Contam. Toxicol. 7(4):200-201.
Rec #: 412
KEYWORDS: organochlorines; toxaphene; double-crested cormorant; *Phalacrocorax auritus*; herring gull; *Larus argentatus*; fat; egg; Bay of Fundy; Canada; North America
NOTES: FIGURES ONLY
ABSTRACT: The level of polychlorinated terphenyls (PCT) in the subcutaneous fat and eggs of herring gulls (*Larus argentatus*) from Bay of Fundy (Canada) was 1.4 and 0.1 µg/g, respectively, when expressed as Aroclor 5460 on wet weight basis. PCT was absent from the eggs and fatty tissue of double-crested cormorants (*Phalacrocorax auritus*). PCT was determined by gas chromatography using OV-210 on Chromosorb WAW, at a column temp. of 200.deg. (Zitko, V., 1971).

Zunk, B. 1984. Cadmium loading of some types of sea gulls from the Bay of Helgoland. Fortschritte in Der Atomspektrometrischen Spurenanalytik. 1:597-607.
Rec #: 460
KEYWORDS: metals; Cd; liver; kidney; great black-backed gull; *Larus marinus*; herring gull; *Larus argentatus*; black-legged kittiwake; *Rissa tridactyla*; Germany; North Sea; Europe; biomagnification; gender differences
ABSTRACT: Cd concentration in the liver and kidney increased in series *Larus marinus* < *L. argentatus* < *Rissa tridactyla*. In the kidney, the Cd concentrations were 2-7-fold higher than in the liver. In all 3 species, the renal Cd concentrations increased during 1st 3 years of life and remained const. thereafter. In females of *L. argentatus*, the hepatic Cd concentrations were above those of males; this may be due to higher Ca requirement for egg shell formation in the females. Sea gull feeding habits and Cd transport in food chain are discussed in relation to environmental pollution with Cd.

Ólafsdóttir, K.; Petersen, Ć; Magnúsdóttir, E. V.; Björnsson, T.; and Jóhannesson, T. 2001. Persistent organochlorine levels in six prey species of the gyrfalcon *Falco rusticolus* in Iceland. Environ. Pollut. 112(2):245-251.

Annotated Bibliography (Cont.).

Rec #: 319
KEYWORDS: mallard; *Anas platyrhynchos;* tufted duck; *Aythya fuligula;* golden plover; *Pluvialis apricaria*; purple sandpiper; *Calidris maritima;* black guillemot; *Cepphus grylle;* rock ptarmigan; *Lagopus mutus;* gyrfalcon; *Falco rusticolus*; organochlorines; HCHs; HCB; PCBs; DDTs; chlordanes; Iceland; Europe; biomagnification; gender differences
ABSTRACT: Our previous investigations have revealed very high levels of organochlorines (OCs) in the Icelandic gyrfalcon *Falco rusticolus*, a resident top predator. We now examine six potential prey species of birds, both resident and migratory, in order to elucidate the most likely route of the OCs to the gyrfalcon. The ptarmigan Lagopus mutus, the most important prey of the gyrfalcon, contained very low levels of OCs. Bioaccumulation of polychlorinated biphenyls (PCBs) and DDTs in mallards *Anas platyrhynchos*, tufted ducks *Aythya fuligula*, golden plovers *Pluvialis apricaria*, purple sandpipers *Calidris maritima*, and black guillemots *Cepphus grylle* reflected their position in the foodchain. The differences in OC-levels seem nevertheless too high just to reflect the different food-chain levels of these species in Iceland. The winter grounds of the migratory golden plovers and tufted ducks appear to be more contaminated than the Icelandic terrestrial habitat of ptarmigans or the freshwater habitat as reflected in mallards, both resident species. However, spending the winter on the coast in Iceland, results in high levels of contaminants in purple sandpipers and black guillemots. Our results indicate OC contamination of the marine ecosystem in Iceland while the terrestrial and freshwater ecosystems are little affected. It is postulated that gyrfalcons receive the major part of the observed contamination from prey other than ptarmigan, especially birds associated with the marine ecosystem and also from migratory birds.

Ólafsdóttir, K.; Skirnisson, K.; Gylfadottir, G.; and Jóhannesson, T. 1998. Seasonal fluctuations or organochlorine levels in the common eider (*Somateria mollissima*) in Iceland. Environ. Pollut. 103(2-3):153-158.
Rec #: 314
KEYWORDS: common eider; *Somateria mollissima*; Iceland; Europe; organochlorines; PCBs; DDTs; HCB; HCHs; seasonal variations; muscle; liver; gender differences
ABSTRACT: Breast muscle of 55 common eiders (*Somateria mollissima*) and liver samples of 12 birds, caught at Skerjafjørour in SW-Iceland in February, May, June and November 1993 were analysed for organochlorine contamination (10-30 different congeners of PCBs, pp'-DDT, -DDE, -DDD, HCB, alpha -, beta -, and gamma-HCH). The levels of the contaminants were similar in both tissues and were at their lowest in February. A substantial increase (up to 10-fold) in the levels of all substances was observed in June, in the females, which at that point had lost about one-third of their late winter body weight. The increase may be due to relocation to other tissues of organochlorines stored in the shrinking bodyfat. During this period the birds must be vulnerable to the toxic effects of these chemicals as they can transiently reach high concentrations in the blood. The levels found were similar or higher than those recently reported for eiders from Spitsbergen, the NWT of Canada and Frans Josefs Land of Russia, especially the levels of PCBs.

Author Index. Numbers in parentheses are the sums of papers authored.

Aarkrog, A. (1) 111
Addison, R. (1) 153
Adelung, Dieter (2) 216, 215
Ainley, D.G. (1) 165
Allen, Janette R. (1) 18
Allen, P. David (2) 61, 61
Alsberg, Tomas (1) 26
Anderka, F.W. (1) 84
Anderson, Daniel W. (4) 19, 152, 52, 152
Anderson, Paul (1) 174
Andreev, A. V. (1) 129
Anker-Nilssen, T. (1) 142
Ankley, Gerald T. (2) 204, 205
Antoine, Nathalie (1) 19
Appelquist, Helge (2) 19, 20
Arcos, J. M. (1) 20
Arnold, M. A. (1) 53
Arsenault, T. (1) 131
Asbirk, Sten (2) 19, 20
Åslund, Kerttu (1) 161
Asmund, G. (1) 176
Asplund, Lillemor (5) 115, 121, 127, 161, 120
Atrashkevich, Gennady I. (2) 130, 129
Atwell, Lisa (1) 20
Aulerich, Richard (1) 143
Auman, Heidi J. (3) 143, 143, 119
Ayotte, Pierre (1) 180
Baars, A. J. (1) 113
Baba, Norihisa (1) 115
Bacon, Corinne E. (1) 116
Bailey, Edgar P. (1) 21
Bakken, Vidar (5) 50, 51, 51, 91, 50
Barbeau, J. M. (1) 193
Barrett, G. C. (2) 172, 118
Barrett, M. W. (1) 193
Barrett, R. T. (3) 22, 22, 203
Barreveld, Hein (1) 55
Bartle, J. A. (1) 141
Bartonek, James C. (1) 162
Bastien, L. J. (1) 126
Baxter, M. S. (1) 63
Bearhop, S. (2) 20, 202
Bech, C. (1) 155

Becker, Paul R. (4) 133, 208, 209, 58
Becker, Peter H. (11) 23, 23, 23, 24, 121, 200, 22, 72, 90, 140, 202
Bell, A. A. (2) 153, 156
Belles-Isles, Jean-Claude (1) 38
Bellward, Gail D. (2) 184, 76
Bend, J. R. (1) 166
Berg, Ole (2) 34, 59
Berge, John Arthur (1) 24
Bergqvist, Per-Anders (2) 80, 117
Bergström, Rune (1) 25
Bernhoft, Aksel (1) 192
Best, Dave (1) 219
Beyerbach, M. (2) 72, 103
Bignert, Anders (6) 26, 127, 161, 189, 25, 188
Bishop, Christine A. (8) 86, 146, 205, 218, 26, 27, 172, 173
Bishop, Paul (1) 52
Bjerk, John Erik (3) 27, 81, 81
Bjørge, Arne (1) 24
Björnsson, T. (1) 229
Blight, Louise K. (1) 28
Blomkvist, Gun (1) 116
Blus, Lawrence J. (1) 105
Böckelmann, W. (1) 103
Boersma, D. (1) 167
Bogan, J. A. (4) 28, 34, 63, 140
Böler, J. B. (1) 55
Bollinger, T. (1) 213
Bondarev, A. Ya. (1) 29
Borch-Iohnsen, Berit (1) 157
Borgå, Katrine (4) 30, 30, 113, 29
Borlakoglu, J.T. (4) 30, 31, 31, 32
Born, E. W. (1) 72
Borthwick, J. (1) 52
Bouquegneau, Jean Marie (8) 19, 32, 32, 66, 67, 67, 68, 68
Bourbonnais, Diane H. (1) 193
Bourne, W. R. P. (5) 28, 32, 34, 63, 140
Bowerman, William W. (2) 123, 143
Braestrup, Liselotte (1) 34

Author Index (Cont.). Numbers in parentheses are the sums of papers authored.

Braune, Birgit Margret (17) 39, 36, 36, 37, 37, 37, 114, 145, 146, 34, 35, 36, 74, 101, 153, 38, 226
Brevik, Einar M. (1) 24
Briggs, Kenneth T. (1) 40
Bright, D. (1) 153
Broman, Dag (2) 41, 41
Brooks, R. J. (1) 26
Brown, A. W. A. (1) 41
Brown, N. J. (1) 41
Brown, Robert L. (1) 110
Brun, Einar (2) 81, 81
Brunström, Björn (3) 42, 78, 108
Buckman, Andrea H. (1) 42
Bull, K. R. (1) 43
Burger, Alan E. (1) 28
Burger, Joanna (14) 44, 45, 45, 46, 46, 47, 47, 48, 48, 49, 49, 43, 44, 44
Burgess, Neil M. (2) 135, 27
Burkow, Ivan C. (1) 109
Bursian, Steven J. (2) 143, 147
Bustnes, Jan Ove (4) 51, 50, 51, 50
Büthe, Annegret (3) 23, 72, 103
Butler, Ronald G. (7) 52, 138, 207, 52, 148, 166, 168
Cahill, T. M. (1) 52
Calambokidas, John (1) 194
Calambokidis, John (1) 53
Caldwell, C. A. (1) 53
Call, D. J. (1) 96
Cameron, Marjorie (1) 54
Camphuysen, Kees C.J. (1) 55
Carlberg, G. E. (1) 55
Carreno, R. A. (1) 213
Carroll, T. R. (1) 93
Carter, I. C. (1) 224
Catry, P. (1) 195
Champoux, Louise (1) 55
Chan, H.M. (2) 56, 133
Cheng, Kimberly M. (2) 109, 184
Cheng, Lana (1) 43
Choi, Jae-Won (3) 57, 122, 56
Choi, P. M. K. (2) 229, 229
Christensen, Guttorm (1) 121

Christensen, Jan H. (1) 211
Christensen, Robert E. (1) 19
Christopher, Steven J. (2) 209, 58
Claeys, Robert R. (1) 187
Clark, Thomas P. (3) 160, 58, 58
Clausen, B. (1) 124
Clausen, Jørgen (2) 59, 34
Cleemann, M (1) 59
Cochrane, Jean F. (1) 197
Coignoul, F. (2) 67, 67
Collins, Brian (3) 207, 181, 207
Comba, M. E. (2) 98, 98
Cooke, A. S. (1) 60
Coon, Nancy C. (1) 199
Corsolini, S. (1) 83
Coulson, J. C. (2) 60, 178
Crabtree, A. N. (2) 60, 178
Craven, A. C. C. (1) 212
Crawford, R. (1) 119
Crist, Thomas O. (1) 223
Cubbage, James C. (1) 53
Custer, Christine.M. (3) 61, 62, 61
Custer, Thomas W. (3) 61, 62, 61
Daelemans, F. F. (2) 62, 147
Daemers, C. (1) 68
Dahlgaard, H. (1) 111
Dahlmann, Gerhard (1) 55
Dale, I. M. (1) 63
D'Amico, Christopher (1) 52
D'Amico, Melissa (1) 52
Danesik, Karen L. (1) 32
Das, K. (1) 32
Davenport, Glenn H. (1) 21
Davis, Jay A. (1) 63
Dawson, P. (1) 140
Day, Robert H. (4) 64, 65, 154, 223
Day, Russell D. (2) 209, 58
De Pauw, E. (1) 66
De Voogt, Pim (1) 66
de Wit, Cynthia A. (7) 189, 121, 127, 161, 189, 120, 188
Deans, I. R. (1) 60
Debacker, Virginie (8) 69, 32, 32, 66, 67, 67, 68, 68

232

Author Index (Cont.). Numbers in parentheses are the sums of papers authored.

Decadt, G. (1) 70
Delbeke, Katrien (2) 71, 70
DeMarch, B. (1) 153
Denisova, A. V. (1) 29
Denker, Eckhard (2) 71, 72
Derocher, Andrew (1) 192
DeWailly, Eric (1) 180
DeWitt, Jamie C. (1) 109
Dickson, D. L. (1) 213
Dieter, Michael P. (1) 196
Dietz, Rune (7) 156, 59, 72, 72, 73, 176, 176
Dils, R. R. (2) 31, 32
Divoky, George J. (1) 162
Docherty, D. (1) 132
Doi, Rikuo (1) 73
Donaldson, G. M. (3) 34, 35, 74
Drabaek, Iver (2) 19, 20
Duffe, Jason (1) 42
Earley, John J. (1) 225
Echols, K. R. (1) 147
Egebäck, A.-L. (1) 120
Elbert, R. A. (1) 52
Elliott, John E. (15) 77, 77, 100, 184, 206, 206, 109, 74, 75, 75, 76, 76, 170, 198, 220
Elvestad, K. (1) 124
Emerick, R. J. (1) 97
Engwall, Magnus (1) 78
Eppe, G. (1) 66
Erdman, T. C (1) 132
Eriksson, Ulla (2) 26, 161
Erikstad, Kjell Einar (4) 51, 50, 51, 50
Eroschenko, Victor P. (1) 175
Esra, Gerald N. (2) 113, 227
Evenset, Anita (2) 109, 121
Ewins, Peter J. (3) 217, 219, 181
Falandysz, Jerzy (7) 79, 80, 80, 80, 81, 123, 79
Falk-Petersen, Stig (1) 186
Fimreite, Norvald (3) 81, 81, 81
Fisk, Aaron T. (4) 42, 150, 205, 82
Flint, Paul L. (1) 94
Focardi, Silvano (4) 140, 83, 83, 139
Fog, Mette (1) 83

Følsvik, Norunn (1) 24
Ford, R. G. (2) 221, 220
Fossi, Cristina (4) 140, 83, 83, 139
Foureman, G. L. (1) 166
Fox, Glen A. (22) 85, 86, 86, 93, 95, 32, 135, 142, 170, 175, 58, 84, 84, 84, 85, 93, 95, 126, 158, 159, 181
Frank, Adrian (2) 87, 87
Frank, Richard (1) 88
Franson, J. Christian (5) 88, 89, 94, 123, 111
Frederichsen, Per (1) 81
Frøslie, Arne (3) 81, 112, 22
Fry, Brian (1) 41
Fry, D. Michael (4) 63, 89, 175, 194
Fukuda, Kayo (1) 134
Furness, Robert W. (17) 23, 91, 204, 20, 22, 90, 90, 140, 141, 151, 195, 196, 202, 202, 203, 203, 203
Gabrielsen, Geir Wing (25) 24, 30, 91, 91, 108, 109, 109, 113, 182, 183, 185, 186, 192, 211, 30, 108, 183, 184, 22, 29, 107, 107, 186, 186, 216
Gale, R. W. (1) 147
Garcia-Fernandez, A. J. (1) 213
Garrett, C. (1) 159
Gaskin, David E. (2) 36, 37
Gaston, Anthony J. (2) 36, 74
Gätzschmann, P. (1) 110
Gershwin, M. Eric (1) 40
Giesy, John P (11) 92, 122, 123, 123, 136, 143, 143, 204, 205, 119, 224
Gilbertson, Michael (2) 93, 143
Gilchrist, H. Grant (4) 215, 213, 213, 214
Gilman, Andrew P. (7) 93, 160, 84, 93, 158, 158, 180
Glaser, Vadim M. (1) 195
Glasius, Marianne (1) 211
Gobert, S. (3) 32, 32, 67
Gochfeld, Michael (10) 46, 46, 47, 47, 48, 48, 49, 49, 44, 94
Gøksyr, Anders (1) 192
Gorsline, Jane (1) 175
Goski, B. C. (1) 193

Author Index (Cont.). Numbers in parentheses are the sums of papers authored.

Goto, R. (1) 128
Gould, W. R. (1) 53
Grafström, Anna Karin (1) 121
Grand, James B. (1) 94
Grasman, Keith.A. (4) 85, 95, 85, 95
Green, Nick (1) 66
Green, S. (1) 148
Greichus, A. (1) 97
Greichus, Yvonne A. (3) 97, 97, 96
Grove, Robert A. (1) 105
Gruchy, I. M. (1) 125
Guiney, Patrick D. (1) 190
Gundersen, Nil (1) 81
Gutreuter, Steve (1) 61
Gylfadottir, G. (1) 230
Haas, Margaret B. (1) 156
Haegele, K.D. (1) 30
Haffner, G. Douglas (2) 97, 131
Häggberg, Lisbeth (3) 26, 161, 189
Haglund, Peter (1) 121
Halldin, Krister (1) 42
Hallett, Douglas J. (9) 93, 166, 98, 98, 98, 148, 158, 166, 198
Hallstadius, L. (1) 111
Hamilton, E. A. (1) 146
Hamr, Premek (2) 218, 181
Hanbidge, Barbara A. (2) 136, 85
Hannon, Michael Robert. (2) 99, 97
Hansen, C. T. (1) 72
Hansen, Kris J. (1) 123
Hansen, M. M. (1) 72
Hanson, H. (1) 153
Harada, Masazumi (1) 73
Harfenist, A. (1) 220
Hargrave, B. (1) 153
Hario, Martti (3) 88, 100, 111
Harmon, R. S. (1) 52
Harner, E. James (2) 64, 65
Harris, H. J. (1) 132
Harris, M. P. (2) 164, 164
Harris, Megan L. (1) 100
Harrison, N. (1) 196
Hart, Leslie E. (1) 184
Hattula, Marja-Liisa (1) 199

Hauteclair, P. (1) 66
Havelange, S. (1) 32
Havelka, T. (1) 118
Haymes, G. T. (1) 93
Hayward, Gregory D. (3) 64, 65, 223
Headley, Alistair D. (1) 101
Hebert, Craig E. (5) 102, 103, 101, 102, 159
Heesen, Theadore C. (1) 227
Heidmann, W. A. (3) 72, 103, 23
Heinemann, D. (2) 220, 221
Heinz, Gary H. (1) 104
Heisinger, James F. (1) 110
HELCOM (1) 104
Henning, Diana (2) 23, 22
Henny, Charles J. (3) 105, 105, 106
Henriksen, Espen O. (9) 91, 108, 108, 109, 183, 183, 107, 107, 192
Henshel, Diane S. (2) 109, 61
Herman, S. G. (1) 177
Herzke, Dorte (2) 109, 121
Hesse, Larry W. (1) 110
Hickey, Joseph J. (1) 19
Hidaka, Hideo (1) 125
Hill, Elwood F. (1) 105
Hilscherova, Klara (1) 123
Hines, Randy K. (2) 61, 61
Hirvi, Juha-Pekka (1) 100
Hobbie, John (1) 41
Hobson, Keith A. (1) 20, 42, 85, 116, 117, 150, 205, 214, 34, 35, 82
Hoekstra, Paul F. (1) 145
Hoffman, David J. (1) 105
Hogstad, O. (1) 110
Hollmén, Tuula (3) 88, 100, 111
Holm, E. (1) 111
Holsbeek, Ludo (3) 19, 119, 67
Holt, Gunnar (2) 27, 112
Honda, Katsuhisa (1) 112
Hontelez, L. C. M. P. (1) 113
Hop, Haakon (2) 30, 113
Horne, Janice A. (1) 191
Howard, Edwin B. (2) 113, 227
Huart, P. (1) 66

Author Index (Cont.). Numbers in parentheses are the sums of papers authored.

Hughes, Donald F. (1) 19
Hughes, Kimberley D. (2) 219, 114
Hühnerfuss, Heinrich (2) 122, 174
Humppi, Tarmo (1) 165
Hutton, M. (1) 114
Hutzinger, O. (1) 229
Ichihashi, Hideki (2) 129, 130
Iida, Tetsuji (1) 115
Imagawa, Takashi (1) 123
Inoue, Tsuyoshi (2) 125, 125
Iseki, Naomasa (1) 122
Iwata, Hisato (1) 115
Jakobsson, Eva (2) 78, 120
James, C. A. (2) 127, 213
Jamieson, W. D. (1) 229
Janatuinen, Jorma (1) 199
Janes-Butler, S. (1) 148
Jansegers, Isabelle (1) 19
Jansson, Bo (8) 115, 116, 121, 127, 189, 189, 220, 188
Jarman, Walter M. (4) 116, 117, 198, 206
Järnberg, Ulf (2) 121, 120
Jauniaux, T. (5) 66, 67, 67, 68, 68
Jefferies, D. J. (1) 165
Jeffrey, Deborah A. (5) 160, 85, 158, 167, 167
Jensen, Sören (4) 26, 116, 117, 163
Jenssen, Bjørn Munro (3) 211, 107, 155
Jermyn-Gee, K. (1) 118
Jóhannesson, T. (2) 229, 230
Johansen, P. (2) 73, 176
Johnels, A. G. (1) 117
Johnson, D. R. (1) 52
Joiris, Claude R. (6) 119, 70, 67, 118, 19, 71
Jones, A. M. (1) 119
Jones, Paul D. (5) 123, 136, 143, 143, 119
Jones, S. P (1) 126
Jones, Yvonne (1) 119
Jonsson, Björn (1) 121
Jurzysta, S. (1) 224
Kageyama, Takae (1) 57
Kahle, Silke (1) 121
Kallenborn, Roland (3) 121, 122, 174

Kan, Shinya (1) 112
Kannan, Kurunthachalam (5) 92, 122, 123, 123, 123
Karakhanova, N. K. (1) 137
Karasov, William H. (1) 136
Karlin, Antti (2) 124, 139
Karlog, O. (1) 124
Karnovsky, Nina J. (2) 42, 82
Kawano, Masahide (4) 57, 125, 125, 56
Keating, Jonathan (3) 215, 213, 214
Keith, J. A. (1) 125
Kennedy, Sean W. (7) 86, 86, 142, 159, 84, 126, 127
Kettrup, A. (1) 164
Khan, R. A. (1) 127
Kierkegaard, Amelie (4) 127, 189, 189, 188
Kilpi, Mikael (2) 88, 111
Kim, Dong Hoon (1) 122
Kim, Eun-young (5) 129, 130, 128, 128, 129
Kinter, William B. (6) 149, 149, 149, 166, 167, 168
Kirven, M. N. (1) 177
Kishchinskii, A. A. (1) 29
Kjos-Hanssen, Bjørn (1) 157
Klass, Erwin E. (1) 162
Kleivane, Lars (1) 113
Knüwer, H. (1) 103
Köhler-Guenther, Angela (1) 174
Koistinen J. (1) 130
Koivusaari, J. (1) 130
Kolbe, Elizabeth (1) 199
König, Wilfried A. (2) 122, 174
Koslowski, Susan E. (1) 131
Koster, M. D. (1) 131
Kotelevtsev, Sergey V. (1) 195
Kraul, Inge (1) 83
Krol, W. J. (1) 131
Krüner, Günter (1) 174
Krynitsky, Alexander J. (1) 162
Kubiak, Timothy J. (4) 205, 219, 119, 132
Kucklick, John R. (3) 133, 208, 209
Kuhnlein, H. V. (1) 133

Author Index (Cont.). Numbers in parentheses are the sums of papers authored.

Kuiken, Thijs (1) 32
Kunisue, Tatsuya (1) 134
Kurita, Hiroko (1) 227
Kurita-Matsuba, Hiroko (2) 143, 205
Kurukgy, M. (1) 212
Kury, Channing R. (1) 134
Kussat, R. (1) 159
Kuzyk, Zou Zou A. (1) 135
Kveseth, N. (1) 81
Laliberte, Claire (1) 180
Lambert, Gabrielle (2) 52, 148
Lande, Eirik (1) 135
Langis, René (1) 136
Langlois, Claude (1) 136
Larsen, Hans Jørgen (1) 192
Larson, Jill M. (2) 136, 205
Lazar, Rodica (2) 97, 131
Lebedev, A. T. (1) 137
Lee, Y-Z. (2) 137, 138
Legierse, Karin (1) 146
Leighton, Frederick A. (6) 138, 52, 76, 137, 138, 167
Lemmetyinen, Risto (3) 124, 138, 139
Leonzio, Claudio (4) 140, 83, 83, 139
Lepper, Peter (1) 182
Lewis, S. A. (1) 140
Lexén, Karin (1) 121
Lie, Elisabet (1) 192
Lierhagen, S. (1) 110
Litzén, Kerstin (3) 26, 161, 25
Lloyd, C. (1) 140
Lock, J. W. (1) 141
Lockhart, L. (1) 153
Logan, K. A. (1) 27
Lorentsen, S-H. (1) 142
Lorenzen Angela (2) 142, 127
Ludwig, James P. (7) 95, 115, 143, 143, 205, 227, 119
Ludwig, Matthew E. (3) 143, 227, 119
Lundbergh, Ivar (1) 41
Mabury, Scott A. (1) 145
Macdonald, Colin R. (3) 103, 101, 144
Mackay, D. (1) 168
Magnúsdóttir, E. V. (1) 229

Malcolm, H. M. (1) 145
Marchant, Tracy (2) 215, 214
Marcovecchio, Jorge Eduardo (1) 112
Markova, Lubov (1) 147
Marshall, W. Keith (2) 38, 226
Martin, D. James (1) 177
Martin, J. William (1) 109
Martin, Jonathan W. (1) 145
Martin, Pamela A. (3) 77, 146, 76
Massart, A.-C. (1) 66
Masunaga, Shigeki (1) 122
Matheson, R. A. F. (1) 146
Matsuda, Muneaki (2) 57, 56
Matsushita, Sanae (1) 125
Matter, John M. (1) 190
Mattina, M. J. I. (1) 131
McLaren, Elizabeth B. (1) 116
Meadows, J. C. (1) 147
Medvedev, Nikolai (1) 147
Mehlum, Fridtjof (2) 147, 188, 62
Mehlum, Fritjof (2) 50, 51, 50
Meire, P. (1) 189
Melancon, M. J. (1) 61
Melancon, Mark J. (1) 61
Menzel, Daniel B. (1) 177
Metcalfe, Christopher D. (1) 131
Meteyer, Carol U. (1) 89
Miller, David S. (10) 149, 149, 149, 207, 52, 148, 148, 166, 166, 167
Min, Byung-Yoon (2) 57, 56
Mineau, Pierre (3) 217, 98, 198
Minh, Tu Binh (1) 134
Moisey, John (3) 150, 160, 82
Moksnes, Milica T. (1) 150
Monteiro, L. R. (1) 151
Moon, Thomas W. (2) 142, 193
Moore, N. W. (1) 151
Mora, Miguel A. (2) 152, 152
Moriarty, F. (1) 153
Muir, Derek C. G. (2) 145, 153
Muirhead, S. J. (1) 90
Murakami, T. (1) 128
Murphy, Stephen M. (4) 64, 65, 154, 223
Murton, R. K. (1) 43

Author Index (Cont.). Numbers in parentheses are the sums of papers authored.

Murvoll, K. M. (1) 155
Näf, Carina (2) 41, 41
Nellissen, J. P. (1) 32
Nettleship, D. N. (1) 155
Neugebauer, E. (1) 213
Newsted, J. L. (1) 119
Newton, I. (1) 156
Ng, P. (1) 26
Nicholson, J. K. (1) 164
Nielsen, Christian Overgaard (2) 156, 72
Nilssen, V. H. (1) 155
Noble, D. G. (3) 74, 74, 75
Norheim, Gunnar (7) 25, 112, 150, 157, 157, 157, 22
Norstrom, Ross J. (48) 37, 42, 86, 93, 100, 102, 103, 109, 146, 150, 160, 160, 184, 193, 205, 206, 206, 217, 218, 153, 26, 27, 58, 58, 74, 75, 75, 76, 82, 98, 98, 101, 102, 127, 137, 144, 158, 158, 159, 159, 167, 168, 170, 173, 181, 194, 207, 220
Nounou, Pierre (1) 161
Nuuja, I. (1) 130
Nygård, T. (1) 110
O'Brien, P. J. (2) 137, 138
Odsjö, Tjelvar (4) 161, 189, 220, 25
Ogi, Haruo (1) 112
Ohlendorf, Harry M. (1) 162
Ohno, Hideki (1) 73
Ólafsdóttir, K. (2) 229, 230
Olcott, Harold S. (1) 177
Olsson, Mats (14) 26, 115, 116, 121, 127, 161, 163, 189, 189, 220, 25, 117, 163, 188
Onuska, F. I. (2) 98, 98
Osborn, D. (4) 43, 145, 164, 164
Ostnes, J. E. (1) 155
Otterlind G. (1) 117
Oxynos, K. (1) 164
Paasivirta, Jaakko (3) 165, 199, 130
Page, G. W. (1) 165
Palokangas, Risto (1) 199
Parker, Keith R. (1) 154
Parra, O. (1) 83
Parslow, J. L. F. (2) 140, 165
Paulsen, G. B (1) 59

Payne, J. F. (2) 137, 153
Peakall, B. (1) 148
Peakall, David B. (32) 93, 138, 149, 149, 149, 170, 171, 207, 52, 169, 52, 75, 84, 93, 137, 138, 148, 148, 155, 166, 166, 166, 167, 167, 167, 168, 168, 168, 170, 177, 210
Pearce, Peter A. (4) 171, 75, 76, 170
Peard, John (2) 53, 194
Pearson, M. A. (1) 96
Pekarik, C. (4) 118, 171, 172, 172
Pellinen, Jukka (1) 165
Perley, B. P. (1) 52
Persson, B. R. R. (1) 111
Persson, Wawa (2) 26, 25
Peterat, B. (1) 103
Petersen, Ć (1) 229
Petersen, Margaret R. (3) 89, 94, 225
Petrosyan, V. S. (1) 137
Pettit, K. E. (3) 26, 172, 173
Pfaffenberger, Bernd (1) 174
Phillips, R. A. (1) 195
Piatt, John F. (1) 174
Pietruszka, C. (1) 220
Polder, Anuschka (2) 91, 186
Poliakova, O. V. (1) 137
Poppenga, Robert H. (2) 88, 111
Potts, G. R. (3) 60, 175, 178
Powell, D. C. (1) 147
Pugh, Rebecca S. (4) 133, 208, 209, 58
Quakenbush, Lori T. (1) 197
Quick, M. P. (1) 175
Rahimtula, A. D. (3) 137, 138, 168
Rantamäki, Pirjo (3) 124, 138, 139
Rappe, Christoffer (2) 80, 161
Rattner, Barnett A. (2) 175, 190
Reiche, P. (1) 177
Reilly, Stephanie M. (1) 49
Reimer, K. (1) 153
Renber, L. (1) 163
Renzoni, A. (2) 137, 139
Reutergårdh, Lars (3) 163, 220, 25
Reynolds, Lincoln M. (2) 171, 210
Richardson, A. (1) 178

Author Index (Cont.). Numbers in parentheses are the sums of papers authored.

Riget, Frank F. (6) 211, 59, 72, 73, 176, 176
Risebrough, K. W. (1) 19
Risebrough, Robert W. (3) 149, 177, 177
Robinson, J. (2) 60, 178
Robinson-Wilson, Everett (1) 106
Rocque, Deborah A. (1) 178
Rodrigue, Jean (1) 180
Roffe, Thomas J. (1) 106
Rolff, Carl (1) 41
Roseneau, David G. (3) 133, 208, 209
Ross, Peter S. (1) 205
Rozemeijer, Marcellino (1) 66
Rudbäck, Eeva (1) 100
Rüdel, Heinz (1) 182
Rudis, Deborah D. (1) 106
Ruiz, X. (1) 20
Russel, H. A. (1) 201
Rüssel-Sinn, H. A. (1) 103
Rutten, A. (1) 68
Ruus, Anders (2) 179, 180
Ryan, John J, (1) 180
Ryan, P. (1) 127
Ryazhenov, N. I. (1) 29
Ryckman, D. P. (2) 131, 181
Ryder, John P. (1) 181
Ryssdal, Jostein (1) 188
Saeki, Kazutoshi (3) 130, 128, 129
Sagerup, Kjetil (4) 182, 183, 183, 107
Sandersen, Erik (1) 188
Sanderson, J. Thomas (1) 184
Sandvik, Morten (1) 179
Särkkä, Jukka (2) 165, 199
Savinov, Vladimir M. (2) 184, 185
Savinova, Tatiana I. (1) 195
Savinova, Tatiana N. (7) 184, 185, 186, 185, 185, 186, 186
Savva, Demetris. (1) 195
Scanlon, Patrick F. (2) 95, 95
Scharenberg, W. (1) 187
Schepens, P. J. C. (1) 62
Scheuhammer, Anton M. (2) 77, 76
Schiettecatte, L.-S. (1) 68
Schlabach, Martin. (1) 121

Schmitzer, J. (1) 164
Schroeder, D. J. (1) 147
Schröter-Kermani, Christa (1) 182
Schwartz, T. R. (2) 132, 147
Scott, J. Michael (1) 187
Seip, Knut Lehre (1) 188
Sellström, Ulla (4) 127, 189, 189, 188
Senthilkumar, Kurunthachalam (1) 122
Seys, J. (1) 189
Shea, David W. (1) 194
Shear, H. (1) 98
Sheffield, Steven R. (1) 190
Shieldcastle, Mark C. (1) 219
Shore, Richard F. (1) 191
Shutt, J. L. (2) 102, 191
Sileo, Louis (3) 136, 205, 132
Simac, Kristin S. (3) 208, 209, 133
Simon, Mary (8) 37, 160, 206, 75, 101, 159, 172, 194
Sirois, Jacques (1) 214
Skaare Janneche Utne (26) 30, 51, 91, 108, 108, 109, 113, 180, 182, 183, 192, 192, 211, 22, 22, 29, 50, 107, 107, 155, 186, 30, 50, 51, 179, 183
Skirnisson, K. (1) 230
Skjegstad, N. (1) 107
Skorping, Arne (2) 182, 183
Smith, L. M. (2) 132, 194
Smith, Louise N. (2) 64, 65
Smith, Milton R. (2) 88, 89
Smithwick, Marla M. (1) 145
Smits, Judit J. E. G. (2) 215, 214
Somers, J. D. (1) 193
Spalding, Marilyn G. (1) 105
Sparks, Timothy H. (2) 191, 145
Speakman, J. R. (1) 202
Spear, Philip A. (1) 193
Speich, Steven M. (2) 53, 194
Sperveslage, Hans (1) 24
Staats de Yanes, G. (1) 72
Stalling, D. L. (2) 132, 194
Steiger, Gretchen H. (1) 53
Steinhanses, Jürgen (1) 182
Stenzel, L. E. (1) 165

Author Index (Cont.). Numbers in parentheses are the sums of papers authored.

Stepanova, Ludmila I. (1) 195
Stewart, A. G. (1) 140
Stewart, Fiona M. (3) 204, 195, 196
Stewart, W. D. P. (1) 119
Stickel, Lucille F. (1) 196
Stout, Jordan H. (1) 197
Stow, Craig A. (1) 197
Stow, Jason P. (1) 135
Strandell, Michael (1) 121
Strauch, J.G., Jr. (1) 197
Straughan, Cameron A. (1) 97
Stromborg, Kenneth L. (6) 61, 62, 136, 205, 61, 224
Struger, John (6) 217, 27, 131, 198, 198, 217
Summer, Cheryl L. (2) 143, 143
Suydam, Robert S. (1) 197
Sydeman, William J. (3) 116, 117, 198
Szaro, Robert C. (1) 199
Szefer, Piotr (2) 81, 79
Tanabe, Shinsuke (7) 115, 129, 130, 134, 227, 128, 129
Tanaka, Hiroyuki (4) 125, 200, 128, 129
Tapia, German (2) 119, 67
Tasker, Mark L. (2) 200, 224
Tatsukawa, Ryo (10) 112, 115, 125, 125, 129, 130, 227, 128, 128, 129
Tatton, J. O. G. (1) 151
Teeple, Stanley M. (2) 201, 93
Ternes, W. (1) 201
Theede, Hans (1) 216
Thingstad, P. G. (1) 110
Thompson, Anu (1) 18
Thompson, David R. (10) 91, 204, 90, 141, 196, 202, 202, 203, 203, 203
Thompson, Steven P. (1) 105
Tillitt, Donald E. (7) 136, 143, 143, 204, 205, 119, 147
Titenko, Alexei M. (1) 134
Tittlemier, Sheryl A. (3) 205, 206, 206
Tremblay, J. (1) 168
Trick, J. A. (1) 132
Trites, A. (1) 146
Trivelpiece, Wayne Z. (3) 207, 52, 148

Troy, Declan (1) 225
Trudeau, Suzanne (5) 86, 84, 85, 107, 126
Trust, Kimberly A. (1) 197
Turle, Richard (4) 207, 27, 144, 207
Ugland, Karl Inne (2) 180, 179
Uvemo, Ulla-Britt (1) 220
Vader, W. (1) 22
Van den Dungen, H. M. (1) 113
Van Franeker, Jan Andries (1) 55
Van Hove Holdrinet, Micheline (1) 88
Van Raat, Patrick (1) 66
Vander Pol, Stacy S. (5) 133, 208, 208, 209, 58
Vaz, Reggie (1) 116
Verbrugge, David A. (3) 136, 119, 224
Vermeer, Kees (2) 210, 210
Verreault Jonathan (1) 211
Vorkamp, Katrin (1) 211
Wada, Toyohito (1) 125
Wagemann, R. (1) 153
Wakeford, Brian (1) 36
Wakeford, Bryan (1) 160
Wakimoto, Tadaaki (2) 57, 56
Walker, C. H. (4) 31, 31, 32, 212
Walsh, Paul M. (3) 91, 203, 212
Ward, P. (1) 43
Watanabe, Mafumi (1) 134
Wayland, Mark (6) 86, 214, 215, 213, 213, 214
Webb, A. (1) 224
Weis, I. Michael (1) 54
Welch, Harold E. (1) 20
Wells, P. G. (1) 168
Wenzel, Christine (3) 216, 215, 216
Weseloh, D. V. Chip (6) 97, 102, 160, 217, 218, 131
Weseloh, D. Vaughn (17) 114, 146, 205, 219, 27, 98, 101, 118, 159, 171, 172, 172, 173, 181, 198, 198, 217
Whitehead, D. (1) 146
Whitehead, Philip E. (6) 77, 74, 75, 76, 159, 220
Widequist, Ulla (2) 220, 120
Wienburg, C. L. (1) 145

Author Index (Cont.). Numbers in parentheses are the sums of papers authored.

Wiens, John A. (9) 64, 65, 154, 187, 221, 222, 223, 220, 221
Wigfield, Donald C. (1) 86
Wiig, Øystein (1) 192
Wilkins, J. P. G. (3) 31, 31, 32
Williams, J. M. (1) 224
Williams, Kim (1) 85
Williams, Lisa L. (3) 123, 143, 224
Wilson, Barry W. (1) 63
Wilson, Heather M. (1) 225
Wilson, James G. (1) 225
Wilson, Laurie K. (1) 100
Wilson, Ulrich W. (1) 105
Winker, Kevin (1) 178
Witter, Michael (1) 194
Wolkers, Hans (1) 24
Wolkers, J. (1) 107
Won, Henry T. (3) 160, 58, 85
Wong, Michael P. (3) 226, 38, 226
Woodburn, M. (1) 90
Worman, J. J. (1) 96
Wright, Julian (2) 191, 145
Xhrouet, C. (1) 66
Yamashita, Nobuyoshi (2) 123, 227
York, Geoff W. (2) 133, 209
Yoshida, Steven H. (1) 40
Young, David R. (2) 227, 113
Young, W. J. (1) 164
Zakrisson, Susanne (1) 26
Zebühr, Yngve (2) 41, 41
Zhu, Jiping (1) 103
Zitko, V. (6) 227, 228, 228, 228, 229, 229
Zunk, B. (1) 229

Publication Index. Numbers in parentheses are the sums of references published.

Acta Vet. Scand. (2) 27, 83
Acta Veterinaria Scandinavica Suppl. (1) 112
Am. Fish Soc. Symp. (1) 174
Am. J. Physiol. (1) 149
Ambio (2) 163, 166
Angewandte Chemie (1) 122
Animals as Monitors of Environmental Pollutants. Symposium on Pathobiology of Environmental Pollutants: Animal Models and Wildlife as Monitors, University of Connecticut, 1977. (3) 93, 113, 149
Annales Zoologici Fennici (1) 138
Annu. Rev. Nutr. (1) 133
Apatity, USSR: Kol. Nauch. Tsentr, Akad. Nauk SSSR. (1) 185
Aquat. Toxicol. (1) 179
Arch. Environ. Contam. Toxicol. (33) 38, 22, 26, 36, 37, 44, 52, 53, 56, 61, 74, 76, 88, 94, 95, 107, 112, 113, 120, 128, 128, 132, 145, 147, 152, 155, 187, 196, 203, 213, 215, 216, 224
Arch. Environ. Health (1) 180
Arch. Toxicol. (1) 78
Arctic (2) 54, 214
Astarte (3) 81, 81, 81
Biol. Conserv. (1) 221
Biol. Ser. Program Fish. Wildl. Serv. (U.S.) (1) 196
Biological Trace Element Research (1) 68
Bioscience (1) 221
Bromatologia i Chemia Toksykologiczna (2) 80, 80
Brookings, SD, USA: South Dakota State Univ. (1) 99
Bull. Environ. Contam. Toxicol. (14) 23, 24, 62, 72, 96, 97, 127, 131, 152, 227, 227, 228, 229, 229
Bull. Environ. Contamin. Toxicol. (2) 34, 199
Bull. Mt. Desert Isl. Biol. Lab. (2) 52, 148

Bulletin de la Societe Royale des Sciences de Liege (3) 19, 32, 118
Can. Field Nat. (4) 19, 181, 201, 210
Can. J. of Fish. Aquat. Sci. (1) 20
Canadian Wildlife Service Analytical Report CRD-88-5 (1) 159
Canadian Wildlife Service, Ontario Region Technical Reports ??? (1) 118
Charleston, SC, USA: College of Charleston, Masters thesis (1) 208
Chemosphere (19) 30, 108, 115, 116, 121, 122, 127, 130, 139, 144, 161, 165, 174, 189, 194, 207, 207, 220, 98
Colonial Waterbirds (2) 105, 194
Comp. Biochem. Physiol. (1) 164
Comp. Biochem. Physiol. C (7) 30, 31, 32, 32, 164, 167, 167
Condor (4) 21, 52, 154, 207
Die Vogelwarte (1) 103
Dis. Aquat. Org. (1) 67
DOE(EV(10254-T1. NTIS Order No.: DE82016736; Contract AC03-79EV10254 (1) 165
Ecol. Appl. (2) 64, 223
Ecol. Model. (1) 188
Ecotoxicol. Environ. Saf. (1) 119
Ecotoxicol. Ser., Vol. 1, Chapman and Hall, New York, NY (1) 169
Ecotoxicology (5) 105, 135, 183, 195, 215
Environ. Chem. Toxicol. (1) 193
Environ. Health Perspect. (6) 89, 95, 106, 142, 170, 211
Environ. Monit. Assess. (12) 23, 44, 45, 48, 74, 83, 85, 110, 172, 181, 213, 217
Environ. Physiol. Biochem. (1) 212
Environ. Pollut. (30) 22, 22, 25, 26, 29, 34, 35, 42, 60, 82, 100, 109, 109, 129, 130, 131, 135, 140, 141, 147, 157, 230, 165, 170, 193, 197, 202, 205, 227, 229
Environ. Pollut. (A Ecol. Biol.) (3) 114, 153, 157
Environ. Pollut. (B. Chem. Phys.) (3) 70, 150, 225

Publication Index (Cont.). Numbers in parentheses are the sums of references published.

Environ. Res. (6) 44, 67, 137, 166, 168, 214
Environ. Rev. (2) 93, 102
Environ. Sci. Technol. (21) 37, 92, 100, 101, 102, 113, 115, 116, 121, 123, 123, 125, 134, 145, 150, 160, 182, 189, 197, 206, 208
Environ. Toxicol. Chem. (24) 37, 41, 41, 49, 51, 51, 58, 62, 61, 63, 86, 88, 94, 117, 32, 136, 148, 160, 168, 178, 180, 205, 219, 225
Environmental Assessment of the Alaskan Continental Shelf. Annual Reports of Principal Inverstigators for the Year Ending March 1980. NOAA(OMPA, Boulder, CO. NOAA-OMPA-AR-80-1. (1) 220
FAO Fish. Rep. (1) 139
Fauna (Blindern)(1) 25
Fortschritte in der Atomspektrometrischen Spurenanalytik (2) 201, 229
Fundam. Appl. Toxicol. (2) 46, 48
Gaithersburg, MD: National Institute of Standards and Technology. NISTIR 7029 (1) 209
Halifax, Nova Scotia, Canada: Environmental Protection Service, Atlantic Region. EPS-5-AR-80-1. (1) 146
IAGLR-86 Program. International Association for Great Lakes Research 29th Conference, May 26-29, 1986 (2) 58, 217
Ibis (1) 224
In. Theoretical approaches to the study of ecosystmes at Arctic and Subarctic seas. Apatity, The Kola Scientific Centre Publishing House (1) 185
In: Environmental assessment of the Alaskan continental shelf. Northeast Gulf of Alaska interim synthesis report. Science Applications, Inc., Boulder, CO 80301, USA (1) 197
In: Extended Abstracts of the International Symposium on the Analysis of Hydrocarbons and Halogenated Hydrocarbons, #40, University of Toronto, Ontario (1) 98
In: Furness, Robert W. and Rainbow, Philip S. (Eds.). Heavy Metals in the Marine Environment. Boca Raton, FL: CRC Press. (2) 91, 212
In: Jauniaux, Thierry; Bouquegneau, Jean Marie; Coignoul, Freddy (Eds.). Marine Mammals, Seabirds and Pollution of Marine Systems. Bulletin de la Societe royale des sciences de Li. (1) 189
In: Johnston, R. (Ed.) Marine pollution. London: Academic Press. (1) 32
In: Kendall, Ronald J., Richard L. Dickerson, John P. Giesy, William P. Suk (Eds.). Principles and Processes for Evaluating Endocrine Disruption in Wildlife, Proceedings from Principles and Processes for Evaluating Endocrine Disruption in Wildlife; March (2) 104, 190
In: Levings, C. D.; Turner, R. B.; Ricketts, B. (Eds.) Proceedings of the Howe Sound Environmental Science Workshop., Can. Tech. Rep. Fish. Aquat. Sci. No. 1879 (1) 220
In: Låg, J. (Ed.) Excess and deficiency of trace elements in relation to human and animal health in Arctic and Subarctic regions. Norwegian Academy of Science and Letters. (1) 157
In: Proceedings of the 38th Conference of the International Association for Great Lakes Research, East Lansing, MI (USA), 28 May-1 Jun 1995 (1) 159
In: Schmidtke, Norbert W. (Ed.), Toxic Contamination in Large Lakes. Volume 1. Chronic Effects of Toxic Contaminants in Large Lakes (1) 166
In: Sellström, U. Determination of some polybrominated flame retardants in

Publication Index (Cont.). Numbers in parentheses are the sums of references published.

biota, sediment, and sewage sludge. Ph.D. Dissertation, Stockholm University. (1) 188
In: Servos, Mark R., Munkittrick, Kelly R., Carey, John H., van der Kraak, Glen J. (Eds.), Environmental Fate and Effects of Pulp and Paper Mill Effluents. Delray Beach, FL: St. Lucie Press. (1) 76
In: Walker, Colin H.; Livingstone, David R. Persistent Pollutants in Marine Ecosystems. New York: Pergamon Press. (1) 75
In: Wells, P.G.; Butler, J.N.; Hughes, J.S. (Eds.). Exxon Valdez Oil Spill: Fate and Effects in Alaskan Waters. Philadelphia, PA (2) 65, 222
Institute of Terrestrial Ecology Annual Report 1981 (1) 156
Intern. J. Environ. Anal. Chem. (1) 228
J Exp Zool (1) 175
J. Anal. At. Spectrom. (1) 58
J. Appl. Ecol. (4) 50, 183, 202, 203
J. Envrion. Monit. (1) 24
J. Forensic Sci. Soc. (1) 175
J. Great Lakes Res. (10) 85, 97, 103, 109, 114, 131, 143, 143, 198, 218
J. Ornithol. (1) 23
J. Toxicol. Environ. Health (3) 45, 47, 184
J. Wildl. Dis. (1) 89
J. Wildl. Manage. (6) 41, 84, 93, 110, 134, 187
Kankyo Kagaku (2) 129, 200
Liege, Belgium: Universite de l'Etat a Liege, Faculty of Sciences, Oceanolgy (1) 69
Mar. Ecol. Prog. Ser. (2) 20, 195
Mar. Environ. Res. (8) 31, 68, 84, 107, 126, 167, 168, 204
Mar. Pollut. Bull. (23) 18, 19, 20, 28, 34, 55, 63, 77, 79, 79, 90, 119, 123, 125, 140, 142, 155, 191, 198, 203, 204, 210, 211

Meddelelser Om Grønland, Bioscience 29 (1) 156
Memoirs of National Institute of Polar Research (1) 91
Nature (10) 28, 43, 60, 117, 119, 151, 175, 177, 177, 178
Nauchnye Osnovy Okhrany Prirody (1) 29
Neth. J. Aquat. Ecol. (1) 200
NOAA Tech. Memo. NOS OMA 113 (1) 53
Nordisk Veterinaermedicin (1) 124
Occas. Pap. Can. Wildl. Serv. (1) 77
Oceanis (1) 71
Oceanologica Acta (1) 111
Oebalia (1) 83
OME 36th Conference of the International Association for Great Lakes Research, June 4-10, 1993. Program and Abstracts (1) 84
Organohalogen Compounds (14) 30, 50, 57, 66, 66, 71, 86, 108, 121, 127, 133, 182, 192, 206
Ornis Fennica (1) 124
Ornis Scand. (1) 36
PCB Conf. II, 1972, suppl. 4E. Solina, Sweden: Swed. Environ. Prot. Board. (1) 163
Pestic. Monit. J. (2) 171, 199
Pharmacol. Biochem. Behav. (4) 43, 46, 47, 49
Preprint, Kol'skij Nauchn. Tsentr Ran, Apatity (Russia) (1) 186
Proc. Inter. Ornithol. Congr. (1) 125
Proceedings International Conference on Transport of Persistent Chemicals in Aquatic Ecosystems. Ottawa, Canada, May 1-3, 1974. (1) 228
Proceedings of the 23rd Annual Aquatic Toxicity Workshop: October 7-9, 1996, Calgary, Alberta. Can. Tech. Rep. Fish. Aquat. Sci. (1) 56
Proceedings of the 26th Conference on Great Lakes Research. May 23-27, 1983,

Publication Index (Cont.). Numbers in parentheses are the sums of references published.

State University of New York at Oswego (2) 158, 198

Proceedings of the 38th Conference of the International Association of Great Lakes Research., International Association for Great Lakes Research, 2200 Bonisteel Boulevard, Ann Arbor, Mi 48109-2099 (USA) (1) 171

Programs and Abstracts of the 28th Confefence on Great Lakes Research University of Wisconsin-Milwaukee (1) 75

Pure. Appl. Chem. (1) 59

Recherche (1) 161

Regul. Toxicol. Pharmacol. (1) 40

Sci. Total Environ. (22) 59, 72, 72, 73, 73, 81, 87, 87, 91, 101, 136, 137, 140, 147, 153, 158, 164, 176, 176, 184, 186, 216

Science (2) 138, 149

Senter for Industriforskning, Report No. 83/11/01-1 (1) 55

Stud. Mater. Ocenaol. (1) 80

Synopsis of Research Conducted under the Northern Contaminants Program 1998-1999. Ottawa: Department of Indian Affairs and Northern Development. (1) 36

Tech. Rep. Ser. No. 187. Canadian Wildl. Serv. St. Foy, PQ, (1) 55

Third periodic assessment of the state of the marine environment of the Baltic Sea, 1989-93 Helsinki : HELCOM (1) 104

Thirteenth International Symposium on Chlorinated Dioxins and Related Compounds. Vienna, Austria, September, 1993. (1) 191

Toxic Chemical Division Reoprt, Canadian Wildlife Service (1) 226

Toxicol. Appl. Pharmacol. (2) 97, 138

Toxicol. Lett. (2) 42, 192

United States Department of the Interior, Fish and Wildlife Service, Special Scientific Report - Wildlife No. 245, Washington, DC (1) 162

University of Guelph. Dept. of Zoology. Ph.D. thesis. (1) 39

Verh. Internat. Verein. Limnol. (1) 98

Water Air Soil Pollut. (1) 151

Water Qual. Res. J. Canada (1) 146

Wildl. Biol. (1) 111

Keyword Index. Numbers in parentheses are the sums of references containing the keyword.

abnormalities (20) 18, 32, 53, 61, 84, 84, 93, 93, 104, 109, 111, 133, 136, 138, 143, 166, 181, 196, 224, 227
Accipiter cooperii (1) 125
Accipiter gentilis (1) 110
Accipiter nisus (1) 169
Accipiter striatus (3) 65, 125, 223
Actitis macularia (4) 64, 65, 125, 223
Adelie penguin (1) 32
adrenal cortex (1) 175
Aechmophorus occidentalis (5) 32, 52, 125, 210, 220
Aethia cristatella (1) 112
Aethia psittacula (2) 112, 223
Aethia pusilla (1) 112
Ag (7) 32, 38, 128, 135, 198, 198, 226
age variations (35) 24, 37, 39, 45, 51, 52, 59, 61, 68, 69, 74, 105, 111, 114, 115, 138, 139, 141, 153, 156, 169, 176, 176, 178, 179, 185, 187, 195, 198, 201, 211, 212, 215, 216, 226
Agelaius phoeniceus (4) 146, 152, 152, 169
AHH (5) 62, 78, 132, 190, 200
Aix sponsa (2) 38, 226
AK (26) 21, 58, 64, 65, 89, 94, 106, 112, 125, 133, 133, 154, 162, 174, 178, 197, 197, 208, 208, 209, 220, 221, 221, 222, 223, 225
Al (7) 85, 105, 106, 110, 198, 198, 226
ALAD (1) 88
albatross (1) 212
Alberta (3) 38, 193, 226
Alca torda (26) 22, 22, 31, 31, 32, 32, 34, 38, 43, 79, 80, 80, 81, 81, 123, 140, 145, 151, 171, 178, 184, 185, 186, 200, 203, 203
Alcedo atthis (1) 70
aldrin (6) 18, 19, 34, 38, 133, 212
Alectoris graeca (2) 38, 169
Aleutian tern (1) 162
Alle alle (19) 20, 32, 34, 42, 72, 72, 73, 79, 82, 125, 150, 156, 157, 157, 184, 185, 186, 205, 222
American avocet (2) 125, 210
American bittern (1) 48
American black duck (4) 125, 136, 226, 229
American coot (2) 38, 48
American kestrel (3) 125, 169, 175
American wigeon (6) 38, 64, 65, 210, 223, 226
American woodcock (2) 38, 125
Anas acuta (9) 38, 129, 130, 136, 137, 195, 210, 223, 226
Anas americana (6) 38, 64, 65, 210, 223, 226
Anas clypeata (5) 38, 137, 195, 223, 226
Anas crecca (7) 38, 64, 65, 87, 136, 223, 226
Anas discors (3) 38, 210, 226
Anas penelope (1) 226
Anas platyrhynchos (21) 38, 52, 64, 65, 87, 125, 137, 139, 146, 149, 149, 151, 168, 169, 175, 195, 196, 210, 218, 223, 229
Anas poecilorhyncha (1) 122
Anas rubripes (4) 125, 136, 226, 229
Anas strepera (5) 38, 210, 218, 223, 226
ancient murrelet (10) 38, 65, 74, 77, 77, 112, 125, 162, 177, 223
Anous tenuirostris (1) 141
Anser albifrons (2) 169, 226
Anser anser (1) 137
Anser fabalis (1) 29
Antarctic petrel (1) 141
Antarctic prion (1) 141
Antarctic tern (1) 32
Aphriza virgata (1) 223
Aquila chrysaetos (1) 125
Arctic loon (6) 32, 73, 79, 129, 130, 199
Arctic Ocean (2) 92, 115
Arctic skua   see parasitic jaeger
Arctic tern (22) 20, 34, 36, 39, 41, 64, 65, 128, 129, 130, 136, 138, 139, 141, 162, 184, 185, 187, 195, 200, 221, 223
Ardea alba (2) 123, 169
Ardea cinerea (9) 70, 87, 112, 113, 117, 122, 137, 153, 195
Ardea herodias (13) 52, 53, 64, 65, 76, 123, 125, 127, 159, 184, 194, 210, 223
Arenaria melanocephala (1) 223
As (20) 32, 38, 44, 52, 85, 94, 105, 110, 111, 133, 135, 136, 178, 184, 185, 186, 197, 198, 225, 226

Keyword Index (Cont.). Numbers in parentheses are the sums of references containing the keyword.

ashy petrel (2) 32, 125
Asia (20) 22, 29, 30, 32, 56, 57, 73, 112, 122, 128, 129, 130, 134, 137, 147, 185, 186, 195, 197, 200
Asio flammeus (1) 125
Asio otus (1) 125
Athene cunicularia (2) 125, 169
Atlantic Ocean (19) 28, 32, 34, 42, 82, 83, 90, 91, 111, 125, 133, 150, 151, 170, 186, 203, 205, 206, 206
Atlantic petrel (3) 32, 90, 204
Atlantic puffin (34) 22, 22, 28, 31, 31, 32, 32, 34, 38, 43, 75, 76, 125, 138, 145, 149, 151, 164, 164, 167, 169, 170, 171, 178, 184, 185, 186, 200, 203, 203, 203, 206, 206, 222
Audouin's gull (3) 20, 83, 139
Audubon's shearwater (1) 32
Aukland Island shag (1) 32
Australasian gannet (2) 32, 141
Australia (1) 32
avocet (3) 72, 83, 139
Aythya affinis (5) 38, 146, 210, 218, 226
Aythya americana (2) 38, 226
Aythya collaris (4) 38, 65, 223, 226
Aythya ferina (2) 137, 195
Aythya fuligula (4) 81, 137, 195, 229
Aythya marila (6) 32, 38, 81, 218, 223, 226
Aythya valisineria (3) 38, 169, 226
AZ (1) 133
Azores (1) 151
B (3) 152, 197, 226
Ba (3) 85, 198, 225
Baja (1) 152
bald eagle (12) 64, 65, 104, 123, 125, 154, 166, 169, 190, 206, 219, 223
Baltic Sea (39) 19, 20, 25, 26, 30, 32, 41, 41, 78, 79, 79, 80, 80, 81, 88, 91, 100, 104, 115, 116, 117, 120, 122, 123, 127, 130, 133, 139, 151, 163, 163, 185, 185, 186, 188, 189, 189, 220, 226
bar-tailed godwit (3) 56, 57, 122
Barents Sea (19) 22, 24, 28, 29, 30, 42, 50, 91, 111, 113, 183, 183, 184, 185, 186, 186, 186, 211, 185

barn swallow (1) 169
Barrow's goldeneye (4) 64, 65, 223, 226
Bartram's sandpiper (1) 32
Bartramia longicauda (1) 32
Bay of Fundy (12) 36, 37, 76, 85, 171, 185, 206, 227, 228, 228, 229, 229
bean goose (1) 29
Belgium (12) 32, 32, 32, 66, 67, 67, 68, 68, 69, 70, 118, 119
belted kingfisher (3) 64, 65, 223
Bengalese finch (1) 169
benzenes (24) 27, 34, 35, 36, 38, 42, 74, 85, 86, 98, 98, 102, 118, 133, 146, 146, 153, 172, 172, 173, 197, 207, 217, 218
Bering Sea (9) 58, 112, 125, 125, 133, 133, 208, 208, 209
Bermuda petrel (2) 32, 125
bile (1) 41
bioaccumulation (69) 22, 22, 23, 24, 30, 31, 36, 37, 39, 45, 49, 51, 52, 53, 58, 59, 61, 62, 68, 69, 70, 71, 74, 76, 81, 81, 83, 85, 111, 114, 114, 115, 116, 119, 124, 130, 134, 139, 140, 141, 151, 153, 155, 156, 159, 160, 176, 176, 178, 179, 180, 185, 186, 187, 193, 193, 195, 196, 198, 201, 202, 203, 204, 206, 211, 212, 215, 216, 216
biomagnification (121) 18, 19, 20, 20, 23, 26, 29, 30, 30, 32, 34, 34, 35, 36, 37, 41, 41, 41, 42, 44, 48, 52, 53, 54, 56, 59, 61, 62, 63, 71, 72, 73, 76, 76, 79, 79, 80, 80, 82, 83, 85, 87, 88, 91, 93, 97, 100, 101, 102, 103, 108, 110, 111, 112, 113, 113, 114, 116, 116, 117, 117, 119, 119, 120, 121, 122, 125, 125, 127, 131, 134, 135, 136, 137, 138, 139, 141, 144, 145, 145, 146, 147, 150, 151, 152, 153, 157, 159, 162, 163, 169, 176, 176, 177, 177, 178, 178, 179, 180, 183, 187, 189, 195, 195, 196, 198, 199, 199, 201, 202, 203, 205, 206, 210, 214, 216, 218, 220, 227, 229, 229
biomarkers (7) 67, 69, 76, 84, 95, 214, 215
black-bellied plover (1) 125
black-bellied storm-petrel (1) 141
black-billed magpie (3) 64, 65, 223
black-browed albatross (3) 32, 128, 141

Keyword Index (Cont.). Numbers in parentheses are the sums of references containing the keyword.

black cormorant   see cormorant
black-crowned night-heron   (14)   27, 48, 73, 88, 104, 105, 118, 123, 166, 169, 173, 190, 210, 218
black-eared kite   (1)   122
black-footed albatross   (9)   28, 32, 112, 123, 128, 128, 129, 206, 206
black guillemot   (41)   19, 20, 20, 29, 32, 34, 34, 36, 36, 38, 39, 42, 62, 72, 72, 73, 79, 82, 91, 113, 130, 135, 145, 147, 149, 150, 153, 156, 163, 166, 168, 169, 171, 184, 185, 186, 200, 205, 211, 221, 229
black-headed gull   (24)   19, 22, 23, 23, 25, 56, 57, 72, 79, 83, 87, 103, 118, 122, 134, 137, 139, 140, 140, 147, 151, 151, 195, 200
black-legged kittiwake   (68)   19, 20, 22, 22, 28, 29, 30, 32, 34, 34, 35, 36, 36, 37, 39, 42, 64, 65, 71, 72, 72, 73, 81, 81, 81, 82, 90, 90, 101, 103, 109, 111, 112, 113, 140, 145, 150, 151, 153, 154, 155, 156, 162, 177, 178, 184, 185, 185, 186, 186, 188, 195, 195, 200, 200, 201, 203, 203, 205, 209, 216, 216, 220, 221, 221, 222, 223, 229
black-necked grebe   (5)   56, 57, 83, 122, 139
black-noddy   see white-capped noddy
black oystercatcher   (5)   64, 65, 154, 221, 223
black petrel   (2)   32, 125
black scoter   (10)   19, 32, 38, 55, 64, 65, 81, 106, 223, 226
black skimmer   (2)   44, 49
black-tailed gull   (3)   56, 57, 122
black tern   (3)   88, 118, 214
black-throated diver   see Arctic loon
black turnstone   (1)   223
black-winged petrel   (1)   141
black-winged stilt   (2)   83, 139
blood   (33)   50, 50, 51, 51, 53, 53, 58, 84, 88, 89, 92, 94, 95, 99, 105, 107, 108, 111, 121, 123, 138, 138, 142, 149, 151, 155, 158, 160, 161, 164, 211, 213, 225
blue-eyed shag   (1)   32
blue grouse   (1)   38
blue penguin   (2)   32, 141

blue petrel   (1)   141
blue shag   (1)   141
blue-winged teal   (3)   38, 210, 226
bobwhite quail   (2)   169, 175
body burden   (5)   37, 58, 128, 151, 212
Bonaparte's gull   (7)   36, 36, 37, 39, 41, 154, 223
Bonasa umbellus   (1)   38
bone   (12)   32, 38, 49, 76, 77, 113, 115, 128, 141, 186, 198, 212
Botaurus lentiginosus   (1)   48
Bounty Island shag   (1)   141
Br   (2)   52, 226
Brachyramphus brevirostris   (1)   223
Brachyramphus marmoratus   (6)   38, 64, 65, 125, 221, 223
brain   (32)   32, 36, 37, 38, 39, 41, 91, 97, 99, 105, 108, 109, 109, 113, 115, 117, 128, 131, 134, 140, 160, 164, 177, 177, 186, 186, 187, 210, 212, 216, 226, 227
Brandt's cormorant   (8)   32, 113, 116, 117, 123, 177, 198, 227
brant   (1)   226
Branta bernicla   (1)   226
Branta canadensis   (13)   38, 48, 64, 65, 87, 133, 136, 162, 169, 210, 218, 223, 226
British Columbia   (16)   28, 32, 38, 74, 75, 76, 77, 77, 100, 127, 133, 159, 184, 206, 220, 226
Brittany   (2)   66, 215
broad-billed prion   (2)   32, 212
brown booby   (5)   32, 125, 128, 129, 141
brown pelican   (6)   32, 123, 125, 169, 177, 212
brown skua   (2)   141, 157
Brünnich's guillemot   see thick-billed murre
Bubo bubo   (3)   110, 112, 163
Bubo virginianus   (1)   125
Bubulcus ibis   (2)   152, 152
Bucephala albeola   (6)   38, 64, 65, 218, 223, 226
Bucephala clangula   (10)   32, 38, 64, 65, 80, 80, 81, 163, 220, 226
Bucephala islandica   (4)   64, 65, 223, 226

Keyword Index (Cont.). Numbers in parentheses are the sums of references containing the keyword.

bufflehead (6) 38, 64, 65, 218, 223, 226
Buller's albatross (1) 141
burrowing owl (2) 125, 169
Buteo buteo (1) 113
Buteo jamaicensis (1) 125
Buteo regalis (1) 125
Buteo swainsoni (1) 125
buzzard (1) 113
$C_{10}H_6N_2Br_4Cl_2$ (2) 206, 206
CA (13) 52, 63, 116, 117, 165, 177, 177, 185, 186, 190, 198, 226, 227
Ca (7) 32, 52, 76, 77, 99, 198, 198
Calidris alpina (2) 125, 223
Calidris maritima (4) 32, 34, 59, 229
Calidris mauri (1) 223
Calidris minutilla (3) 65, 125, 223
Calidris ptilocnemis (2) 178, 223
Calidris tenuirostris (3) 56, 57, 122
California gull (6) 113, 125, 169, 175, 210, 227
California quail (1) 169
Callipepla californica (1) 169
Calonectris (1) 212
Calonectris diomedea (4) 32, 83, 139, 151
Canada (89) 19, 20, 26, 28, 32, 34, 35, 36, 36, 36, 37, 37, 38, 39, 41, 42, 42, 54, 55, 56, 58, 74, 74, 75, 75, 76, 76, 77, 77, 82, 84, 85, 85, 86, 88, 92, 93, 95, 95, 100, 102, 125, 126, 127, 127, 133, 133, 135, 136, 136, 145, 146, 146, 149, 150, 153, 155, 159, 168, 170, 171, 180, 181, 181, 184, 185, 186, 193, 205, 206, 206, 210, 210, 213, 213, 214, 214, 215, 218, 219, 220, 226, 226, 227, 228, 228, 228, 229, 229
Canada goose (13) 38, 48, 64, 65, 87, 133, 136, 162, 169, 210, 218, 223, 226
canvasback duck (3) 38, 169, 226
cape gannet (1) 32
cape pigeon (2) 32, 141
Capella gallinago see *Gallinago gallinago*
carcass (39) 24, 32, 36, 37, 37, 38, 39, 41, 54, 55, 61, 61, 62, 71, 81, 96, 97, 97, 99, 111, 115, 119, 125, 125, 127, 134, 151, 152, 153, 158, 160, 165, 165, 177, 177, 178, 187, 191, 194
carotene (2) 97, 99
Casmerodius alba see *Ardea alba*
Caspian tern (13) 27, 95, 104, 143, 143, 146, 169, 173, 190, 204, 214, 218, 227
Cassin's auklet (6) 38, 77, 77, 112, 125, 177
Catharacta (1) 212
Catharacta hamiltoni (1) 204
Catharacta lonnbergi (2) 141, 157
Catharacta skua (11) 32, 34, 90, 114, 151, 195, 200, 202, 203, 203, 204
Cathartes aura (2) 32, 52
Catoptrophorus semipalmatus (1) 125
cattle egret (2) 152, 152
CBC (2) 97, 99
Cd (72) 19, 24, 32, 32, 38, 43, 44, 44, 44, 45, 45, 48, 49, 67, 67, 68, 69, 73, 74, 76, 85, 87, 91, 94, 94, 103, 105, 106, 110, 111, 112, 113, 114, 118, 124, 128, 129, 130, 133, 133, 135, 136, 140, 141, 152, 153, 156, 157, 164, 164, 176, 176, 178, 184, 185, 186, 196, 197, 198, 198, 201, 212, 213, 213, 214, 215, 215, 216, 216, 225, 226, 229
Cepphus columba (12) 38, 53, 64, 65, 116, 117, 154, 162, 194, 198, 221, 223
Cepphus grylle (41) 19, 20, 20, 29, 32, 34, 34, 36, 36, 38, 39, 42, 62, 72, 72, 73, 79, 82, 91, 113, 130, 135, 145, 147, 149, 150, 153, 156, 163, 166, 168, 169, 171, 184, 185, 186, 200, 205, 211, 221, 229
Cerorhinca monocerata (16) 28, 38, 65, 74, 75, 77, 77, 112, 116, 117, 125, 177, 198, 206, 206, 223
Ceryle alcyon (3) 64, 65, 223
Charadrius hiaticula (2) 23, 72
Charadrius semipalmatus (1) 125
Charadrius vociferus (1) 125
Chatham Island petrel (1) 141
Chen caerulescens (2) 38, 226
Chen rossii (1) 226
chicken (2) 78, 169
Chile (1) 83
China (1) 112
China Sea (1) 112

Keyword Index (Cont.). Numbers in parentheses are the sums of references containing the keyword.

chinstrap penguin (1) 32
Chlidonias niger (3) 88, 118, 214
chlordanes (96) 19, 22, 24, 27, 29, 30, 30, 34, 35, 36, 37, 38, 42, 50, 51, 51, 54, 55, 56, 56, 58, 59, 61, 61, 62, 74, 74, 75, 77, 82, 85, 86, 89, 91, 93, 93, 95, 100, 107, 108, 109, 113, 116, 116, 118, 125, 125, 130, 131, 133, 133, 134, 135, 136, 144, 146, 151, 152, 153, 158, 158, 162, 164, 168, 170, 172, 172, 173, 178, 179, 180, 181, 183, 183, 186, 190, 192, 193, 197, 199, 207, 208, 208, 209, 210, 211, 211, 214, 217, 217, 218, 219, 225, 226, 227, 229
chough (1) 191
chukar (2) 38, 169
Circus cyaneus (1) 125
Clangula hyemalis (23) 29, 32, 34, 38, 59, 64, 65, 72, 81, 87, 87, 106, 123, 128, 129, 130, 156, 163, 185, 186, 220, 223, 226
CO (1) 226
Co (6) 32, 85, 128, 152, 185, 198
Colinus virginianus (2) 169, 175
Columba livia (4) 38, 110, 169, 195
common cormorant see cormorant
common eider (62) 20, 22, 32, 34, 36, 38, 39, 41, 41, 54, 59, 63, 72, 72, 72, 73, 78, 80, 81, 83, 87, 87, 88, 89, 94, 103, 110, 111, 111, 113, 119, 122, 124, 135, 136, 140, 147, 153, 156, 157, 157, 157, 171, 174, 178, 180, 184, 185, 186, 186, 188, 196, 197, 199, 201, 212, 213, 213, 214, 215, 226, 230
common goldeneye (10) 32, 38, 64, 65, 80, 80, 81, 163, 220, 226
common grackle (1) 169
common gull (25) 27, 56, 57, 64, 65, 71, 72, 73, 79, 80, 103, 121, 122, 134, 137, 140, 147, 154, 162, 165, 195, 200, 214, 221, 223
common loon (8) 32, 64, 65, 123, 136, 145, 162, 223
common merganser (13) 32, 38, 55, 64, 65, 79, 87, 136, 154, 163, 220, 223, 226
common moorhen (1) 38
common murre (99) 19, 19, 20, 21, 22, 22, 25, 26, 28, 28, 31, 31, 32, 32, 32, 32, 34, 38, 55, 58, 63, 64, 65, 66, 67, 67, 68, 68, 69, 70, 71, 79, 80, 80, 81, 81, 81, 90, 91, 103, 104, 112, 115, 116, 116, 117, 117, 118, 119, 120, 121, 123, 125, 127, 127, 133, 140, 145, 151, 151, 156, 161, 162, 163, 163, 165, 167, 169, 171, 174, 175, 177, 178, 184, 185, 186, 187, 188, 188, 189, 189, 195, 196, 198, 200, 201, 203, 203, 204, 208, 208, 209, 215, 216, 220, 220, 221, 222, 223
common pigeon (4) 38, 110, 169, 195
common puffin see Atlantic puffin
common quail see Japanese quail
common raven (6) 32, 34, 59, 64, 65, 223
common scoter see black scoter
common snipe (2) 38, 125
common teal see green-winged teal
common tern (39) 20, 22, 23, 23, 23, 27, 32, 36, 39, 44, 49, 56, 57, 72, 83, 88, 90, 104, 122, 125, 137, 139, 146, 147, 151, 151, 166, 169, 171, 173, 178, 187, 190, 195, 200, 202, 210, 210, 218
Cook's petrel (1) 141
Cooper's hawk (1) 125
coot (5) 32, 80, 83, 139, 210
cormorant (21) 31, 31, 32, 32, 34, 59, 70, 72, 83, 122, 123, 139, 140, 141, 151, 156, 163, 178, 191, 200, 225
cormorants (1) 212
Corvus caurinus (3) 64, 65, 223
Corvus corax (6) 32, 34, 59, 64, 65, 223
Corvus cornix (1) 147
Cory's shearwater (4) 32, 83, 139, 151
Coturnix coturnix (4) 68, 169, 175, 190
Cr (24) 19, 32, 38, 44, 44, 44, 45, 45, 47, 48, 52, 76, 85, 94, 111, 135, 152, 178, 185, 186, 197, 198, 198, 226
crane (1) 87
Craveri's murrelet (1) 125
crested auklet (1) 112
Crocethia alba (3) 56, 57, 122
crow (1) 147
Cs (2) 38, 128
CT (2) 131, 226
Cu (43) 19, 32, 32, 32, 38, 67, 67, 68, 68,

249

Keyword Index (Cont.). Numbers in parentheses are the sums of references containing the keyword.

69, 76, 77, 85, 101, 105, 106, 110, 111, 112, 118, 124, 128, 129, 130, 135, 135, 141, 157, 178, 184, 185, 185, 186, 196, 197, 198, 198, 212, 213, 215, 216, 216, 226
Cyanocitta stelleri (3) 64, 65, 223
Cyclorrhynchus psittacula see *Aethia psittacula*
Cygnus cygnus (1) 87
Cygnus olor (3) 87, 140, 169
CYP1A (7) 76, 91, 107, 153, 179, 192, 192
cytochrome P_{450} (9) 91, 107, 137, 153, 168, 184, 190, 200, 204
Danube Delta (1) 83
Daption capense (2) 32, 141
Daurian redstart (1) 195
DDTs (185) 18, 19, 22, 22, 23, 24, 25, 25, 26, 26, 27, 28, 29, 29, 30, 30, 31, 32, 34, 34, 35, 36, 37, 38, 41, 42, 50, 51, 51, 54, 55, 56, 58, 58, 59, 59, 60, 60, 61, 61, 62, 74, 74, 75, 77, 79, 79, 80, 81, 81, 83, 83, 83, 84, 85, 85, 86, 88, 88, 89, 91, 91, 93, 93, 95, 95, 97, 97, 99, 100, 102, 103, 104, 105, 107, 108, 109, 112, 113, 113, 114, 116, 116, 117, 118, 119, 124, 125, 125, 125, 130, 131, 133, 133, 133, 134, 134, 135, 136, 138, 139, 140, 146, 147, 149, 150, 151, 152, 153, 153, 155, 156, 157, 158, 158, 159, 160, 161, 162, 163, 164, 164, 165, 165, 166, 168, 169, 170, 170, 171, 171, 172, 172, 173, 175, 175, 177, 177, 178, 178, 179, 180, 181, 183, 183, 185, 185, 186, 186, 187, 187, 190, 191, 192, 193, 194, 197, 198, 199, 199, 200, 201, 207, 208, 208, 209, 210, 210, 211, 211, 214, 217, 217, 218, 219, 225, 226, 227, 227, 227, 228, 228, 229, 229, 230
Dendragapus canadensis (1) 38
Dendragapus obscurus (1) 38
Denmark (4) 19, 32, 83, 124
dieldrin (85) 18, 19, 24, 26, 32, 34, 35, 36, 37, 38, 42, 54, 55, 58, 58, 60, 60, 61, 62, 74, 74, 75, 77, 83, 85, 85, 86, 88, 89, 91, 93, 93, 95, 97, 103, 125, 131, 133, 133, 135, 136, 140, 144, 146, 151, 152, 153, 156, 158, 158, 162, 164, 164, 165, 168, 169, 170, 171, 172, 172, 173, 175, 177, 178, 178, 181, 190, 191, 193, 197, 198, 199, 207, 208, 208, 209, 210, 212, 217, 217, 218, 219, 225, 226, 228
Diomedea bulleri (1) 141
Diomedea cauta (3) 128, 129, 141
Diomedea cauta salvini (1) 141
Diomedea chlororhynchos (3) 128, 129, 204
Diomedea chrysostoma (2) 128, 141
Diomedea epomophora (4) 128, 128, 129, 141
Diomedea exulans (3) 32, 141, 204
Diomedea immutabilis (5) 32, 112, 123, 128, 200
Diomedea melanophris (3) 32, 128, 141
Diomedea nigripes (9) 28, 32, 112, 123, 128, 128, 129, 206, 206
disease (3) 93, 100, 106
diver see red-throated loon
diving petrel (2) 32, 141
domestic fowl (1) 137
dosing study (32) 43, 46, 46, 47, 47, 48, 49, 52, 52, 58, 68, 78, 97, 99, 137, 138, 138, 148, 148, 149, 149, 149, 166, 166, 167, 167, 167, 168, 168, 175, 204, 207
double-crested cormorant (76) 19, 27, 32, 36, 39, 48, 52, 53, 61, 61, 62, 63, 64, 65, 66, 74, 75, 76, 76, 84, 86, 92, 96, 97, 97, 97, 99, 100, 104, 105, 105, 109, 110, 115, 119, 123, 123, 125, 133, 134, 136, 143, 143, 146, 147, 152, 152, 159, 162, 166, 170, 171, 172, 173, 180, 181, 184, 190, 193, 194, 194, 204, 205, 210, 210, 218, 220, 223, 224, 227, 227, 228, 228, 228, 229, 229
dovekie see little auk
dunlin (2) 125, 223
eagle owl (3) 110, 112, 163
eared grebe (2) 48, 210
EDCs (2) 104, 190
effects (173) 18, 19, 21, 23, 25, 27, 32, 32, 32, 40, 42, 43, 45, 45, 46, 46, 47, 47, 48, 50, 50, 52, 52, 53, 55, 55, 60, 61, 61, 64, 65, 66, 67, 67, 68, 68, 69, 70, 76, 78, 81, 83, 84, 84, 85, 86, 88, 89, 89, 91, 91, 91, 93, 93, 93, 94, 97, 99, 100, 102, 104, 105, 105, 106, 107,

Keyword Index (Cont.). Numbers in parentheses are the sums of references containing the keyword.

107, 108, 109, 111, 112, 113, 115, 125, 126, 127, 127, 131, 132, 133, 133, 135, 135, 136, 137, 138, 138, 138, 140, 140, 142, 143, 143, 145, 147, 148, 148, 149, 149, 149, 151, 151, 153, 153, 154, 159, 161, 161, 165, 165, 166, 166, 166, 167, 167, 167, 168, 168, 168, 169, 170, 174, 175, 175, 177, 179, 181, 181, 182, 183, 184, 186, 187, 187, 188, 189, 190, 191, 192, 192, 193, 194, 195, 196, 196, 197, 198, 199, 200, 200, 203, 204, 205, 207, 210, 210, 211, 212, 213, 214, 215, 216, 217, 219, 220, 221, 221, 222, 223, 224, 224, 225, 227, 227

egg (190) 18, 22, 22, 23, 23, 23, 24, 25, 25, 26, 26, 27, 27, 32, 34, 35, 36, 37, 37, 38, 41, 42, 44, 44, 45, 48, 49, 53, 53, 55, 58, 58, 60, 60, 61, 61, 62, 72, 74, 75, 75, 76, 77, 79, 81, 81, 81, 83, 83, 85, 88, 88, 91, 92, 93, 93, 93, 94, 94, 95, 96, 97, 97, 98, 98, 98, 100, 101, 102, 102, 103, 105, 105, 114, 116, 116, 117, 117, 118, 119, 120, 121, 123, 123, 124, 125, 127, 127, 130, 131, 131, 132, 133, 134, 135, 136, 136, 137, 138, 139, 139, 140, 142, 143, 143, 144, 146, 146, 147, 147, 150, 151, 151, 153, 153, 155, 155, 156, 158, 158, 159, 159, 160, 160, 161, 162, 163, 164, 165, 166, 169, 170, 170, 171, 171, 172, 172, 173, 175, 178, 180, 181, 181, 182, 184, 186, 188, 189, 189, 191, 191, 193, 193, 194, 194, 195, 197, 198, 198, 199, 200, 201, 204, 205, 206, 206, 207, 207, 208, 208, 209, 210, 210, 211, 214, 217, 217, 218, 219, 220, 220, 224, 225, 227, 227, 228, 228, 228, 229, 229

eggshell (1) 45

eggshell thickness (26) 19, 25, 27, 53, 60, 61, 61, 83, 104, 105, 125, 147, 153, 166, 169, 171, 177, 181, 181, 190, 193, 194, 199, 210, 217, 225

Egretta alba (1) 151

Egretta garzetta (3) 83, 122, 139

Egretta thula (2) 105, 123

Egretta tricolor (1) 196

elegant tern (1) 125

elimination (18) 23, 36, 37, 39, 41, 45, 109, 112, 122, 129, 140, 141, 158, 160, 178, 196, 212, 226

embryo (1) 62

endrin (7) 18, 19, 38, 152, 193, 217, 225

Endomychura craveri see *Synthliboramphus craveri*

energetics (3) 36, 220, 221

enzymes (10) 55, 137, 167, 167, 168, 169, 179, 195, 200, 212

erect-crested penguin (1) 141

Eremophila alpestris (1) 125

EROD (30) 55, 61, 61, 62, 63, 76, 78, 91, 95, 100, 107, 107, 108, 127, 135, 136, 137, 143, 153, 167, 167, 168, 179, 184, 190, 191, 192, 192, 200, 204

Erolia alpina see *Calidris alpina*

Erolia minutilla see *Calidris minutilla*

esophagus (1) 128

ethylenes (1) 98

Eudyptes chrysocome (2) 32, 141

Eudyptes chrysolophus (1) 157

Eudyptes pachyrhynchus (1) 141

Eudyptes sclateri (1) 141

Eudyptula albosignata (1) 32

Eudyptula minor (2) 32, 141

Eurasian wigeon (1) 226

Europe (184) 18, 19, 19, 20, 20, 22, 22, 22, 23, 23, 23, 24, 24, 25, 25, 26, 27, 28, 29, 30, 30, 30, 31, 32, 32, 32, 32, 34, 41, 41, 42, 43, 50, 50, 51, 51, 55, 55, 60, 60, 62, 63, 66, 66, 67, 67, 68, 68, 69, 70, 71, 71, 72, 78, 79, 79, 80, 80, 80, 81, 81, 81, 81, 83, 83, 87, 87, 88, 90, 90, 91, 91, 91, 100, 101, 103, 104, 107, 107, 108, 108, 109, 109, 110, 111, 111, 112, 113, 113, 115, 116, 117, 118, 119, 119, 120, 121, 121, 121, 122, 123, 124, 124, 127, 130, 133, 135, 138, 139, 139, 140, 140, 140, 142, 145, 147, 150, 151, 151, 153, 155, 156, 157, 157, 157, 161, 163, 163, 164, 164, 164, 165, 165, 174, 175, 175, 177, 178, 179, 180, 182, 182, 183, 183, 184, 185, 185, 186, 186, 186, 187, 188, 189, 189, 189, 191, 192, 192, 195, 195, 196, 199, 200, 201, 202, 202, 203, 203, 203, 204, 211, 212, 212, 215, 216, 216, 220, 224, 225, 226, 229, 229, 230, 185

Keyword Index (Cont.). Numbers in parentheses are the sums of references containing the keyword.

European shag see shag
European storm-petrel see storm-petrel
European wigeon see Eurasian wigeon
eyeball (1) 128
fairy prion (2) 32, 141
Falcipennis canadensis see *Dendragapus canadensis*
Falco columbarius (2) 125, 169
Falco mexicanus (1) 125
Falco peregrinus (5) 32, 74, 169, 177, 223
Falco rusticolus (1) 229
Falco sparverius (3) 125, 169, 175
Faroe Islands (2) 20, 186
fat (42) 28, 29, 31, 31, 32, 32, 34, 34, 38, 41, 41, 41, 42, 56, 57, 58, 59, 66, 79, 79, 80, 80, 81, 82, 96, 108, 109, 125, 125, 125, 128, 140, 150, 157, 164, 177, 186, 186, 193, 210, 215, 229
Fe (24) 19, 32, 38, 52, 67, 67, 69, 76, 77, 85, 101, 105, 106, 110, 111, 112, 118, 129, 130, 135, 197, 198, 198, 226
feather (51) 19, 20, 20, 22, 23, 32, 36, 36, 37, 38, 39, 44, 45, 48, 49, 49, 52, 53, 73, 90, 90, 91, 99, 105, 110, 121, 125, 128, 128, 129, 130, 136, 140, 141, 151, 164, 193, 195, 198, 202, 202, 203, 203, 203, 204, 207, 212, 215, 216, 216, 226
feces (5) 36, 41, 99, 101, 151
feral pigeon see common pigeon
Ferruginous hawk (1) 125
Finland (13) 79, 88, 100, 111, 124, 130, 133, 138, 139, 165, 185, 186, 199
Fiordland crested penguin (1) 141
FL (1) 226
flesh-footed shearwater (1) 141
fork-tailed storm petrel (9) 28, 32, 64, 65, 74, 77, 77, 162, 223
Forster's tern (11) 27, 44, 49, 104, 118, 132, 166, 169, 173, 190, 194
France (2) 32, 69
Franklin's Gull (3) 48, 123, 210
Fratercula arctica (34) 22, 22, 28, 31, 31, 32, 32, 34, 38, 43, 75, 76, 125, 138, 145, 149, 151, 164, 164, 167, 169, 170, 171, 178, 184, 185, 186, 200, 203, 203, 203, 206, 206, 222
Fratercula cirrhata (9) 28, 38, 64, 65, 112, 162, 220, 221, 223
Fratercula corniculata (5) 28, 65, 112, 162, 223
Fregata magnificens (1) 32
Fregata minor (1) 141
Fregetta tropica (1) 141
Fulica americana (2) 38, 48
Fulica atra (5) 32, 80, 83, 139, 210
fulmar prion (1) 141
fulmars (1) 212
Fulmarus (1) 212
Fulmarus glacialis (50) 20, 28, 28, 32, 34, 35, 36, 37, 42, 43, 55, 65, 70, 71, 72, 72, 73, 75, 82, 90, 111, 112, 125, 128, 128, 129, 142, 145, 145, 150, 153, 155, 156, 157, 157, 162, 164, 164, 177, 184, 185, 186, 200, 200, 202, 203, 203, 203, 204, 222
GA (1) 226
gadwall (5) 38, 210, 218, 223, 226
gallbladder (2) 41, 128
Gallinago gallinago (2) 38, 125
Gallinula chloropus (1) 38
Gallinula comeri (1) 32
Gallus domesticus (2) 78, 169
Gallus gallus (1) 137
gannet (14) 32, 34, 60, 81, 81, 123, 125, 145, 153, 178, 186, 200, 203, 212
Gavia (1) 221
Gavia adamsii (3) 32, 65, 223
Gavia arctica (6) 32, 73, 79, 129, 130, 199
Gavia immer (8) 32, 64, 65, 123, 136, 145, 162, 223
Gavia pacifica (1) 223
Gavia stellata (10) 32, 79, 80, 103, 123, 123, 129, 130, 200, 223
gender differences (37) 31, 37, 39, 42, 45, 50, 59, 68, 69, 79, 81, 97, 99, 109, 111, 114, 115, 124, 134, 138, 140, 153, 156, 157, 160, 169, 185, 196, 197, 199, 211, 214, 225, 226, 229, 229, 230
Germany (22) 22, 23, 23, 23, 24, 71, 72, 90,

Keyword Index (Cont.). Numbers in parentheses are the sums of references containing the keyword.

103, 121, 122, 140, 151, 164, 174, 182, 187, 201, 202, 215, 216, 229
gizzard (1) 131
gland (11) 32, 38, 85, 119, 125, 128, 140, 149, 168, 175, 226
glaucous gull (52) 20, 22, 24, 28, 29, 32, 34, 34, 42, 42, 50, 50, 51, 51, 59, 62, 65, 72, 72, 73, 82, 91, 91, 101, 107, 107, 108, 108, 109, 113, 121, 129, 130, 147, 150, 156, 157, 157, 176, 176, 182, 183, 183, 184, 185, 186, 186, 192, 192, 205, 211, 223
glaucous-winged gull (18) 30, 36, 53, 64, 65, 74, 75, 112, 153, 154, 162, 169, 190, 194, 206, 206, 220, 223
golden eagle (1) 125
golden plover (1) 229
gonad (6) 38, 128, 134, 164, 216, 226
goosander see common merganser
goose (1) 137
goshawk (1) 110
gough gallinule (1) 32
great black-backed gull (16) 32, 34, 44, 60, 70, 118, 123, 136, 172, 185, 186, 196, 199, 200, 219, 229
great blue heron (13) 52, 53, 64, 65, 76, 123, 125, 127, 159, 184, 194, 210, 223
Great Britain (26) 31, 32, 32, 34, 60, 60, 63, 79, 90, 145, 151, 153, 156, 175, 175, 177, 178, 185, 186, 189, 191, 195, 202, 203, 225, 226
great cormorant see cormorant
great crested grebe (5) 32, 70, 79, 87, 123
great egret (2) 123, 169
great frigatebird (1) 141
great horned owl (1) 125
great knot (3) 56, 57, 122
Great Lakes (91) 26, 27, 30, 37, 58, 58, 61, 61, 62, 75, 76, 84, 84, 84, 85, 85, 86, 86, 92, 93, 93, 93, 95, 95, 97, 98, 98, 98, 101, 102, 102, 103, 104, 109, 114, 115, 118, 119, 123, 123, 126, 127, 131, 131, 132, 133, 136, 142, 143, 143, 144, 146, 147, 158, 158, 159, 160, 160, 166, 169, 170, 171, 172, 172, 173, 175, 181, 184, 185, 186, 190, 191, 193, 194, 197, 198, 198, 201, 204, 205, 206, 207, 207, 210, 217, 217, 218, 219, 224, 226, 227
great cormorant see cormorant
great shearwater (4) 32, 34, 90, 125
great skua (11) 32, 34, 90, 114, 151, 195, 200, 202, 203, 203, 204
great-tailed grackle (2) 152, 152
great white egret (1) 151
great white pelican (2) 83, 139
great-winged petrel see grey-faced petrel
greater scaup (6) 32, 38, 81, 218, 223, 226
greater yellowlegs (1) 223
green-winged teal (7) 38, 64, 65, 87, 136, 223, 226
Greenland (19) 19, 20, 34, 42, 59, 59, 72, 72, 73, 82, 111, 133, 150, 156, 176, 176, 186, 205, 211
greenshank (3) 56, 57, 122
grey-backed storm-petrel (1) 141
grey-faced petrel (1) 141
grey-headed albatross (2) 128, 141
grey heron (9) 70, 87, 112, 113, 117, 122, 137, 153, 195
grey partridge (2) 38, 169
grey petrel (4) 32, 128, 129, 141
Grus canadensis (1) 223
Grus grus (1) 87
Guanay cormorant (2) 113, 227
guillemot see common murre
Gulf of Alaska (7) 58, 112, 133, 133, 208, 208, 209
gull (1) 181
gull-billed tern (3) 83, 139, 200
gulls (1) 212
gut (2) 121, 164
gyrfalcon (1) 229
Haematopus bachmani (5) 64, 65, 154, 221, 223
Haematopus ostralegus (8) 23, 23, 32, 72, 103, 114, 140, 151
Haliaeetus albicilla (9) 32, 80, 112, 115, 117, 120, 123, 130, 163
Haliaeetus leucocephalus (12) 64, 65, 104, 123, 125, 154, 166, 169, 190, 206, 219, 223

Keyword Index (Cont.). Numbers in parentheses are the sums of references containing the keyword.

Halobaena caerulea (1) 141
Halocyptena microsoma see *Oceanodroma microsoma*
harlequin duck (9) 34, 59, 64, 65, 154, 220, 221, 223, 226
HBCD (1) 127
HCB (95) 18, 22, 23, 24, 25, 26, 26, 29, 30, 30, 37, 38, 42, 50, 51, 51, 56, 58, 59, 61, 61, 74, 75, 77, 79, 79, 80, 81, 83, 86, 89, 91, 93, 93, 95, 98, 98, 100, 103, 107, 108, 109, 113, 116, 118, 130, 133, 135, 146, 146, 150, 152, 157, 158, 158, 159, 162, 164, 165, 168, 169, 170, 170, 172, 172, 173, 178, 179, 180, 181, 183, 183, 186, 187, 192, 193, 197, 198, 199, 200, 207, 208, 208, 209, 211, 211, 217, 217, 218, 219, 226, 227, 228, 229, 230
HCHs (79) 18, 19, 22, 22, 23, 24, 26, 29, 29, 30, 30, 34, 34, 35, 36, 37, 38, 42, 51, 54, 55, 56, 58, 59, 59, 74, 74, 75, 77, 79, 80, 81, 83, 89, 91, 100, 108, 113, 115, 116, 122, 125, 125, 130, 133, 134, 136, 146, 147, 150, 151, 152, 153, 158, 164, 170, 172, 172, 173, 174, 179, 180, 186, 187, 192, 193, 199, 200, 207, 208, 210, 211, 217, 218, 225, 226, 227, 229, 230
HDBPs (1) 205
heart (7) 39, 70, 97, 119, 128, 134, 186
herring gull (183) 22, 22, 22, 23, 23, 23, 24, 25, 26, 27, 27, 32, 34, 36, 37, 42, 43, 44, 44, 44, 45, 45, 46, 46, 47, 47, 48, 49, 49, 52, 55, 56, 57, 58, 58, 64, 65, 72, 75, 75, 76, 76, 79, 81, 81, 81, 83, 84, 84, 84, 85, 85, 86, 86, 87, 88, 90, 92, 93, 93, 93, 94, 95, 95, 97, 98, 98, 98, 100, 101, 102, 102, 103, 103, 104, 114, 114, 118, 119, 122, 123, 123, 124, 125, 126, 127, 128, 129, 130, 131, 131, 131, 134, 136, 137, 137, 138, 138, 139, 139, 140, 140, 140, 142, 143, 144, 145, 146, 146, 147, 148, 148, 149, 149, 150, 151, 151, 158, 158, 159, 160, 160, 163, 164, 165, 166, 166, 167, 167, 167, 168, 168, 168, 169, 170, 171, 172, 172, 173, 175, 178, 179, 180, 180, 181, 182, 184, 185, 185, 185, 186, 190, 191, 191, 193, 194, 195, 197, 198, 198, 199, 200, 201, 202, 204, 206, 206, 207, 207, 210, 210, 214, 217, 217, 218, 219, 223, 227, 228, 228, 229, 229, 229
Heteroscelus incanus (3) 64, 65, 223
Hg (133) 19, 20, 20, 20, 22, 22, 22, 23, 23, 24, 32, 32, 34, 35, 36, 36, 36, 37, 38, 39, 44, 44, 44, 45, 45, 48, 52, 53, 56, 58, 63, 67, 69, 70, 72, 72, 73, 73, 74, 76, 77, 80, 81, 85, 88, 88, 90, 90, 91, 93, 93, 94, 94, 97, 103, 105, 105, 106, 110, 110, 111, 112, 112, 114, 116, 118, 119, 119, 121, 124, 125, 128, 128, 129, 129, 130, 131, 133, 133, 135, 135, 136, 139, 140, 140, 141, 151, 153, 156, 156, 157, 157, 161, 164, 164, 165, 171, 176, 176, 178, 181, 184, 185, 186, 195, 196, 197, 198, 198, 199, 200, 201, 202, 202, 203, 203, 203, 204, 209, 212, 213, 213, 214, 214, 215, 215, 216, 216, 217, 218, 225, 226, 227
Himantopus himantopus (2) 83, 139
Hirundo rustica (1) 169
histopathology (5) 67, 105, 106, 157, 213
Histrionicus histrionicus (9) 34, 59, 64, 65, 154, 220, 221, 223, 226
hooded merganser (3) 38, 55, 226
horned grebe (9) 32, 64, 65, 79, 80, 137, 195, 210, 223
horned lark (1) 125
horned puffin (5) 28, 65, 112, 162, 223
hudsonian godwit (1) 223
Hutton's shearwater (1) 141
Hydrobates pelagicus (5) 28, 32, 34, 43, 200
hydrocarbons (11) 41, 89, 102, 133, 137, 167, 175, 184, 190, 191, 193
IA (1) 226
Hydroprogne caspia see *Sterna caspia*
Iceland (4) 133, 203, 229, 230
Iceland gull (6) 59, 65, 72, 72, 73, 156
ID (1) 52
IL (1) 226
immunology (6) 40, 95, 192, 211, 214, 215
imperial shag see blue-eyed shag
Indian Ocean (3) 128, 128, 129
intestine (8) 38, 41, 109, 127, 128, 182, 192, 192

254

Keyword Index (Cont.). Numbers in parentheses are the sums of references containing the keyword.

Ireland (6)   31, 32, 151, 151, 185, 225
Irish Sea (4)   18, 32, 140, 165
Italy (3)   83, 139, 140
ivory gull (7)   42, 72, 73, 75, 82, 150, 156
jackass penguin (1)   32
Japan (5)   30, 32, 73, 112, 122
Japanese cormorant (1)   112
Japanese quail (4)   68, 169, 175, 190
K (1)   198
Kattegat (3)   20, 32, 91
Kerguelen petrel (3)   32, 90, 141
kidney (65)   20, 32, 36, 37, 38, 39, 41, 43, 67, 67, 68, 68, 69, 70, 72, 72, 73, 76, 77, 80, 85, 87, 87, 91, 97, 105, 105, 106, 110, 112, 113, 114, 119, 119, 122, 127, 128, 128, 129, 130, 133, 135, 137, 140, 141, 142, 149, 151, 156, 157, 164, 164, 196, 197, 198, 212, 213, 213, 214, 215, 215, 216, 216, 226, 229
killdeer (1)   125
king eider (16)   34, 59, 72, 72, 73, 129, 130, 153, 156, 184, 185, 186, 197, 213, 225, 226
king shag (1)   141
kingfisher (1)   70
kittiwakes (1)   212
Kittlitz's murrelet (1)   223
Korea (3)   56, 57, 122
LA (1)   226
Labrador (1)   135
Lagopus lagopus (3)   38, 41, 136
Lagopus mutus (3)   34, 59, 229
lapwing (1)   137
Larus (1)   212
Larus argentatus (183)   22, 22, 22, 23, 23, 23, 24, 25, 26, 27, 27, 32, 34, 36, 37, 42, 43, 44, 44, 44, 45, 45, 46, 46, 47, 47, 48, 49, 49, 52, 55, 56, 57, 58, 58, 64, 65, 72, 75, 75, 76, 76, 79, 81, 81, 81, 83, 84, 84, 84, 85, 85, 86, 86, 87, 88, 90, 92, 93, 93, 93, 94, 95, 95, 97, 98, 98, 98, 100, 101, 102, 102, 103, 103, 104, 114, 114, 118, 119, 122, 123, 123, 124, 125, 126, 127, 128, 129, 130, 131, 131, 131, 134, 136, 137, 137, 138, 138, 139, 139, 140, 140, 140, 142, 143, 144, 145, 146, 146, 147, 148, 148, 149, 149, 150, 151, 151, 158, 158, 159, 160, 160, 163, 164, 165, 166, 166, 167, 167, 167, 168, 168, 168, 169, 170, 171, 172, 172, 173, 175, 178, 179, 180, 180, 181, 182, 184, 185, 185, 185, 186, 190, 191, 191, 193, 194, 195, 197, 198, 198, 199, 200, 201, 202, 204, 206, 206, 207, 207, 210, 210, 214, 217, 217, 218, 219, 223, 227, 228, 228, 229, 229, 229
Larus atricilla (2)   169, 196
Larus audouinii (3)   20, 83, 139
Larus brevirostris   see *Rissa brevirostris*
Larus cachinnans (1)   20
Larus californicus (6)   113, 125, 169, 175, 210, 227
Larus canus (25)   27, 56, 57, 64, 65, 71, 72, 73, 79, 80, 103, 121, 122, 134, 137, 140, 147, 154, 162, 165, 195, 200, 214, 221, 223
Larus crassirostris (3)   56, 57, 122
Larus delawarensis (9)   27, 104, 123, 169, 172, 173, 181, 190, 218
Larus dominicanus (1)   141
Larus fuscus (12)   25, 32, 34, 90, 100, 135, 169, 178, 185, 186, 200, 204
Larus genei (2)   83, 139
Larus glaucescens (18)   30, 36, 53, 64, 65, 74, 75, 112, 153, 154, 162, 169, 190, 194, 206, 206, 220, 223
Larus glaucoides (6)   59, 65, 72, 72, 73, 156
Larus hyperboreus (52)   20, 22, 24, 28, 29, 32, 34, 34, 42, 42, 50, 50, 51, 51, 59, 62, 65, 72, 72, 73, 82, 91, 91, 101, 107, 107, 108, 108, 109, 113, 121, 129, 130, 147, 150, 156, 157, 157, 176, 176, 182, 183, 183, 184, 185, 186, 186, 192, 192, 205, 211, 223
Larus marinus (16)   32, 34, 44, 60, 70, 118, 123, 136, 172, 185, 186, 196, 199, 200, 219, 229
Larus melanocephalus (1)   200
Larus minutus (2)   32, 79
Larus novaehollandiae (3)   141, 151, 169
Larus occidentalis (4)   125, 169, 177, 190
Larus philadelphia (7)   36, 36, 37, 39, 41, 154, 223
Larus pipixcan (3)   48, 123, 210

Keyword Index (Cont.). Numbers in parentheses are the sums of references containing the keyword.

Larus ridibundus (24) 19, 22, 23, 23, 25, 56, 57, 72, 79, 83, 87, 103, 118, 122, 134, 137, 139, 140, 140, 147, 151, 151, 195, 200
Larus sabini (2) 129, 130
laughing gull (2) 169, 196
Laysan albatross (5) 32, 112, 123, 128, 200
Leach's storm-petrel (23) 28, 32, 43, 52, 52, 74, 75, 76, 77, 77, 125, 162, 167, 168, 168, 169, 170, 171, 175, 200, 206, 206, 207
least auklet (1) 112
least petrel (2) 32, 125
least sandpiper (3) 65, 125, 223
lesser black-backed gull (12) 25, 32, 34, 90, 100, 135, 169, 178, 185, 186, 200, 204
lesser noddy see white-capped noddy
lesser scaup (5) 38, 146, 210, 218, 226
lesser yellowlegs (2) 125, 223
Leucocarbo atriceps (1) 141
Leucocarbo ranfurlyi (1) 141
Li (2) 128, 226
light-mantled sooty albatross (3) 128, 129, 141
Limnodromus griseus (1) 125
Limosa fedoa (1) 125
Limosa haemastica (1) 223
Limosa lapponica (3) 56, 57, 122
little auk (19) 20, 32, 34, 42, 72, 72, 73, 79, 82, 125, 150, 156, 157, 157, 184, 185, 186, 205, 222
little egret (3) 83, 122, 139
little grebe (1) 70
little gull (2) 32, 79
little pied cormorant see litte shag
little shag (1) 141
little shearwater (1) 141
little tern (7) 32, 56, 57, 83, 139, 151, 200
liver (174) 19, 20, 24, 28, 29, 30, 32, 32, 32, 34, 36, 37, 37, 37, 38, 39, 41, 41, 42, 42, 43, 49, 53, 53, 59, 61, 62, 63, 63, 66, 67, 67, 68, 68, 69, 70, 72, 72, 73, 76, 76, 77, 78, 79, 79, 80, 80, 80, 82, 84, 84, 85, 85, 86, 86, 87, 87, 89, 91, 91, 92, 95, 97, 97, 100, 103, 105, 105, 106, 107, 107, 108, 108, 109, 109, 110, 111, 112, 113, 113, 113, 114, 115, 118, 119, 119, 122, 122, 123, 126, 127, 128, 128, 129, 129, 130, 131, 133, 135, 135, 136, 137, 138, 139, 140, 140, 140, 141, 142, 145, 145, 147, 149, 151, 152, 155, 156, 157, 157, 157, 160, 164, 164, 165, 167, 167, 168, 169, 174, 175, 176, 176, 177, 178, 178, 179, 180, 182, 183, 183, 184, 185, 185, 186, 186, 192, 192, 195, 196, 197, 198, 199, 200, 201, 204, 205, 206, 210, 211, 212, 212, 213, 213, 214, 214, 215, 215, 216, 216, 220, 226, 227, 229, 230
Lobipes lobatus see *Phalaropus lobatus*
Lonchura domestica (1) 169
long-billed curlew (1) 125
long-eared owl (1) 125
long-tailed duck (23) 29, 32, 34, 38, 59, 64, 65, 72, 81, 87, 87, 106, 123, 128, 129, 130, 156, 163, 185, 186, 220, 223, 226
long-tailed jaeger (4) 129, 130, 141, 223
Loomelania melania see *Oceanodroma melania*
loons (1) 221
Lophodytes cucullatus (3) 38, 55, 226
Lophortyx californica see *Callipepla californica*
Louisiana heron (1) 196
Lugensa brevirostris see *Pterodroma brevirostris*
lung (4) 119, 128, 215, 216
MA (3) 32, 186, 226
macaroni penguin (1) 157
Macronectes giganteus (2) 32, 141
Macronectes halli (3) 128, 129, 141
magellanic penguin (1) 32
magnificent frigatebird (1) 32
malathion (1) 169
mallard (21) 38, 52, 64, 65, 87, 125, 137, 139, 146, 149, 149, 151, 168, 169, 175, 195, 196, 210, 218, 223, 229
Manitoba (5) 19, 38, 41, 136, 226
Manx shearwater (8) 32, 34, 43, 90, 164, 164, 200, 203
marbled godwit (1) 125
marbled murrelet (6) 38, 64, 65, 125, 221, 223

Keyword Index (Cont.). Numbers in parentheses are the sums of references containing the keyword.

marsh hawk (1) 125
marsh sandpiper (1) 137
masked booby (1) 32
ME (5) 134, 149, 185, 199, 226
Mediterranean gull (1) 200
Mediterranean Sea (4) 20, 83, 139, 151
Megadyptes antipodes (1) 141
Melanitta (1) 221
Melanitta fusca (10) 32, 38, 64, 65, 80, 81, 106, 133, 223, 226
Melanitta nigra (10) 19, 32, 38, 55, 64, 65, 81, 106, 223, 226
Melanitta perspicillata (8) 38, 64, 65, 106, 133, 220, 223, 226
Meleagris gallopavo (1) 169
Mergus merganser (13) 32, 38, 55, 64, 65, 79, 87, 136, 154, 163, 220, 223, 226
Mergus serrator (14) 32, 38, 63, 64, 65, 72, 79, 87, 136, 140, 156, 163, 223, 226
merlin see pigeon hawk
metabolism (18) 30, 32, 41, 41, 52, 72, 97, 109, 114, 115, 122, 125, 137, 142, 148, 174, 195, 212
metallothionein (8) 32, 32, 68, 69, 76, 77, 114, 153
metals (159) 19, 19, 20, 20, 20, 22, 22, 22, 23, 23, 24, 32, 32, 32, 34, 35, 36, 36, 36, 37, 38, 39, 43, 43, 44, 44, 44, 45, 45, 46, 46, 47, 47, 48, 48, 49, 49, 52, 53, 55, 56, 58, 63, 67, 67, 68, 68, 69, 70, 72, 72, 73, 73, 74, 76, 77, 80, 81, 85, 87, 87, 88, 88, 89, 90, 90, 91, 93, 93, 94, 94, 97, 99, 101, 103, 105, 105, 106, 110, 110, 111, 112, 112, 113, 114, 116, 118, 119, 119, 121, 124, 125, 128, 128, 129, 129, 130, 131, 133, 133, 135, 135, 136, 139, 140, 140, 141, 151, 152, 153, 156, 156, 157, 157, 157, 161, 164, 164, 165, 171, 176, 176, 178, 181, 184, 185, 185, 186, 186, 195, 196, 197, 198, 198, 199, 200, 201, 202, 202, 203, 203, 203, 204, 209, 212, 213, 213, 214, 214, 215, 215, 216, 216, 217, 218, 225, 226, 227, 229
methylparathion (1) 169
Mexico (3) 125, 152, 152
MFOs (6) 137, 166, 167, 167, 168, 169

Mg (8) 76, 77, 106, 110, 111, 197, 198, 226
MI (2) 123, 226
Milvus lineatus (1) 122
mirex (56) 26, 29, 34, 35, 36, 37, 38, 42, 55, 58, 58, 61, 61, 74, 74, 77, 85, 85, 86, 91, 93, 93, 95, 102, 107, 108, 109, 133, 136, 144, 158, 158, 162, 170, 172, 173, 175, 179, 180, 181, 183, 183, 191, 193, 198, 199, 207, 208, 208, 209, 217, 217, 218, 219, 226, 228
MN (4) 19, 48, 61, 226
Mn (24) 44, 44, 44, 45, 45, 47, 48, 76, 77, 85, 94, 101, 105, 106, 110, 111, 112, 128, 129, 130, 185, 185, 197, 198
MO (1) 226
Mo (4) 85, 111, 197, 198
moorhen see gough gallinule
Morus serrator (2) 32, 141
mottled petrel (2) 32, 141
mourning dove (3) 38, 152, 152
MT (1) 226
muscle (81) 19, 20, 28, 32, 34, 36, 37, 38, 39, 41, 41, 42, 53, 67, 67, 68, 69, 70, 72, 72, 73, 74, 76, 79, 79, 80, 80, 80, 91, 97, 105, 110, 112, 113, 115, 115, 116, 117, 118, 119, 119, 122, 128, 128, 130, 134, 135, 136, 138, 139, 140, 140, 140, 146, 151, 156, 160, 163, 164, 164, 168, 177, 178, 180, 183, 184, 185, 185, 186, 186, 187, 196, 199, 200, 204, 212, 215, 216, 220, 226, 230
mute swan (3) 87, 140, 169
Mycteria americana (1) 123
Na (3) 76, 99, 198
NC (2) 32, 226
ND (1) 19
neotropic cormorant see olivaceous cormorant
Nesocichla eremita (1) 32
Netherlands (5) 32, 55, 66, 113, 212
New Brunswick (14) 32, 36, 36, 37, 38, 39, 75, 76, 76, 84, 146, 171, 226, 229
New Zealand (2) 32, 141
Newfoundland (13) 38, 58, 74, 75, 76, 127, 133, 149, 168, 170, 171, 206, 226
NH (1) 226

Keyword Index (Cont.). Numbers in parentheses are the sums of references containing the keyword.

Ni (11) 32, 38, 52, 85, 101, 135, 178, 185, 186, 198, 226
NJ (10) 32, 43, 44, 44, 46, 47, 48, 49, 94, 226
NM (2) 53, 133
North America (230) 19, 19, 20, 20, 21, 26, 27, 28, 30, 32, 34, 34, 35, 36, 36, 36, 37, 37, 37, 38, 39, 41, 42, 42, 43, 44, 44, 44, 45, 45, 46, 46, 47, 47, 48, 48, 49, 49, 52, 53, 53, 54, 55, 56, 58, 58, 58, 59, 59, 61, 61, 62, 63, 64, 65, 72, 72, 73, 74, 74, 75, 75, 76, 76, 77, 77, 82, 84, 84, 84, 85, 85, 86, 86, 88, 89, 92, 93, 93, 93, 94, 94, 95, 95, 96, 97, 97, 98, 98, 98, 99, 100, 101, 102, 102, 103, 104, 105, 105, 106, 109, 110, 111, 112, 114, 115, 116, 117, 118, 119, 123, 123, 125, 126, 127, 127, 131, 131, 131, 132, 133, 133, 133, 134, 135, 136, 136, 142, 143, 143, 144, 145, 146, 146, 147, 149, 150, 152, 152, 153, 154, 155, 156, 158, 158, 159, 159, 160, 160, 162, 165, 166, 168, 169, 170, 170, 171, 171, 172, 172, 173, 174, 175, 176, 176, 177, 177, 178, 180, 181, 181, 184, 185, 186, 187, 190, 191, 193, 193, 194, 194, 197, 197, 197, 198, 198, 198, 199, 201, 204, 205, 205, 206, 206, 207, 207, 208, 208, 209, 210, 210, 211, 212, 213, 213, 214, 214, 215, 217, 217, 218, 219, 220, 220, 221, 221, 222, 223, 224, 225, 226, 226, 227, 227, 227, 228, 228, 228, 229, 229
North Sea (29) 19, 20, 23, 23, 24, 32, 32, 55, 69, 71, 71, 90, 103, 115, 118, 119, 133, 151, 151, 174, 182, 185, 186, 189, 200, 202, 216, 224, 229
northern bobwhite see bobwhite quail
northern fulmar (50) 20, 28, 28, 32, 34, 35, 36, 37, 42, 43, 55, 65, 70, 71, 72, 72, 73, 75, 82, 90, 111, 112, 125, 128, 128, 129, 142, 145, 145, 150, 153, 155, 156, 157, 157, 162, 164, 164, 177, 184, 185, 186, 200, 200, 202, 203, 203, 203, 204, 222
northern gannet see gannet
northern giant petrel (3) 128, 129, 141
northern goshawk see goshawk
northern harrier see marsh hawk
northern phalarope see red-necked phalarope
northern pintail see pintail
northern shoveler see shoveler
Northwater Polynya (4) 42, 82, 150, 205
Northwest Territories (10) 20, 38, 54, 56, 133, 145, 155, 213, 214, 226
northwestern crow (3) 64, 65, 223
Norway (57) 22, 22, 24, 25, 27, 28, 30, 32, 34, 42, 50, 50, 51, 51, 55, 62, 81, 81, 81, 91, 91, 101, 107, 107, 108, 108, 109, 109, 110, 111, 112, 113, 121, 133, 135, 142, 147, 150, 155, 157, 157, 179, 180, 182, 183, 183, 184, 185, 186, 186, 186, 192, 192, 195, 203, 211, 216
Nova Scotia (7) 38, 76, 146, 171, 206, 226, 228
Numenius americanus (1) 125
Numenius phaeopus (1) 223
Nunavut (8) 34, 35, 36, 37, 145, 213, 214, 215
NV (1) 105
NY (12) 44, 44, 45, 45, 46, 47, 48, 49, 49, 94, 133, 226
Nycticorax nycticorax (14) 27, 48, 73, 88, 104, 105, 118, 123, 166, 169, 173, 190, 210, 218
Oceanites nereis (1) 141
Oceanites oceanicus (2) 32, 125
Oceanodroma furcata (9) 28, 32, 64, 65, 74, 77, 77, 162, 223
Oceanodroma homochroa (2) 32, 125
Oceanodroma leucorhoa (23) 28, 32, 43, 52, 52, 74, 75, 76, 77, 77, 125, 162, 167, 168, 168, 169, 170, 171, 175, 200, 206, 206, 207
Oceanodroma melania (2) 32, 125
Oceanodroma microsoma (2) 32, 125
Oceanodroma monorhis (1) 112
OCS (14) 25, 38, 42, 55, 58, 86, 146, 150, 172, 173, 187, 200, 207, 218
OH (1) 226
oil (44) 21, 32, 40, 52, 52, 64, 65, 66, 69, 104, 127, 137, 138, 138, 142, 148, 148, 149, 149, 154, 161, 165, 166, 166, 167, 167, 167,

Keyword Index (Cont.). Numbers in parentheses are the sums of references containing the keyword.

168, 168, 169, 174, 175, 185, 188, 189, 196, 197, 207, 220, 221, 221, 222, 223, 224
OK (1) 226
oldsquaw see long-tailed duck
olivaceous cormorant (1) 152
Ontario (8) 32, 38, 76, 88, 133, 181, 210, 226
OR (4) 28, 63, 187, 226
organic Hg (20) 38, 67, 68, 69, 72, 81, 90, 105, 118, 119, 128, 129, 133, 141, 151, 169, 203, 204, 213, 213
organochlorines (274) 18, 19, 22, 22, 23, 24, 25, 25, 26, 26, 27, 27, 28, 29, 29, 30, 30, 30, 31, 31, 32, 32, 34, 34, 34, 35, 36, 37, 37, 38, 41, 41, 42, 42, 50, 50, 51, 51, 54, 55, 55, 56, 56, 57, 58, 58, 59, 59, 60, 60, 61, 61, 62, 62, 63, 66, 66, 67, 71, 71, 72, 74, 74, 75, 76, 77, 78, 79, 79, 80, 80, 81, 81, 81, 82, 83, 83, 83, 84, 84, 85, 85, 86, 86, 88, 88, 89, 91, 91, 93, 93, 93, 95, 95, 96, 97, 97, 97, 98, 98, 98, 99, 100, 101, 102, 102, 103, 103, 104, 104, 105, 107, 107, 108, 108, 109, 109, 109, 112, 113, 113, 114, 115, 116, 116, 117, 117, 118, 118, 119, 119, 120, 121, 121, 122, 123, 124, 125, 125, 125, 126, 127, 127, 130, 131, 131, 132, 133, 133, 133, 134, 134, 135, 136, 136, 137, 138, 139, 140, 142, 143, 143, 144, 145, 146, 146, 147, 147, 147, 149, 150, 150, 151, 152, 153, 153, 155, 155, 156, 157, 158, 158, 159, 159, 160, 161, 162, 163, 163, 164, 164, 165, 165, 166, 168, 169, 170, 170, 171, 171, 172, 172, 173, 174, 175, 175, 175, 177, 177, 178, 178, 179, 180, 180, 181, 181, 183, 183, 184, 185, 185, 186, 186, 186, 187, 187, 190, 191, 192, 193, 193, 194, 194, 197, 197, 198, 199, 199, 200, 200, 201, 204, 205, 206, 206, 207, 207, 208, 208, 209, 210, 210, 211, 211, 212, 214, 217, 217, 218, 219, 220, 220, 224, 225, 226, 227, 227, 227, 228, 228, 229, 229, 229, 230
organohalogens (1) 205
organolead (1) 38
organometallics (23) 24, 38, 67, 68, 69, 72, 81, 90, 105, 118, 119, 123, 128, 129, 133, 141, 151, 169, 182, 203, 204, 213, 213
organophosphates (1) 169
organotins (3) 24, 123, 182
osmoregulation (1) 149
osprey (6) 32, 52, 112, 123, 166, 169
ovary (2) 140, 210
oxychlordane (1) 191
Oxyura jamaicensis (4) 38, 65, 223, 226
oystercatcher (8) 23, 23, 32, 72, 103, 114, 140, 151
P (5) 76, 77, 99, 198, 226
PA (1) 226
Pachyptila crassirostris (1) 141
Pachyptila desolata (1) 141
Pachyptila turtur (2) 32, 141
Pachyptila vittata (2) 32, 212
Pacific loon (1) 223
Pacific Ocean (21) 28, 32, 92, 112, 123, 125, 125, 125, 128, 128, 129, 133, 162, 177, 178, 185, 186, 200, 206, 206, 208
Pagodroma (1) 212
Pagodroma nivea (1) 32
Pagophila eburnea (7) 42, 72, 73, 75, 82, 150, 156
PAHs (10) 41, 89, 102, 133, 137, 167, 184, 190, 191, 193
pancreas (3) 128, 164, 226
Pandion haliaetus (6) 32, 52, 112, 123, 166, 169
parakeet auklet (2) 112, 223
parasites (4) 53, 182, 183, 213
parasitic jaeger (9) 32, 34, 129, 130, 141, 195, 200, 203, 223
Passer montanus (1) 195
Pb (52) 19, 24, 32, 38, 43, 44, 44, 44, 45, 45, 46, 46, 47, 48, 48, 49, 49, 52, 73, 74, 76, 77, 85, 87, 88, 89, 91, 94, 94, 101, 103, 110, 111, 113, 114, 116, 124, 128, 133, 133, 136, 140, 141, 153, 157, 176, 186, 198, 201, 212, 225, 226
PBBs (1) 115
PBDEs (10) 102, 104, 109, 115, 127, 160, 188, 189, 189, 211
PCBs (226) 18, 19, 22, 22, 23, 24, 25, 25,

Keyword Index (Cont.). Numbers in parentheses are the sums of references containing the keyword.

26, 26, 27, 27, 28, 29, 30, 30, 30, 31, 31, 32, 32, 34, 34, 34, 35, 36, 37, 37, 38, 42, 42, 50, 50, 51, 51, 54, 55, 56, 56, 57, 59, 59, 60, 61, 61, 62, 62, 66, 67, 71, 71, 72, 74, 74, 75, 77, 79, 79, 80, 81, 81, 81, 83, 83, 83, 84, 84, 85, 85, 86, 86, 88, 89, 91, 91, 93, 93, 95, 95, 96, 97, 97, 98, 98, 100, 100, 102, 102, 103, 103, 105, 107, 107, 108, 108, 109, 109, 109, 112, 113, 113, 114, 116, 116, 117, 118, 118, 119, 119, 120, 121, 123, 124, 125, 125, 126, 127, 127, 130, 131, 132, 133, 133, 133, 134, 135, 136, 136, 138, 139, 140, 142, 143, 143, 144, 145, 146, 147, 147, 150, 152, 153, 155, 155, 156, 157, 158, 158, 159, 161, 162, 163, 163, 164, 164, 165, 165, 168, 169, 170, 170, 171, 171, 172, 172, 173, 175, 175, 177, 178, 179, 180, 180, 181, 182, 183, 183, 184, 185, 185, 186, 186, 186, 187, 187, 190, 191, 192, 192, 193, 194, 194, 197, 197, 198, 199, 199, 200, 200, 204, 205, 207, 208, 208, 209, 210, 210, 211, 211, 214, 217, 217, 218, 219, 220, 224, 225, 226, 227, 227, 227, 228, 228, 229, 229, 230

PCDDs (64) 26, 27, 32, 37, 37, 38, 41, 55, 56, 57, 61, 63, 66, 75, 76, 80, 84, 85, 86, 89, 91, 93, 95, 100, 101, 102, 104, 109, 117, 118, 119, 121, 123, 127, 130, 132, 135, 136, 142, 143, 143, 144, 146, 159, 159, 161, 169, 170, 172, 173, 180, 184, 190, 191, 193, 194, 204, 205, 217, 218, 220, 224, 227, 227

PCDEs (1) 130

PCDFs (43) 26, 27, 32, 37, 37, 38, 41, 55, 56, 57, 66, 76, 80, 85, 86, 89, 100, 101, 104, 109, 117, 118, 123, 130, 135, 137, 142, 144, 146, 159, 161, 172, 173, 180, 184, 191, 193, 194, 204, 218, 220, 227, 227

PCNs (3) 78, 120, 121

Pedioecetes phasianellus see *Tympanuchus phasianellus*

pekin duck see mallard

pelagic cormorant (12) 32, 64, 65, 73, 74, 100, 154, 162, 177, 178, 194, 223

Pelecanoides georgicus (1) 141

Pelecanoides urinatrix (2) 32, 141

Pelecanus erythrorhynchos (7) 19, 32, 96, 97, 123, 125, 210

Pelecanus occidentalis (6) 32, 123, 125, 169, 177, 212

Pelecanus onocrotalus (2) 83, 139

penguins (1) 212

Perdix perdix (2) 38, 169

peregrine falcon (5) 32, 74, 169, 177, 223

perfluorinated acids (2) 122, 145

petrels (1) 212

PFOS (4) 92, 122, 123, 145

Phaethon lepturus (2) 32, 141

Phaethon rubricauda (1) 141

Phalacrocorax (1) 212

Phalacrocorax aristotelis (25) 18, 20, 22, 22, 31, 31, 32, 32, 34, 60, 60, 145, 151, 153, 155, 175, 178, 191, 200, 203, 203, 212, 221, 222, 225

Phalacrocorax atriceps (1) 32

Phalacrocorax auritus (76) 19, 27, 32, 36, 39, 48, 52, 53, 61, 61, 62, 63, 64, 65, 66, 74, 75, 76, 76, 84, 86, 92, 96, 97, 97, 97, 99, 100, 104, 105, 105, 109, 110, 115, 119, 123, 123, 125, 133, 134, 136, 143, 143, 146, 147, 152, 152, 159, 162, 166, 170, 171, 172, 173, 180, 181, 184, 190, 193, 194, 194, 204, 205, 210, 210, 218, 220, 223, 224, 227, 227, 228, 228, 228, 229, 229

Phalacrocorax bougainvillii (2) 113, 227

Phalacrocorax capillatus (1) 112

Phalacrocorax carbo (21) 31, 31, 32, 32, 34, 59, 70, 72, 83, 122, 123, 139, 140, 141, 151, 156, 163, 178, 191, 200, 225

Phalacrocorax colensoi (1) 32

Phalacrocorax filamentosus (1) 73

Phalacrocorax melanoleucos (1) 141

Phalacrocorax nigrogularis (1) 32

Phalacrocorax olivaceus (1) 152

Phalacrocorax pelagicus (12) 32, 64, 65, 73, 74, 100, 154, 162, 177, 178, 194, 223

Phalacrocorax penicillatus (8) 32, 113, 116, 117, 123, 177, 198, 227

Phalacrocorax pygmeus (2) 83, 139

Phalacrocorax urile (5) 64, 65, 162, 178,

Keyword Index (Cont.). Numbers in parentheses are the sums of references containing the keyword.

223
Phalaropus fulicarius (3) 32, 125, 177
Phalaropus lobatus (6) 36, 39, 64, 65, 125, 223
pharmacokinetics (1) 58
Phasianus colchicus (2) 38, 169
pheasant (2) 38, 169
phenols (2) 38, 137
Philomachus pugnax (1) 29
Phoebetria fusca (2) 32, 204
Phoebetria palpebrata (3) 128, 129, 141
Phoenicurus auroreus (1) 195
Phoenicurus phoenicurus (1) 169
phosmethylan (1) 169
phthalates (1) 228
Pica pica (3) 64, 65, 223
pigeon guillemot (12) 38, 53, 64, 65, 116, 117, 154, 162, 194, 198, 221, 223
pigeon hawk (2) 125, 169
pink-footed shearwater (1) 125
pintail (9) 38, 129, 130, 136, 137, 195, 210, 223, 226
plastic (2) 28, 32
Platalea leucorodia (1) 32
Plautus alle see *Alle alle*
Plegadis chihi (1) 123
Pluvialis apricaria (1) 229
Pluvialis squatarola (1) 125
pochard (2) 137, 195
Podiceps auritus (9) 32, 64, 65, 79, 80, 137, 195, 210, 223
Podiceps caspicus (2) 48, 210
Podiceps cristatus (5) 32, 70, 79, 87, 123
Podiceps grisegena (7) 64, 65, 70, 79, 80, 154, 223
Podiceps nigricollis (5) 56, 57, 83, 122, 139
Poland (6) 32, 79, 80, 80, 123, 186
polyisobutylene (1) 55
Polysticta stelleri (3) 29, 106, 197
pomarine jaeger (7) 64, 65, 72, 72, 156, 200, 223
porphyrin (15) 55, 84, 85, 86, 86, 91, 95, 107, 126, 127, 135, 166, 169, 190, 191
prairie falcon (1) 125

Procellaria aequinoctialis (4) 128, 128, 129, 141
Procellaria cinerea (4) 32, 128, 129, 141
Procellaria westlandica (1) 141
Pterodroma (1) 212
Pterodroma axillaris (1) 141
Pterodroma brevirostris (3) 32, 90, 141
Pterodroma cahow (2) 32, 125
Pterodroma cookii (1) 141
Pterodroma incerta (3) 32, 90, 204
Pterodroma inexpectata (2) 32, 141
Pterodroma lessonii (1) 141
Pterodroma longirostris (1) 28
Pterodroma macroptera (1) 141
Pterodroma mollis (2) 90, 204
Pterodroma nigripennis (1) 141
Pterodroma pycrofti (1) 141
Ptychoramphus aleuticus (6) 38, 77, 77, 112, 125, 177
Puffinus (1) 212
Puffinus assimilis (1) 141
Puffinus carneipes (1) 141
Puffinus creatopus (1) 125
Puffinus gravis (4) 32, 34, 90, 125
Puffinus griseus (8) 28, 32, 65, 112, 125, 177, 200, 220
Puffinus huttoni (1) 141
Puffinus lherminieri (1) 32
Puffinus pacificus (3) 169, 175, 190
Puffinus puffinus (8) 32, 34, 43, 90, 164, 164, 200, 203
Puffinus tenuirostris (6) 32, 112, 125, 141, 177, 200
pulp mill (4) 55, 76, 159, 191
purple sandpiper (4) 32, 34, 59, 229
Pycroft's petrel (1) 141
pygmy cormorant (2) 83, 139
Pygoscelis adeliae (1) 32
Pygoscelis antarctica (1) 32
Pyrrhocorax pyrrhocorax (1) 191
Quebec (12) 26, 38, 55, 56, 76, 76, 133, 136, 145, 171, 180, 226
Quelea quelea (1) 169
Quiscalus mexicanus (2) 152, 152

Keyword Index (Cont.). Numbers in parentheses are the sums of references containing the keyword.

Quiscalus quiscalus (1) 169
radionuclides (3) 111, 133, 161
razorbill (26) 22, 22, 31, 31, 32, 32, 34, 38, 43, 79, 80, 80, 81, 81, 123, 140, 145, 151, 171, 178, 184, 185, 186, 200, 203, 203
Rb (2) 52, 128
Recurvirostra americana (2) 125, 210
Recurvirostra avosetta (3) 72, 83, 139
red-billed gull (3) 141, 151, 169
red-billed quelea (1) 169
red-breasted merganser (14) 32, 38, 63, 64, 65, 72, 79, 87, 136, 140, 156, 163, 223, 226
red-faced cormorant (5) 64, 65, 162, 178, 223
red-legged kittiwake (1) 200
red-necked grebe (7) 64, 65, 70, 79, 80, 154, 223
red-necked phalarope (6) 36, 39, 64, 65, 125, 223
red phalarope (3) 32, 125, 177
red-tailed hawk (1) 125
red-tailed tropicbird (1) 141
red-throated loon (10) 32, 79, 80, 103, 123, 123, 129, 130, 200, 223
red-winged blackbird (4) 146, 152, 152, 169
redhead duck (2) 38, 226
redshank (2) 72, 140
redstart (1) 169
reproductive success (55) 19, 23, 32, 42, 50, 50, 52, 53, 55, 61, 76, 84, 93, 93, 94, 100, 104, 105, 109, 131, 132, 134, 136, 143, 147, 148, 151, 154, 159, 166, 168, 169, 170, 171, 175, 175, 177, 190, 191, 193, 196, 199, 200, 201, 204, 205, 207, 210, 210, 217, 219, 220, 221, 221, 222
retinol (14) 55, 84, 107, 107, 135, 155, 161, 169, 190, 191, 192, 192, 193, 215
review (31) 32, 32, 38, 40, 56, 76, 89, 91, 91, 93, 93, 104, 133, 143, 151, 153, 161, 166, 168, 169, 175, 185, 186, 189, 190, 200, 212, 220, 221, 222, 226
rhinoceros auklet (16) 28, 38, 65, 74, 75, 77, 77, 112, 116, 117, 125, 177, 198, 206, 206, 223

ring-billed gull (9) 27, 104, 123, 169, 172, 173, 181, 190, 218
ring dove (2) 169, 175
ring-necked duck (4) 38, 65, 223, 226
ringed plover (2) 23, 72
Rissa (1) 212
Rissa brevirostris (1) 200
Rissa tridactyla (68) 19, 20, 22, 22, 28, 29, 30, 32, 34, 34, 35, 36, 36, 37, 39, 42, 64, 65, 71, 72, 72, 73, 81, 81, 81, 82, 90, 90, 101, 103, 109, 111, 112, 113, 140, 145, 150, 151, 153, 154, 155, 156, 162, 177, 178, 184, 185, 185, 186, 186, 188, 195, 195, 200, 200, 201, 203, 203, 205, 209, 216, 216, 220, 221, 221, 222, 223, 229
rock dove see common pigeon
rock ptarmigan (3) 34, 59, 229
rock sandpiper (2) 178, 223
rockhopper penguin (2) 32, 141
roseate tern (5) 45, 49, 151, 178, 200
Ross' goose (1) 226
royal albatross (4) 128, 128, 129, 141
ruddy duck (4) 38, 65, 223, 226
ruff (1) 29
ruffed grouse (1) 38
Russia (12) 22, 29, 128, 129, 130, 134, 137, 147, 185, 186, 195, 197
Rynchops niger (2) 44, 49
S (2) 52, 226
Sabine's gull (2) 129, 130
Salvin's albatross (1) 141
sanderling (3) 56, 57, 122
sandhill crane (1) 223
sandwich tern (7) 23, 32, 72, 151, 178, 196, 200
Saskatchewan (6) 19, 38, 133, 133, 193, 226
Sb (1) 105
SC (3) 52, 212, 226
Scolopax minor (2) 38, 125
scoters (1) 221
Scotland (18) 28, 32, 32, 34, 43, 79, 90, 119, 140, 164, 164, 186, 196, 203, 203, 203, 204, 212

Keyword Index (Cont.). Numbers in parentheses are the sums of references containing the keyword.

SD (5) 61, 96, 97, 99, 110
Se (51) 34, 35, 38, 44, 44, 44, 45, 45, 48, 52, 72, 73, 74, 76, 77, 85, 88, 94, 94, 105, 106, 111, 114, 116, 128, 129, 133, 135, 136, 140, 152, 153, 156, 157, 176, 178, 184, 185, 186, 197, 198, 213, 213, 214, 214, 215, 215, 216, 216, 225, 226
Sea of Japan (1) 112
seasonal variations (20) 32, 36, 51, 67, 70, 79, 109, 119, 134, 139, 152, 152, 169, 196, 201, 210, 212, 214, 225, 230
semipalmated plover (1) 125
shag (25) 18, 20, 22, 22, 31, 31, 32, 32, 34, 60, 60, 145, 151, 153, 155, 175, 178, 191, 200, 203, 203, 212, 221, 222, 225
sharp-shinned hawk (3) 65, 125, 223
sharp-tailed grouse (2) 38, 169
shearwaters (1) 212
shelduck (5) 23, 32, 72, 103, 151
Shetland (4) 32, 90, 195, 202
short-billed dowitcher (1) 125
short-eared owl (1) 125
short-tailed shearwater (6) 32, 112, 125, 141, 177, 200
shoveler (5) 38, 137, 195, 223, 226
shy albatross   see white-capped albatross
silver gull   see red-billed gull
skin (4) 115, 128, 164, 226
skua   see long-tailed jaeger or pomarine jaeger
skuas (1) 212
Slavonian grebe   see horned grebe
slender-billed gull (2) 83, 139
slender-billed shearwater   see short-tailed shearwater
Sn (2) 32, 85
snow goose (2) 38, 226
snow petrel (1) 32
snowy egret (2) 105, 123
socotra cormorant (1) 32
soft-plumaged petrel (2) 90, 204
Somateria fischeri (6) 29, 89, 94, 106, 197, 225
Somateria mollissima (62) 20, 22, 32, 34, 36, 38, 39, 41, 41, 54, 59, 63, 72, 72, 72, 73, 78, 80, 81, 83, 87, 87, 88, 89, 94, 103, 110, 111, 111, 113, 119, 122, 124, 135, 136, 140, 147, 153, 156, 157, 157, 157, 171, 174, 178, 180, 184, 185, 186, 186, 188, 196, 197, 199, 201, 212, 213, 213, 214, 215, 226, 230
Somateria spectabilis (16) 34, 59, 72, 72, 73, 129, 130, 153, 156, 184, 185, 186, 197, 213, 225, 226
sooty albatross (2) 32, 204
sooty shearwater (8) 28, 32, 65, 112, 125, 177, 200, 220
sooty tern (2) 32, 141
South America (1) 83
South Georgian diving petrel (1) 141
southern black-backed gull (1) 141
southern giant petrel (2) 32, 141
sparrowhawk (1) 169
spatial variations (117) 20, 22, 22, 23, 26, 27, 27, 32, 34, 34, 43, 44, 49, 52, 53, 54, 58, 59, 61, 63, 66, 69, 71, 74, 76, 77, 77, 79, 80, 81, 81, 81, 83, 83, 84, 85, 86, 86, 90, 92, 93, 94, 98, 101, 102, 103, 105, 111, 112, 114, 115, 117, 118, 119, 121, 122, 123, 133, 135, 136, 140, 145, 145, 146, 147, 151, 152, 153, 156, 156, 157, 158, 159, 160, 162, 164, 165, 169, 170, 170, 171, 171, 172, 172, 173, 176, 176, 178, 181, 184, 185, 186, 186, 189, 193, 194, 194, 198, 199, 202, 203, 204, 206, 206, 208, 209, 210, 210, 211, 211, 212, 214, 215, 217, 218, 225, 226
spectacled eider (6) 29, 89, 94, 106, 197, 225
Speotyto cunicularia   see *Athene cunicularia*
Spheniscus demersus (1) 32
Spheniscus magellanicus (1) 32
spleen (2) 128, 186
spoonbill (1) 32
spot-billed duck (1) 122
spotted sandpiper (4) 64, 65, 125, 223
spotted shag (2) 32, 141
spruce grouse (1) 38
Squatarola squatarola   see *Pluvialis squatarola*

Keyword Index (Cont.). Numbers in parentheses are the sums of references containing the keyword.

Sr (6) 38, 52, 128, 197, 198, 226
stable isotopes (18) 20, 30, 34, 35, 41, 42, 82, 102, 107, 113, 116, 117, 178, 180, 183, 202, 205, 214
starling (2) 125, 169
Stejneger's petrel (1) 28
Steller's eider (3) 29, 106, 197
Steller's jay (3) 64, 65, 223
Stercorarius longicaudus (4) 129, 130, 141, 223
Stercorarius parasiticus (9) 32, 34, 129, 130, 141, 195, 200, 203, 223
Stercorarius pomarinus (7) 64, 65, 72, 72, 156, 200, 223
Sterna (1) 212
Sterna albifrons (7) 32, 56, 57, 83, 139, 151, 200
Sterna aleutica (1) 162
Sterna caspia (13) 27, 95, 104, 143, 143, 146, 169, 173, 190, 204, 214, 218, 227
Sterna dougallii (5) 45, 49, 151, 178, 200
Sterna elegans (1) 125
Sterna forsteri (11) 27, 44, 49, 104, 118, 132, 166, 169, 173, 190, 194
Sterna fuscata (2) 32, 141
Sterna hirundo (39) 20, 22, 23, 23, 23, 27, 32, 36, 39, 44, 49, 56, 57, 72, 83, 88, 90, 104, 122, 125, 137, 139, 146, 147, 151, 151, 166, 169, 171, 173, 178, 187, 190, 195, 200, 202, 210, 210, 218
Sterna nilotica (3) 83, 139, 200
Sterna paradisaea (22) 20, 34, 36, 39, 41, 64, 65, 128, 129, 130, 136, 138, 139, 141, 162, 184, 185, 187, 195, 200, 221, 223
Sterna sandvicensis (7) 23, 32, 72, 151, 178, 196, 200
Sterna striata (1) 141
Sterna vittata (1) 32
Stictocarbo punctatus punctatus (1) 141
Stictocarbo punctatus steadi (2) 32, 141
stomach (1) 128
stomach contents (2) 30, 105
stomach oil (2) 28, 34
storm-petrel (5) 28, 32, 34, 43, 200

storm petrels (1) 212
Streptopelia risoria (2) 169, 175
Sturnus vulgaris (2) 125, 169
Sula bassana (14) 32, 34, 60, 81, 81, 123, 125, 145, 153, 178, 186, 200, 203, 212
Sula capensis (1) 32
Sula dactylatra (1) 32
Sula leucogaster (5) 32, 125, 128, 129, 141
Sula serrator see *Morus serrator*
surf scoter (8) 38, 64, 65, 106, 133, 220, 223, 226
surfbird (1) 223
Svalbard (21) 22, 24, 30, 42, 50, 50, 55, 62, 91, 107, 108, 108, 109, 111, 121, 133, 147, 157, 157, 182, 192
Swainson's hawk (1) 125
Sweden (20) 25, 26, 32, 41, 78, 87, 87, 111, 116, 117, 120, 121, 127, 133, 161, 163, 186, 189, 189, 220
Swinhoe's storm petrel (1) 112
Switzerland (1) 79
Synthliboramphus antiquus (10) 38, 65, 74, 77, 77, 112, 125, 162, 177, 223
Synthliboramphus craveri (1) 125
Synthliboramphus hypoleucus (1) 28
Tachybaptus ruficollis (1) 70
Tachycineta bicolor (1) 146
Tadorna tadorna (5) 23, 32, 72, 103, 151
Taeniopygia guttata (1) 169
TCDDs (1) 133
Temminck's cormorant (1) 73
temporal trends (90) 20, 22, 25, 25, 26, 26, 27, 35, 36, 37, 42, 44, 46, 51, 60, 61, 73, 74, 75, 75, 77, 79, 80, 83, 84, 84, 85, 85, 90, 91, 91, 97, 98, 100, 101, 102, 102, 103, 104, 118, 121, 127, 131, 134, 144, 145, 150, 151, 153, 155, 156, 159, 160, 161, 163, 164, 165, 166, 170, 170, 171, 171, 172, 172, 173, 176, 181, 182, 186, 186, 188, 189, 189, 194, 197, 198, 199, 202, 203, 207, 208, 209, 212, 214, 217, 218, 220, 226, 228, 228
TEQs (35) 37, 42, 56, 57, 61, 66, 80, 85, 95, 95, 97, 100, 100, 108, 109, 119, 123, 130, 134, 135, 136, 142, 143, 143, 145, 155, 159,

Keyword Index (Cont.). Numbers in parentheses are the sums of references containing the keyword.

   161, 180, 184, 193, 204, 205, 224, 227
terns (1) 212
Th (1) 198
Thalasseus elegans see *Sterna elegans*
Thalassoica antarctica (1) 141
Thayer's gull see Iceland gull
thick-billed murre (52) 20, 20, 22, 29, 30, 32, 34, 34, 34, 35, 36, 37, 38, 42, 58, 59, 72, 72, 73, 74, 75, 82, 91, 91, 111, 112, 113, 115, 125, 125, 133, 133, 147, 150, 153, 155, 156, 157, 157, 162, 174, 184, 185, 186, 203, 208, 208, 209, 216, 221, 221, 222
thyroid hormones (4) 55, 166, 169, 211
Ti (4) 19, 52, 198, 226
Tl (1) 85
Tokyo Bay (1) 30
Totanus flavipes see *Tringa flavipes*
toxaphene (18) 19, 38, 56, 61, 79, 89, 109, 116, 121, 133, 153, 162, 197, 199, 211, 220, 226, 229
toxicology (111) 23, 32, 37, 42, 52, 53, 55, 55, 56, 57, 61, 61, 62, 63, 66, 67, 68, 68, 69, 76, 78, 80, 84, 84, 85, 85, 86, 86, 88, 91, 91, 93, 94, 95, 95, 97, 97, 99, 100, 100, 102, 105, 107, 107, 108, 109, 111, 114, 123, 124, 124, 126, 127, 127, 130, 132, 133, 134, 135, 136, 137, 138, 138, 141, 142, 143, 143, 145, 148, 148, 149, 151, 153, 155, 157, 161, 166, 166, 166, 167, 167, 167, 168, 168, 168, 169, 170, 175, 177, 179, 184, 187, 190, 191, 192, 192, 193, 195, 196, 196, 200, 204, 205, 207, 212, 213, 215, 216, 224, 225, 227
trachea (1) 128
tree sparrow (1) 195
tree swallow (1) 146
Tringa flavipes (2) 125, 223
Tringa melanoleuca (1) 223
Tringa nebularia (3) 56, 57, 122
Tringa stagnatilis (1) 137
Tringa totanus (2) 72, 140
Tristan skua (1) 204
Tristan thrush (1) 32
tubenoses (1) 212
tufted duck (4) 81, 137, 195, 229

tufted puffin (9) 28, 38, 64, 65, 112, 162, 220, 221, 223
turkey (1) 169
turkey vulture (2) 32, 52
TX (1) 226
Tympanuchus phasianellus (2) 38, 169
ultraviolet radiation (2) 19, 133
upland sandpiper see Bartram's sandpiper
Uria aalge (99) 19, 19, 20, 21, 22, 22, 25, 26, 28, 28, 31, 31, 32, 32, 32, 32, 34, 38, 55, 58, 63, 64, 65, 66, 67, 67, 68, 68, 69, 70, 71, 79, 80, 80, 81, 81, 81, 90, 91, 103, 104, 112, 115, 116, 116, 117, 117, 118, 119, 120, 121, 123, 125, 127, 127, 133, 140, 145, 151, 151, 156, 161, 162, 163, 163, 165, 167, 169, 171, 174, 175, 177, 178, 184, 185, 186, 187, 188, 188, 189, 189, 195, 196, 198, 200, 201, 203, 203, 204, 208, 208, 209, 215, 216, 220, 220, 221, 222, 223
Uria grylle see *Cepphus grylle*
Uria lomvia (52) 20, 20, 22, 29, 30, 32, 34, 34, 34, 35, 36, 37, 38, 42, 58, 59, 72, 72, 73, 74, 75, 82, 91, 91, 111, 112, 113, 115, 125, 125, 133, 133, 147, 150, 153, 155, 156, 157, 157, 162, 174, 184, 185, 186, 203, 208, 208, 209, 216, 221, 221, 222
USA (82) 19, 21, 28, 32, 43, 44, 44, 44, 45, 45, 46, 46, 47, 47, 48, 48, 49, 49, 52, 53, 53, 58, 61, 61, 62, 63, 64, 65, 89, 94, 94, 96, 97, 99, 105, 105, 106, 110, 116, 117, 123, 123, 125, 131, 132, 133, 133, 134, 136, 145, 147, 149, 154, 162, 165, 174, 177, 177, 178, 185, 186, 187, 190, 194, 194, 197, 197, 198, 199, 208, 208, 209, 212, 220, 221, 221, 222, 223, 224, 225, 226, 227
UT (1) 226
V (5) 38, 85, 128, 198, 226
VA (3) 44, 185, 199
Vanellus vanellus (1) 137
velvet scoter see white-winged scoter
viscera (1) 41
vitamin A (13) 55, 85, 91, 95, 97, 99, 155, 161, 169, 192, 192, 193, 215
VT (1) 226

Keyword Index (Cont.). Numbers in parentheses are the sums of references containing the keyword.

WA (5) 28, 53, 105, 190, 194
Wadden Sea (7) 22, 23, 72, 121, 140, 164, 187
wandering albatross (3) 32, 141, 204
wandering tattler (3) 64, 65, 223
wedge-tailed shearwater (3) 169, 175, 190
western grebe (5) 32, 52, 125, 210, 220
western gull (4) 125, 169, 177, 190
western sandpiper (1) 223
westland petrel (1) 141
whimbrel (1) 223
white-billed diver see yellow-billed loon
white-capped albatross (3) 128, 129, 141
white-capped noddy (1) 141
white-chinned petrel (4) 128, 128, 129, 141
white-faced ibis (1) 123
white-flippered penguin (1) 32
white-fronted goose (2) 169, 226
white-fronted tern (1) 141
white-headed petrel (1) 141
white Pekin ducks (1) 149
white pelican (7) 19, 32, 96, 97, 123, 125, 210
white-tailed sea eagle (9) 32, 80, 112, 115, 117, 120, 123, 130, 163
white-tailed tropicbird (2) 32, 141
white-throated sparrow (1) 169
white-winged dove (1) 152
white-winged scoter (10) 32, 38, 64, 65, 80, 81, 106, 133, 223, 226
whooper swan (1) 87
WI (11) 19, 61, 61, 62, 132, 133, 136, 147, 194, 224, 226
willet (1) 125
willow ptarmigan (3) 38, 41, 136
Wilson's petrel (2) 32, 125
wood duck (2) 38, 226
woodstork (1) 123
Xantus' murrelet (1) 28
yellow-billed loon (3) 32, 65, 223
yellow-eyed penguin (1) 141
yellow-legged gull (1) 20
yellow-nosed albatross (3) 128, 129, 204
yolk (1) 193

Yukon (4) 56, 133, 145, 226
zebra finch (1) 169
Zenaida asiatica (1) 152
Zenaida macroura (3) 38, 152, 152
Zn (47) 19, 32, 32, 32, 38, 52, 67, 67, 68, 68, 69, 76, 77, 85, 101, 105, 106, 110, 111, 112, 114, 118, 128, 129, 130, 135, 135, 141, 152, 156, 157, 164, 178, 184, 185, 185, 186, 196, 197, 198, 198, 212, 213, 215, 216, 216, 226
Zonotrichia albicollis (1) 169
Zr (1) 198

www.ingramcontent.com/pod-product-compliance
Lightning Source LLC
Chambersburg PA
CBHW081720170526
45167CB00009B/3645